Smart Manufacturing

Smart Manufacturing

Editors

Zhuming Bi
Li Da Xu
Puren Ouyang

MDPI • Basel • Beijing • Wuhan • Barcelona • Belgrade • Manchester • Tokyo • Cluj • Tianjin

Editors
Zhuming Bi
Purdue University Fort
Wayne
USA

Li Da Xu
Old Dominion University
USA

Puren Ouyang
Ryerson University
Canada

Editorial Office
MDPI
St. Alban-Anlage 66
4052 Basel, Switzerland

This is a reprint of articles from the Special Issue published online in the open access journal *Machines* (ISSN 2075-1702) (available at: https://www.mdpi.com/journal/machines/special_issues/smart_manufacturing).

For citation purposes, cite each article independently as indicated on the article page online and as indicated below:

LastName, A.A.; LastName, B.B.; LastName, C.C. Article Title. *Journal Name* **Year**, *Volume Number*, Page Range.

ISBN 978-3-0365-5309-2 (Hbk)
ISBN 978-3-0365-5310-8 (PDF)

Cover image courtesy of Zhuming Bi

Contents

About the Editors

Zhuming Bi

Zhuming Bi received the Ph.D. degree from the Harbin Institute of Technology, Harbin, China, and the University of Saskatchewan, Saskatoon, SK, Canada, in 1994 and 2002, respectively. Currently, he is a Professor at the Department of Civil and Mechanical Engineering, Purdue University Fort Wayne (PFW). Professor Bi is the Harris Chair of Wireless Communication and Applied Research at PFW; he has authored or co-authored 4 books, including three textbooks on 'Finite Element Analysis Applications', 'Computer Aided Design and Manufacturing', and 'Digital Manufacturing'. In addition, Professor Bi has published 10 book chapters, 150 journal articles, and over 60 articles in conference proceedings. His research interests are enterprise information systems, digital manufacturing, finite element analysis, machine designs, robotics and automation, and enterprise systems.

Li Da Xu

Li Da Xu received the B.S. degree in information science from the University of Science and Technology of China (1978), an M.S. degree in information science from the University of Science and Technology of China (1981), and a Ph.D. degree in systems science and engineering from Portland State University, USA (1986). He is an IEEE Fellow, academician of the European Academy of Sciences (Division of Engineering), academician of the Russian Academy of Engineering (formerly USSR Academy of Engineering), and academician of the Engineering Academy of Armenia. Dr. Xu is a 2016-2021 Highly Cited Researcher in the field of engineering named by Clarivate Analytics (formerly Thomson Reuters Intellectual Property & Science).

Puren Ouyang

Puren Ouyang received his MS and Ph.D. degrees from the University of Saskatchewan of Canada in 2002 and 2005, respectively. Currently, he is an Associate Professor at the Department of Aerospace Engineering, Ryerson University of Canada. His research interests are robotics control, hybrid systems, and mechatronics.

Editorial

Smart Manufacturing—Theories, Methods, and Applications

Zhuming Bi [1,*], **Lida Xu** [2] **and Puren Ouyang** [3]

1 Department of Civil and Mechanical Engineering, Purdue University Fort Wayne, Fort Wayne, IN 46805, USA
2 Department of Information Technology, Old Dominion University, Norfolk, VA 23529, USA
3 Department of Aerospace Engineering, Ryerson University, 350 Victoria St, Toronto, ON M5B 2K3, Canada
* Correspondence: biz@pfw.edu

Citation: Bi, Z.; Xu, L.; Ouyang, P. Smart Manufacturing—Theories, Methods, and Applications. *Machines* 2022, *10*, 742. https://doi.org/10.3390/machines10090742

Received: 5 August 2022
Accepted: 8 August 2022
Published: 29 August 2022

Publisher's Note: MDPI stays neutral with regard to jurisdictional claims in published maps and institutional affiliations.

1. Smart Manufacturing (SM) Theories

Smart manufacturing (SM) distinguishes itself from other system paradigms by introducing '*smartness*' as a measure to a manufacturing system; however, researchers in different domains have different expectations of system smartness from their own perspectives. In this *Special Issue* (SI), SM refers to a system paradigm where digital technologies are deployed to enhance system smartness by (1) empowering physical resources in production, (2) utilizing virtual and dynamic assets over the internet to expand system capabilities, (3) supporting data-driven decision making at all domains and levels of businesses, or (4) reconfiguring systems to adapt changes and uncertainties in dynamic environments. System smartness is measured by one or a combination of system performance metrics, such as the degree of automation, cost-effectiveness, leanness, robustness, flexibility, adaptability, sustainability, and resilience. This SI aims to present the most representative works in advancing the theories, methods, and applications of SM.

Rapidly developed digital technologies have continuously stimulated shifts of manufacturing system paradigms; most recently, the study of SM has attracted numerous researchers in academia and practitioners in industry [1–5]. However, people in different domains have highly diversified expectations of system smartness, leading to the ambiguity, diversity, and inconsistency of SM concepts in terms of system architecture, reference models, enabling technologies, and evaluation matrices. Bi et al. [6] generalized the definition of SM by unifying diversified expectations of system smartness as customizable measures, and they presented two concepts of *digital triad* (DT-II) and the *Internet of Digital Triad Things* (IoDTT) to emphasize the *functional requirements* (FRs) of SM to accommodate changes and uncertainties in sustainable and cost-effective ways. Bányai [7] analyzed the needs of adaptability and flexibility in *matrix production*; he argued that flexible manufacturing systems could be the correct solutions to deal with changes in production. He emphasized the importance of effective models and methods in optimizing system controls. In particular, he proposed a hybrid metaheuristic algorithm based on multiphase black hole and flower pollination to plan and schedule manufacturing resources in material handling systems using robots.

Sahal et al. [8] investigated the roles of *digital twins* (DTs) in modelling physical assets and supporting decision-making activities in decentralized and distributed manufacturing. They found that DTs required collaboration among stakeholders to reach the consensuses of decisions and predict risks; the critical FRs of collaborations were defined in terms of interoperability, authentication, scalability, and the avoidance of single-point failures. A ledger-based collaborative framework was proposed to fulfill the identified FRs in smart transportation systems, and the incorporated technologies included blockchain technologies (BCTs), predictive analysis techniques, and other digital technologies. Ubiquitous smart things in the *Internet of Things* (IoT) make it feasible to collect real-time data of the conditions of any manufacturing resources from anywhere at any time; Tan et al. [9] adopted DTs to synchronize and utilize real-time data in a cyber space; the challenges of integrating DTs with smart things in IoT were explored, and a new scheme and framework were constructed to simulate DTs with real-time data.

2. System Design Methods

SM has benefited greatly from rapidly developed information technologies, such as DTs, BCT, IoT, *cloud computing* (CC), *big data analytics* (BDA), *cyber-physical systems* (CPSs), and *edge-computing*. These technologies have been changing the landscape of the research and development of SM radically, in a sense that (1) solutions of acquiring and transferring data become increasingly more affordable in regards to implementing, deploying, and integrating 'smarter' things in a system; (2) business-relevant data become increasingly bigger in terms of *'volume'*, *'variety'*, and *'velocity'*, where advanced data analytics can be used to capture, store, process, and utilize data to cope with changes in dynamic environments; (3) the system boundary becomes increasingly vaguer, and system architecture has to be dynamically adaptable to physical and virtual collaborations of business partners over time [10–13].

System design methods are used to select system elements, configure these elements into components and systems, and model, evaluate, and compare design options against *key performance indicators* (KPIs) for system optimization. However, traditional system design methods are mostly for the design of static systems with clear system boundaries. There are needs required for the advancement of system design methods so that a smart manufacturing system can be reconfigurable to achieve high-level smartness in its system lifecycle. The configurations of a smart system must be customized to the constraints of manufacturing resources and the prioritized KPIs. Bi et al. [14,15] proposed a systematic design methodology as the guide for designs of smart manufacturing systems in specified applications. The *axiomatic design theory* (ADT) was adopted and expanded to design, analyze, and assess smart manufacturing systems, and the applicability of the proposed methodology was verified using three case studies. Erasmus et al. [16] proposed an information architecture to integrate CC and IoT with smart devices for human–robot collaboration; the architecture was modularized for *small- or medium-sized enterprises* (SMEs) to access extensible cloud services, and it was used as a reference architecture for information management systems in Industry 4.0. The architecture was tested and evaluated with the information systems of ten real-world factories. Kim and Lee [17] extended the SM concept to a maintenance system in ship building and servicing; the framework, procedure, and architecture of a smart maintenance system were developed to systematically design large-scale SM systems.

3. Applications

SM is expected to meet some emerging requirements of automation, adaptability, sustainability, and resilience of modern manufacturing systems in the digital era at numerous aspects, including (1) dealing with any level of system complexity relating to the number and variants of system elements, the interactions of system elements, and anticipated and unanticipated changes over time; (2) maximizing system entropy to adopt changes in a dynamic environment; (3) responding to real-time changes in the shortest possible time; (4) monitoring, diagnosing, and predicting system states and trends, generating preventive solutions for adverse changes, and upgrading systems to adapt preferable changes; (5) supporting the seamless coordination, collaboration, and cooperation of stakeholders; (6) orchestrating manufacturing resources across enterprise bounds to seize novel opportunities; (7) providing generic architecture applicable to different products, functions, and regions [1,18–20].

Cutting-edge digital technologies have been widely explored in regard to solving various engineering problems in real-world applications. For examples, Hou et al. [21] developed a function–structure model to evaluate performance and cost in product development; products were characterized in functional and structural domains, respectively, and an evolutionary algorithm (EA) was used to map functions into corresponding structures for the verification of design constrains and the evaluation of design solutions. Kang et al. [22] discussed various challenges of using vibrioses to protect the environment during fossil fuel exploration; numerical simulation models were developed to analyze the response of a vibriosis subjected to specific boundary conditions and excitations, and simulation results were used to identify the weakest vibriosis junctions. Liu et al. [23] proposed an integrated robotic system for its application in an ill-structured on-site environment with the purpose of cost-efficiency. The proposed system consisted of two-terminal manipulators for parallel sorting processes, and it was seamlessly integrated in an automated assembly system to perform sorting tasks consistently in a shortened cycle time. Yung et al. [24] discussed the challenges in designing and manufacturing highly diversified space instruments. The specifications of space instruments were greatly distinguished from those of products on Earth, and careful considerations had to be determined on the size, weight, cost, complexity, and extreme space environments. A systematic literature search method was used to look into the impact of product design and innovation on the development of space instruments; the survey provided important information and critical considerations for using cutting-edge digital technologies in designing and manufacturing space instruments.

4. Future Research Directions

Increasingly more manufacturing enterprises are ready to incorporate newly developed technologies, such as DTs, CPSs, IoT, BDA, and BCT, with traditional manufacturing technologies, such as *flexible manufacturing systems (FMSs)*, *total quality management (TQM)*, *supply chain management (SCM)*, *enterprise resource planning (ERP)*, and *computer-integrated manufacturing (CIM)*. However, existing theories, methods, and tools still exhibit limitations in supporting cost-effective *vertical integration, decentralization, smart sensing and actuating, autonomy and self-organization*, and uses of *semantic models* [25]. The research of SM in theories, methods, and applications should be advanced to transfer integrated digital technologies into productivity, profitability, and sustainability of systems. This Editorial Team anticipated that future research in SM would mainly incorporate areas of (1) ubiquitous sensing, (2) fusing and integrating data from heterogeneous sources, (3) effective BDA methods, (4) data visualization methods for human interactions, (5) data-driven decision-making supports, (6) workflow composition methods, (7) the standardization and specifications of smart modules, and (8) quantified criteria such as adaptability, sustainability, and resilience for system evaluation [1,6,14].

Acknowledgments: We thank the editors of MDPI for their excellent support to this SI; it would have been impossible without their persistence and great effort. We thank all colleagues who showed an interest in submitting their works and, we especially thank all reviewers for their valuable author recommendations, helping us finalize their high-quality manuscripts; all accepted papers passed a rigorous reviewing process before they were finally accepted and published in the SI. The selected papers represent the quality, breadth, and depth of the study on theories, methods, and applications of SM.

Conflicts of Interest: The authors declare no conflict of interest.

References

1. Bi, Z.M.; Jin, Y.; Maropoulos, P.; Zhang, W.J.; Wang, L. Internet of things (IoT) and big data analytics (BDA) for digital manufacturing. *Int. J. Prod. Res.* **2021**. [CrossRef]
2. Bi, Z.M.; Xu, L.D.; Wang, C. Internet of things for enterprise systems of modern manufacturing. *IEEE Trans. Ind. Inform.* **2014**, *10*, 1537–1546.
3. Bi, Z.M.; Wang, G.P.; Thompson, J.; Ruiz, D.; Rosswurm, J.; Roof, S.; Guandique, C. System framework of adopting additive manufacturing in mass production line. *Enterp. Inf. Syst.* **2021**, *16*, 606–629. [CrossRef]
4. Bi, Z.M.; Zhang, W.J. *Practical Guide to Digital Manufacturing–First Time-Right from Digital Twin to Physical Twin*; Springer International Publishing: New York, NY, USA, 2021; ISBN 978-3-030-70303-5.
5. Bi, Z.M.; Wang, X.Q. *Computer Aided Design and Manufacturing (CAD/CAM)*; Wiley: New York, NY, USA, 2020.
6. Bi, Z.; Zhang, W.-J.; Wu, C.; Luo, C.; Xu, L. Generic design methodology for smart manufacturing systems from a practical Perspective, part I—Digital triad concept and Its application as a system reference model. *Machines* **2021**, *9*, 207. [CrossRef]

7.   Bányai, T. Optimization of material supply in smart manufacturing environment: A metaheuristic approach for matrix production. *Machines* **2021**, *9*, 220. [CrossRef]
8.   Sahal, R.; Alsamhi, S.H.; Brown, K.N.; O'Shea, D.; McCarthy, C.; Guizani, M. Blockchain-Empowered digital twins collaboration: Smart transportation use case. *Machines* **2021**, *9*, 193. [CrossRef]
9.   Tan, Y.; Yang, W.; Yoshida, K.; Takakuwa, S. Application of IoT-Aided simulation to manufacturing systems in cyber-physical system. *Machines* **2019**, *7*, 2. [CrossRef]
10.   Wang, C.; Bi, Z.M.; Xu, L.D. IoT and cloud computing in automation of assembly modeling systems. *IEEE Trans. Ind. Inform.* **2014**, *10*, 1426–1434. [CrossRef]
11.   Wang, L.; Xu, L.D.; Bi, Z.M.; Xu, Y. Data cleaning for RFID and WSN integration. *IEEE Trans. Ind. Inform.* **2014**, *10*, 408–418. [CrossRef]
12.   Viriyasitavat, W.; Xu, L.D.; Bi, Z.M. User-oriented selections of validators for trust of internet-of thing services. *IEEE Trans. Ind. Inform.* **2022**, *18*, 4859–4867. [CrossRef]
13.   Viriyasitavat, W.; Xu, L.D.; Bi, Z.M. Blockchain-based business process management framework for service composition in industry 4.0. *J. Intell. Manuf.* **2020**, *31*, 1737–1748. [CrossRef]
14.   Bi, Z.; Zhang, W.-J.; Wu, C.; Luo, C.; Xu, L. Generic design methodology for smart manufacturing systems from a practical perspective. part II—Systematic designs of smart manufacturing systems. *Machines* **2021**, *9*, 208. [CrossRef]
15.   Bi, Z.M.; Zhang, W.J.; Wu, C.; Li, L. New digital triad (DT-II) concept for lifecycle information integration of sustainable manufacturing systems. *J. Ind. Inf. Integr.* **2022**, *26*, 100316. [CrossRef]
16.   Erasmus, J.; Grefen, P.; Vanderfeesten, I.; Traganos, K. Smart hybrid manufacturing control using cloud computing and the internet-of-things. *Machines* **2018**, *6*, 62. [CrossRef]
17.   Kim, G.S.; Lee, Y.H. Transformation towards a smart maintenance factory: The case of a vessel maintenance depot. *Machines* **2021**, *9*, 267. [CrossRef]
18.   Viriyasitavat, W.; Bi, Z.M.; Hoonsopon, D. Blockchain technologies for interoperation of business processes in smart supply chains. *J. Ind. Inf. Integr.* **2022**, *26*, 100326. [CrossRef]
19.   Viriyasitavat, W.; Xu, L.; Dhiman, G.; Sapsomboon, A.; Pungpapong, V.; Bi, Z.M. Service workflow: State-of-the-Art and future trends. *IEEE Trans. Serv. Comput.* **2022**. [CrossRef]
20.   Bi, Z.M.; Chen, B.; Xu, L.D.; Wu, C.; Malott, C.; Chamberlin, M. Enterline, T. Security and safety assurance of collaborative manufacturing in industry 4.0. *Enterp. Inf. Syst.* **2022**. [CrossRef]
21.   Huo, Y.-L.; Hu, X.-B.; Chen, B.-Y.; Fan, R.-G. A product conceptual design method based on evolutionary game. *Machines* **2019**, *7*, 18. [CrossRef]
22.   Kang, Z.; Li, G.; Wang, F.; Zhang, H.; Su, R. Analysis of vibration plate cracking based on working stress. *Machines* **2018**, *6*, 51. [CrossRef]
23.   Liu, P.; Li, G.; Su, R.; Wen, G. Automatic test and sorting system for the slide valve body of oil control valve based on cartesian coordinate robot. *Machines* **2018**, *6*, 64. [CrossRef]
24.   Yung, K.-L.; Tang, Y.-M.; Ip, W.-H.; Kuo, W.-T. A Systematic review of product design for space instrument innovation, reliability, and manufacturing. *Machines* **2021**, *9*, 244. [CrossRef]
25.   Dotoli, M.; Fay, A.; Miskowicz, M.; Seatzu, C. An overview of current technologies and emerging trends in factory automation. *Int. J. Prod. Res.* **2019**, *57*, 5047–5067. [CrossRef]

Article

Generic Design Methodology for Smart Manufacturing Systems from a Practical Perspective, Part I—Digital Triad Concept and Its Application as a System Reference Model

Zhuming Bi [1,*], Wen-Jun Zhang [2], Chong Wu [3,*], Chaomin Luo [4] and Lida Xu [5]

1   Department of Civil and Mechanical Engineering, Purdue University Fort Wayne, Fort Wayne, IN 46805, USA
2   Department of Mechanical Engineering, University of Saskatchewan, Saskatoon, SK S7N 5A9, Canada; Chris.Zhang@usask.ca
3   School of Economics and Management, Harbin Institute of Technology, Harbin 150001, China
4   Department of Electrical and Computer Engineering, Mississippi State University, Starkville, MS 39762, USA; chaomin.luo@ece.msstate.edu
5   Department of Information Technology, Old Dominion University, Norfolk, VA 23529, USA; lxu@odu.edu
*   Correspondence: biz@pfw.edu (Z.B.); wuchong@hit.edu.cn (C.W.)

Citation: Bi, Z.; Zhang, W.-J.; Wu, C.; Luo, C.; Xu, L. Generic Design Methodology for Smart Manufacturing Systems from a Practical Perspective, Part I—Digital Triad Concept and Its Application as a System Reference Model. *Machines* **2021**, *9*, 207. https://doi.org/10.3390/machines9100207

Academic Editors: Angelos P. Markopoulos and Ibrahim Nur Tansel

Received: 26 July 2021
Accepted: 20 September 2021
Published: 23 September 2021

Publisher's Note: MDPI stays neutral with regard to jurisdictional claims in published maps and institutional affiliations.

Abstract: Rapidly developed information technologies (IT) have continuously empowered manufacturing systems and accelerated the evolution of manufacturing system paradigms, and smart manufacturing (SM) has become one of the most promising paradigms. The study of SM has attracted a great deal of attention for researchers in academia and practitioners in industry. However, an obvious fact is that people with different backgrounds have different expectations for SM, and this has led to high diversity, ambiguity, and inconsistency in terms of definitions, reference models, performance matrices, and system design methodologies. It has been found that the state of the art SM research is limited in two aspects: (1) the highly diversified understandings of SM may lead to overlapped, missed, and non-systematic research efforts in advancing the theory and methodologies in the field of SM; (2) few works have been found that focus on the development of generic design methodologies for smart manufacturing systems from the practice perspective. The novelty of this paper consists of two main aspects which are reported in two parts respectively. In the first part, a simplified definition of SM is proposed to unify the existing diversified expectations, and a newly developed concept named digital triad (*DT-II*) is adopted to define a reference model for SM. The common features of smart manufacturing systems in various applications are identified as functional requirements (*FRs*) in systems design. To model a system that is capable of reconfiguring itself to adapt to changes, the concept of IoDTT is proposed as a reference model for smart manufacturing systems. In the second part, these two concepts are used to formulate a system design problem, and a generic methodology, based on axiomatic design theory (ADT), is proposed for the design of smart manufacturing systems.

Keywords: smart manufacturing; information technologies (IT); system of systems (*SoS*); digital manufacturing (DM); digital twins (DT-I); digital triad (*DT-II*); cyber-physical systems; Internet of Things (IoT); Internet of Digital Triad Things (IoDTT); big data analytics (BDA); cloud computing (CC); axiomatic design theory (ADT)

1. Introduction

Manufacturing creates products for customers, and the demand for new and advanced products has been increasing monotonically due to (1) the rise of the global population, (2) an increase in the standards of living, and (3) the way of consumption in "throwaway" societies. Accordingly, manufacturing technologies have been greatly advanced to meet the needs of consumers' markets [1–5]. In the 2020s, traditional manufacturing systems face strong pressure to enhance their capabilities in handling the growing complexity and scale of—and changes to—the manufacturing environment, and the manufacturing industry has entered the digital era, due to the adoption of advanced information technologies and operation technologies [6].

The historical advancement of manufacturing technologies has been widely discussed [7,8], and the following trends have become common sense to researchers in manufacturing: (1) products become increasingly advanced, diversified, and with fragmented demand; (2) global manufacturing capabilities become saturated in comparison to customers' needs, and manufacturing enterprises face ever-increasing competitions regionally and globally; (3) the scope and complexities of manufacturing businesses are continuously increased, and this forces enterprises to adopt more advanced technologies to automate manufacturing operations and decision making in various domains and levels of business; (4) the boundaries of manufacturing systems become vague and dynamic, since enterprises have to collaborate with others to make complex products or systems; (5) manufacturing systems are expected to be optimized against more performance metrics, including functionality, quality, productivity, cost, lead-time, personalization, adaptability, and sustainability; (6) the businesses of manufacturing systems are gradually extended to cover all the stages of product lifecycles, from raw materials to the disposal of used products; (7) manufacturing technologies have advanced from human operation to automation, from disciplinary to multidisciplinary, from standalone to integrated and comprehensive, and from sophisticated to adaptable and reconfigurable.

Since the advances in manufacturing technologies can be characterized by the degree of automation (of either (1) manufacturing processes or (2) the decision making support at different domains and levels of system operations), information technologies (IT) have played indispensable roles in advancing manufacturing technologies. Every leap in the manufacturing paradigms was triggered by the corresponding IT; to name a few, mechanization and electronic and electrical controls for mass production, numerical controls (NC) or computer NC (CNC), networking, and group technologies (GT) for flexible manufacturing systems (FMS), computer aided design and manufacturing (CADM), material resource planning (MRP), and enterprise resource planning (ERP) for computer integrated manufacturing (CIM), quality control (QC) and total quality management (TQM) for lean production (LP), product data management (PDM), product lifecycle management (PLM), and enterprise systems (ES) for agile manufacturing (AM), business process management (BPM), service-oriented architecture (SoA), and agent-based techniques for virtual manufacturing (VM), and Internet of Things (IoT), radio frequency identification (RFID) [9–11], cyber-physical systems (CPS), human-cyber-physical systems (HCPS), blockchain technology (BCT), big data analytics (BDA), and cloud computing (CC) for digital manufacturing (DM) and smart manufacturing (SM) [3,4]. The studies in DM have attracted a great deal of attention recently. For example, Dey et al. [12,13] investigated the impacts of optimized autonomation policies on the control variables in the inventory management of a smart production system. In the digital era, the digitization of manufacturing businesses has progressed exponentially, and the increasing adoption of digital technologies in our economy has reshaped the way we live and work. The trend of digitization has brought new opportunities and challenges for manufacturing enterprises to gain business competitiveness over their strategical competitors in the globalized market [14,15]. The further development of the digital economy in the near future will be characterized by four dimensions of smartness: smart manufacturing, smart products and services, smart supply chains, and smart processes [16].

SM or DM have become the frontier in advancing manufacturing technologies [17]. However, today's manufacturing systems are, in fact, systems of systems, due to the high-level of system complexity, the large variety of manufacturing assets and enabling technologies, and the dynamics of systems over time. This causes diversity and confusion for practitioners in understanding SM and its relevant concepts, reference models, performance matrices, and the selection of design methodologies. On the one hand, any manufacturing system should be a sophisticated system tailored to the given manufacturing resources, markets, and business environment, and there is no universal SM solution for enterprises. On the other hand, manufacturing businesses are involved with the transformation of materials, values, information, and knowledge of multiple disciplines, and system performances are assessed very differently in these disciplines. In other words, the stakeholders of one manufacturing system have different expectations of SM, which might be correlated, coupled, or even conflicted with each other.

The authors are highly motivated to gain a thorough understanding of the state of the art development of smart manufacturing (SM) systems and to identify the direction of future research in the field of SM. The main contributions from the reported work are as follows:

(1) It is found that the existing works on SM show the limitations of at two aspects, i.e., (a) the highly diversified understandings of the functionalities and expectations of SM that may result in overlapped, missed, or non-systematic research efforts in advancing the theory and methodologies in the field of SM; (b) few works have been published that propose a generic design methodology for the design of smart manufacturing systems in practice.

(2) The definition of SM is simplified to unify the diversified expectations. A newly developed concept, digital triad (*DT-II*), is adopted to define a reference model for SM; it reflects all of the main characteristics of digital solutions at the different levels and domains of system operations.

(3) The common features of various smart manufacturing systems are identified; particularly, the concept of IoDTT is proposed as a reference model to represent the need for system reconfiguration in the event of uncertainties and changes in business environments.

(4) The generality and specialty in designing and implementing various smart manufacturing systems are discussed, to illustrate the need for developing a general design methodology to guide the design of a smart manufacturing system from a practical perspective.

The remainder of paper is organized as follow. In Section 2, an overview of SM and its development is provided, to understand the limitations of the existing works. In Section 3, the concept of SM is refined, and the corresponding reference model is proposed, using the concept of digital triad (*DT-II*). In Section 4, the design of a modular robotic system is used as an example to illustrate the system design problems at different phases, and the generality of designing and implementing tailored smart systems is discussed, to highlight the need for generic design methodologies. In the following part II of this work, two proposed concepts are adopted for formulating system design problems, and ADT is suggested as the generic design methodology for SM. Three case studies are introduced to illustrate the application of the proposed design methodology, and the future research directions regarding SM are discussed as a summary.

2. Overview of Smart Manufacturing (SM)

2.1. Original Definition and Variations

Smart manufacturing (SM) is also commonly referred to as "Industry 4.0" in Europe and "Made in China 2025" in China, as the latest iteration of the industrial revolution that began 260 years ago. The recent development of information technologies (IT) has provided the concrete foundation to revolutionize manufacturing in the form of SM [18]. The concept was coined by the National Science Foundation (NSF) at its workshop on cyberinfrastructure in 2006 [19]. SM was originally defined as a fully integrated and collaborative manufacturing system that responds in real time to meet the changing demands and conditions in the factory, the supply chain, and customer needs [20,21]. The vision and goals of SM were then developed by the Smart Manufacturing Leadership Coalition in 2011 [22]. The roadmaps and standards for Industry 4.0 and SM were firstly developed by the German Commission for Electrical, Electronic & Information Technologies [23] in 2014 and the National Institute of Standards and Technology (NIST) in 2016 [20], respectively.

The clean energy smart manufacturing innovation institute (CESMII) was created to promote SM as a sustainable driving force for manufacturing in the United States to adopt smart sensors, controls, platforms, and models in 2016. The closely related concept, Made in China 2025, was initialized to advance the independent manufacturing technologies in China in 2015 [24]. From the perspective of applications, SM refers to an IoT-based application to automate manufacturing processes and utilize data analytics tools to improve the performances of manufacturing systems. SM uses information technologies, computer-integrated technology, flexible workforces, and digital technology to improve the system adaptability for the changes and uncertainties in system operations [25]. SM was treated as an application of Industry 4.0 in manufacturing, where advanced IT, such as the Internet of Things (IoT), cyber physical systems (CPS), machine learning (ML), additive manufacturing (AM), and robotics, were used in process automation and decision making support [26]. It should be noted that Industry 4.0 refers to a system in which automated facilities are networked, so that data can be collected, processed, and utilized to make smart decisions in system operations.

SM distinguishes itself from other manufacturing paradigms by defining the meaning of "smart", whereas researchers in different disciplines and roles define different expectations of smartness. For example, the smartness of a manufacturing system, according to Romero et al. [27] covers (1) the ability to communicate to exchange data, and to collect and report data regarding the state of manufacturing assets; (2) the embedded knowledge for the representation of human expertise and the understanding of system elements and environments; (3) learning capabilities through the application of diversified algorithms, methods, and tools; (4) reasoning capabilities for data-driven decision making; (5) the perception capability to sense, understand, and respond to environmental changes; (6) control capabilities to ensure smooth manufacturing processes, and to make and deliver products to end-users; (7) self-organization to reconfigure systems to accommodate changes and uncertainties; (8) context awareness that retrieves the information and knowledge that characterizes the state of the systems and environments. Lenz et al. [28] extended the scope of SM to the collection of raw materials; the data in the entire product lifecycle was acquired and processed to improve the accuracy and reliability of the process signatures and to improve manufacturing processes. Filleti et al. [29] used a grinding unit as an example to discuss the impact of manufacturing processes and real-time monitoring on the performance of production and environmental indicators. The manufacturing assets were mostly developed with assured safety and security at the device level; security assurance became a critical challenge when the manufacturing assets were interactive and interoperative. In addition, Maggi et al. [30] and Viriyasitava et al. [31–35] emphasized the importance of addressing cybersecurity awareness regarding the configuring systems, to adjust to changes and uncertainties in dynamic business environments.

Felice et al. [36] proposed a bibliometric model for analyzing the existing works on SM from 2011 to 2018; their objectives were to identify relevant topics and explore the interdependencies of these topics. It was found that researchers generally expected SM to improve the efficiency, cost-effectiveness, safety, and sustainability of systems, and the main enabling technologies for SM were automation, IoT, CPS, BDA, cloud computing (CC), modelling and simulation, and additive manufacturing (AM). SM integrated manufacturing assets, sensors, technological platforms, networks, data-driven modelling, simulations, decision making, and diagnoses and predictions to improve efficiency, flexibility, adaptability, and resilience [37,38]. With the trend of digitization, SM becomes the new frontier of manufacturing systems [17]. SM, as a flexible system, should be able to self-optimize performance across a broader network, self-adapt to change and learn from new conditions in real or near-real time, and autonomously run entire production processes.

2.2. Main Characteristics

The five main characteristics of SM identified by Deloitte (2021) are that they are: (1) connected to all smart things through the internet, including traditional datasets, real-time data enabling collaboration, and the collaborations across departments; (2) optimized for predictable capacities, increased asset uptime and efficiency, highly automated production and material handling, and minimized cost; (3) transparent for live metrics and tools, and have real-time linkages to demand forecasts and order tracking; (4) proactive for anomaly identification and resolutions, restocking and replenishment, and the identification of quality issues; (5) agile for flexible and adaptable scheduling and production changes, with configurable machines and layouts. SM was expected to improve the efficiency of manufacturing assets, improve the quality of products and manufacturing processes, reduce the cost of system operations, and to increase the safety and sustainability of products, manufacturing processes, systems, and human living environments. SM was characterized as networked, big data, digitization, data-driven, resource sharing, connected, sustainable and resilient; in particular, sustainability and resiliency distinguished SM from other manufacturing paradigms [39].

SM, as an engineering system, was compared to a biological system by Byrne et al. [40], and it was suggested that SM should be the convergence of biology and SM, as part of the evolution of the system paradigms involved in bio-inspiration, bio-integration and bio-intelligence. Therefore, SM would incorporate components, features, characteristics, and capabilities to converge a manufacturing system with a biological system. SM is a type of intelligent manufacturing (IM) system, although it is certainly at a more advanced level of intelligence in comparison to traditional IM. Wang et al. [41] elaborated the differences between SM and IM in detail, and they argued that SM was a comprehensive outcome of integrating an increasing number of digital tools with an intelligent manufacturing system. IM and SM were closely correlated; indeed, the differences were mainly in qualitative and quantitative measures, and SM was seen as the advanced stage of IM.

2.3. Technological Drivers

Manufacturing paradigms are enabled by the available technologies, particularly information technologies in the digital era. Kulvatunyou et al. [42] analysed the development of the standards for the integration of semantic data and concluded that the standards for digital manufacturing must be advanced to maintain, represent, and present data in the form of knowledge and insight for collaborative decision making support. It is well known that the digital transformation from Industry 3.0 to 4.0 was enabled by some critical ITs, such as IoT, CC, BDA, and AI [16]. IoT enables the connection of everything in manufacturing, in the same way as Industrial IoT (IIoT); in return, IIoT supports the new and unprecedented interactions among the hardware, software, virtual assets, and humans. In addition, IIoT can be integrated with AI for enterprises to increase their flexibility, agility, efficiency, and resilience [21].

Bi and Zhang [3] classified technical drivers into three types, as shown in Figure 1. From the perspective of system inputs, manufacturing businesses are pushed by humans' expectations for civilization and continuous development. From the perspective of system outputs, manufacturing businesses are pulled by users' demand for better and more products. From the perspective of system transformation, manufacturing processes are gradually advanced by integrating an increasing number of relevant technologies through the iteration of continuous improvement (CI). For system transformation, the trends of the reciprocating drivers were (1) the increasing decentralization of the manufacturing businesses, (2) additive manufacturing using polymeric, metallic, and bio-materials, (3) networks and the integration of supply chains, (4) cyber security assurance, (5) the elimination of latency, (6) high-level automation and self-optimization, (7) the unprecedented scale and level of connections, (8) the adoption of advanced artificial intelligence and machine learning, (9) the sustainability of products, processes, and systems, and (10) the expectations of responsiveness, robustness, and residence [40].

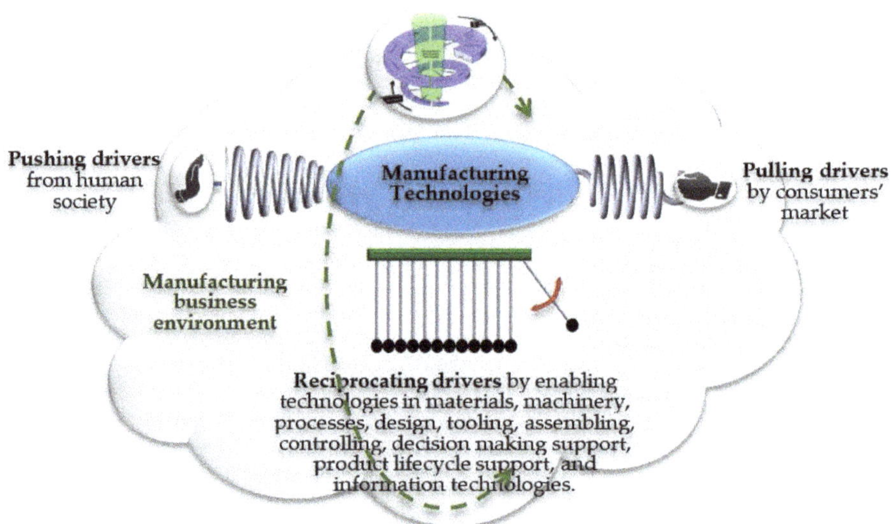

Figure 1. Three types of technological drivers in manufacturing (Reprinted with permission from ref. [3]. 2021, Springer Nature).

SM relies on the integration of various digital technologies. Ghobakhloo [14] argued that digitization was attributed to the information technologies that are used for decentralization, horizontal and vertical integrations, interoperability, virtualization, and modularity. Modularity was viewed as one of the most effective ways to deal with the complexity, changes, and uncertainties in a dynamic environment, and the idea of modularity was applicable to both internal and external manufacturing assets [43]. The virtual assets were used for servitization, and the corresponding manufacturing system was referred to as a product-service system. Brad and Murar [44] discussed the business model, economic impact, and reconfiguring technologies of a product-service system. Feldner and Herber [45] evaluated the communication protocol of IPv6 to support the interactions and interoperations of networked things. Kumar [46] discussed the impact of IoT, CPS, human–robot interaction (HRI), and augmented reality (AR) on the development of new materials and manufacturing processes in SM. Innovation in digital technologies led to advances in new materials and manufacturing processes. SM became practical due to rapidly growing technologies, the ever-increasing complexity of the supply chains, the global fragmentation of production and demands, the growing pressure of competitiveness from unexpected sources, organizational realignment caused by the marriage of information technologies and organization technologies, and ongoing talent challenges [17]. Standardization is powerful in developing and implementing advanced manufacturing technologies. For the seamless integration of various manufacturing assets, the efforts regarding standardization are indispensable. Leading standards organizations, such as the International Organization for Standardization (ISO), National Institute of Standards and Technology (NIST), International Telecommunication Union (ITU), and Institute of Electrical and Electronics Engineers (IEEE), have developed the standards for the architecture, reference models, and frameworks of smart manufacturing [5,18,20,21].

2.4. Applications

The manufacturing industry has a strong impact on the anthropogenic environment; with the growing concern regarding the deterioration of the environment, SM must expand its business scope to the entire product lifecycle by recycling, reusing, and remanufacturing to promote circular economy. Barletta et al. [47] considered the bounds of distinguishing products and services as the environmental breakeven points that could be determined by environmental assessment. Blomeke et al. [5] proposed a recycling 4.0 framework with integrated SM technologies and solutions to support recycling, reusing, and remanufacturing. Jaspert et al. [48] referred to the reconfiguration of SM as smart retrofitting, and an SM could be retrofitted at any functional layer of an enterprise, from the physical, sensor, connectivity, and data layers, to the application layer. Most of the reported applications were in the preliminary stages and met the expectations only for certain aspects. For example, Zenisek et al. [49] introduced an experimental SM system with integrated mixed reality and additive manufacturing for predictive maintenance, and they identified challenges in data merging, online applicability, the conflict of reactivity and false positive rates, and shortfalls in customers' other expectations.

Rodger et al. [50] discussed the need for sustainability in highly automated car body manufacturing; existing methods and algorithms were surveyed to identify the bottlenecks for improving the sustainability of manufacturing systems from the perspective of product lifecycles. Siiskonen et al. [51] explored the potential application of SM in producing personalized medicines; however, they limited this to the synergic outcomes by integrating the platforms for the design of products and manufacturing processes. They found that customer satisfaction was improved by modularized table designs at increased production costs. Ghobakhloo and Ching [15] surveyed the application of digital technologies in small- and medium-sized enterprises (SMEs); they found the use of SM in SMEs was restrained, due to technological, organizational, and environmental factors. Kamble et al. [52] investigated the application of SM in India's automotive industry, and the benefits of SM were assessed across 10 dimensions (i.e., quality, cost, time, flexibility, integration, productivity, computing, sustainability, diagnosis, and prognosis).

2.5. Limitations of Existing Works

Despite its great potential, numerous researchers discussed the limitations of the existing works on SM. For example, Phuyal et al. [53] analyzed the technological gaps in adopting state of art IT in smart manufacturing systems. Their main concerns were the maturity and readiness of these technologies (i.e., AI, CPS, BDA, AR, IoT, and robotic technologies) in achieving tangible expectations in real-world industrial applications. Others discussed the challenges regarding the integration and complexity of systems.

SM integrates many newly developed manufacturing technologies in systems. While advanced technologies are generally complex, a system that is poorly or inappropriately designed for a particular application is not expected to benefit enterprises. Uysal and Mergen [54] expressed concern over the sharing and integration of data across systems and products, and discussed the feasibility of using an intelligent digital mesh (IDM) to form a system made up of dynamically interconnected elements. SM aims for a high level of automation with minimized human effort in collecting, transferring, processing, and mining data. However, the relevant studies were mostly fragmented by focusing on one or a small number of issues, such as decision making, cyber–physical interactions, information infrastructures, digitalization, human–machine interactions, cloud computing, and virtual services [55], and generalizability was lacking from system perspectives. For example, a smart manufacturing system focused on the utilization of hardware assets may undervalue the impact of these assets on system-level performances. To improve the agility and adaptability of SM, advanced methods and tools are needed to connect digital technologies and their business goals for cost effectiveness [56].

The cost of a smart manufacturing system is another significant downside. This is particularly true for small to midsize enterprises (SMEs). Over 95% of enterprises are SMEs that lack manufacturing assets, other than their core competencies for new business opportunities. SMEs were not able to afford the considerable expense for the advanced technology, since short-term benefits were mostly prioritized, and the savings over the long term would outweigh the startup costs. Although the gaps regarding the adoption of digital technologies by SMEs were discussed [37,38], limited works were found that focused on the development of roadmaps and system frameworks for assessing the maturity and readiness of SMEs to adopt digital technologies.

A set of the performance metrics are used to compare and optimize systems. Helu et al. [57] indicate that there are limitations in the assessment of the performances of adopted digital technologies; since the evaluation relies on an appropriate model for the breadth and depth of the technologies that were usually unavailable. Quantitative measures are essential to develop the systematic design methodologies of manufacturing systems. However, the existing assessment models, methods, and criteria are mostly empirical, and built on numerous assumptions and hypotheses; manufacturing systems are viewed as black or grey boxes, and system performances are mainly evaluated based on system inputs and outputs, with the limited context of value-added and non-value-added manufacturing processes in systems [58]. Most models lack consistent data, reliable analysis methods, and user-friendly tools for decision makers to assess system performance in an understandable way [59]. The performance assessment methods fail to match advanced technologies to market demands in manufacturing [60]. Few methods were available to design and evaluate system configurations and implement a smart manufacturing system based on a set of the specified assessment metrics. Despite the large variety of available performance metrics, the importance of selecting the right metrics was overlooked, and no coherent framework was available to adequately measure the effectiveness of system configurations [61].

Despite its attractiveness, the research topic of SM is relatively new, and the relevant studies are mostly preliminary. From the perspective of its applications, the existing works on SM show the following limitations.

SM tends to pursue a full wish list of the functional requirements of the traditional manufacturing paradigms, including automation, productivity, leanness, flexibility, agility, sustainability, adaptability, and resilience. While it is reasonable to represent system expectations from different perspectives, system complexity will easily become unmanageable when designing and controlling a smart manufacturing system. Moreover, the resulting system might not be optimized, since many performance metrics conflict with each other, and the most critical metrics for the weakest aspects must be emphasized.

SM emphasizes (a) the necessities of digital technologies, (b) networking and virtual assets, and (c) adaptability to environmental changes and uncertainties. However, in the existing definitions of SM, the uniqueness of digital technologies was not distinguished from those of other advanced technologies, or from the representations of all advanced technologies in traditional manufacturing paradigms. Additionally, there is no mechanism to distinguish fixed/dynamic, internal/external, and physical/virtual resources, as well as their corresponding roles in systems' lifecycles. Lastly, SM was still modelled as a system with stable structures (commonly known as a hierarchical or multidimensional grid structure), which is ineffective in representing system configurations that are intended to adapt to changes and uncertainties over time. There are some conflicts among steady system models, system reconfiguration, and the adaptability over system lifecycles [3,4].

Other than some discussions on enabling technologies, system architecture, reference models, and performance metrics, no systematic methodology has been explored in designing a smart manufacturing system. The existing examples of smart manufacturing systems were conceptual and lacking in details of system developments.

The available performance metrics were highly diversified. Moreover, most of the metrics limited their applications to system designs only, since the quantifications were data-driven and made for black box or gray box systems. It should be noted that, during the phase of system design or reconfiguration, an evaluation model should be a white box and should represent the dependencies of the performance metrics (such as the level of automation, configurability, and residence) on the design variables (such as the selection of system elements, and the assembly and interaction of system elements) directly.

3. Proposed Definition of SM

In good manufacturing practices, any manufacturing system must be tailored to a specified application, and the same applies for a smart manufacturing system. Therefore, it is our argument that SM is better defined to satisfy the most common system requirements, rather than all requirements, which might be optional, of less importance, and tailored to specific applications. Based on a generic definition of SM focusing on its core values, a reference model can be derived to develop a systematic methodology for the designs of SM. In this section, a new definition of SM is proposed, and the rationales of this proposal are provided by (1) identifying the common features of SM as functional requirements (*FRs*), (2) adopting two new concepts, *DT-II* and *IoDTT*, as the design solutions (DSs) for the given *FRs*, and (3) discussing the solutions for the limitations of the existing works identified in Section 2.

3.1. New Definition of SM

We propose to redefine SM and clarify the relevant concept of smartness as follows:

"Smart manufacturing (SM) is a type of manufacturing paradigm for the enhanced smartness of systems, in which digital technologies are used to empower the physical things in manufacturing products, access virtual assets over networks for expanded manufacturing capabilities, support data-driven decision making in any domain and at any level of manufacturing operations, and reconfigure systems to adapt to the changes in customer needs when making products".

"The smartness of a manufacturing system refers to its ability to (1) offer better manufacturing processes or (2) support better decision making in other manufacturing operations. Smartness can be measured by one, or a combination of, system performance metric(s), such as the degree of automation, cost-effectiveness, leanness, robustness, flexibility, adaptability, sustainability, and resilience".

In comparison to the existing definitions of SM, the proposed definition is generic and applicable to any smart system that adopts digital technologies and is capable of using virtual resources and reconfiguring systems. More importantly, the proposed definition considers the continuous improvement (CI) of system smartness at any increment, domain and level, rather than the overall system-level performance of everything with unmanageable complexity.

3.2. Functional Requirements (FRs) of SM

We are interested in developing a generic model for the system elements in SM. Therefore, the core values of SM with its new definition are treated as functional requirements (*FRs*). Consequently, a new representation of a generic system element must meet the following *FRs*.

FR1—Performing manufacturing processes: a manufacturing system transforms raw materials into the final products through a series of manufacturing processes in the material flows of systems. Manufacturing processes are performed on physical manufacturing assets. Therefore, making products (*FR1*) is the primary *FR* of a manufacturing system, and it should be defined based on the variants and volumes of products that customers need. In addition, since a complex product involves many parts and components that are made using different manufacturing processes, physical manufacturing assets can be defined for the products, components, parts, and processes. For the purpose of generality, one physical entity is involved in one system element; such a physical entity can be present at any level of the product, from a specific process, part, component, product, product family, or product series over its lifecycle.

FR2—Dealing with changes in customer needs in manufacturing processes: a smart manufacturing system is sustainable; it is capable of reconfiguring itself to deal with the changes in customers' needs, as well as the disturbances and uncertainties in manufacturing environments. Customers' needs correspond to the functionalities, variations, volumes, delivery times, and other expectations of products that change in the market over time. The solutions for a system model to adapt to changes are (1) the software flexibility of adjustable assets, (2) the hardware modularity, which is capable of configuring the system by selecting different modules and assembling them in different ways, and (3) the combination of software and hardware flexibility [43].

FR3—Supporting virtual analyses of manufacturing processes: to perform a manufacturing process successfully, engineers must define, plan, program, verify, validate, control, and monitor the process to ensure it is carried out correctly the first time. With the ever-increasing complexity of manufacturing processes, performing these complex tasks is far beyond manual effort. In SM, digital models are developed as digital twins (DT-I) of the corresponding physical assets; digital twins are utilized for the optimization, simulation, and verification of manufacturing processes, and for controlling, monitoring, and diagnosing manufacturing processes in actual operations.

FR4—Acquiring, processing, and mining data for use in digital models: maintaining a manufacturing system involves numerous decision-making activities at all levels, domains, and aspects of businesses. On the other hand, SM emphasizes responsiveness, adaptability, and resilience to the changes and uncertainties in dynamic environments. Therefore, decision-making activities are closed-loop and data-driven processes that rely on reliable and abundant data about everything in the system. Physical assets, in SM, are intended to acquire real-time data; they are networked to collect and share data, and the collected data is processed, mined, and utilized by digital models for decision making support.

FR5—Making decisions for enhanced system smartness: the decisions for manufacturing operations are made based on the data collected from machines, operators, sensors, suppliers, markets, and users. Due to the rapid growth in networked elements in a system, decision-making activities about manufacturing operations usually involve exceptionally large data sets that are characterized as big data in terms of variety, volume, velocity, veracity, and value (5V). Big data is analyzed and mined to allow smarter decisions to be made regarding manufacturing operations, and to achieve better system performance for any aspects of interest, such as agility, robustness, adaptability, flexibility, and resilience.

FR6—Accessing virtual resources: the scope of manufacturing businesses has been continuously increased due to the growing complexity of products and the need for manufacturing businesses to extend over product lifecycles. Manufacturing systems are highly pressured by having to reconfigure themselves to meet the changes in customer needs over time. Virtual manufacturing resources become increasingly important for the host enterprises to cover the increasing scope of manufacturing businesses [8]. The information infrastructure of an enterprise system (ES) should be capable of accessing virtual resources and supporting the interactions and interoperations of internal and external resources seamlessly.

FR7—Supporting decision making for incorporation-level businesses: in the digital era, manufacturing systems become increasingly distributed and decentralized, and manufacturing businesses within a system become more closely related to the stakeholders, such as suppliers, service providers, logistic systems, and users, across the boundaries of the system. Therefore, sustainable manufacturing operations require numerous decision-making processes at the incorporation-level, such as the selection of suppliers or service providers, the composition of workflows for emerging business opportunities, and the reconfiguration of systems or virtual enterprise alliances to adapt to changing customer needs. Decision making support systems should be able to deal with the incorporation-level big data to support the interactions and interoperations of enterprises with assured security, privacy, and responsiveness [31–35].

3.3. Generic Model of System Elements—Digital Triad (DT-II)

A system can generally be modelled by (1) a set of system elements, (2) the relations between system elements, (3) a series of the transformation from the inputs to the outputs of the system, and (4) a set of the performance metrics that are quantified based on (1), (2), and (3). Modelling a system begins with the representation of the system's elements. A smart manufacturing system is, in fact, a system of systems (*SoS*). SM is a multiplicity of technologies and elements, such as IoTs, CPSs, BDA, ML, BCT, CC, and collaborative robots [62]. Here, the newly developed concept, digital triad (*DT-II*), is used to represent an abstract system element of SM [3,4]. *DT-II* is an extension of the concept of digital twins (DT-I), and *DT-II* includes the enablers to meet the functional requirements *FR1, FR2, FR3, FR4,* and *FR5* that were discussed in Section 3.2. As shown in Figure 2, a digital triad (*DT-II*) is a coalition of life models, digital models, and the corresponding physical models. Accordingly, *DT-II* is modelled by three models and their interactions as,

$$\mathrm{DT-II}(t) = \{L_M(t), D_M(t), P_M(t), C_N(L_M(t), D_M(t), P_M(t), t), E_T(t)\}\ t = t_1, t_2, \ldots t_n, \ldots \quad (1)$$

where *DT-II* (t) represents the state of a digital triad at time t, $L_M(t)$, $D_M(t)$, and $P_M(t)$ are the sets of life, digital, and physical models at t, respectively; $C_N(L_M(t), D_M(t), P_M(t), t)$ are the interactions of life, digital, and physical models at t; $E_T(t)$ is the set of enablers for the operations of *DT-II*; $t_1, t_2, t_3, \ldots$ are timeframes to update *DT-II*.

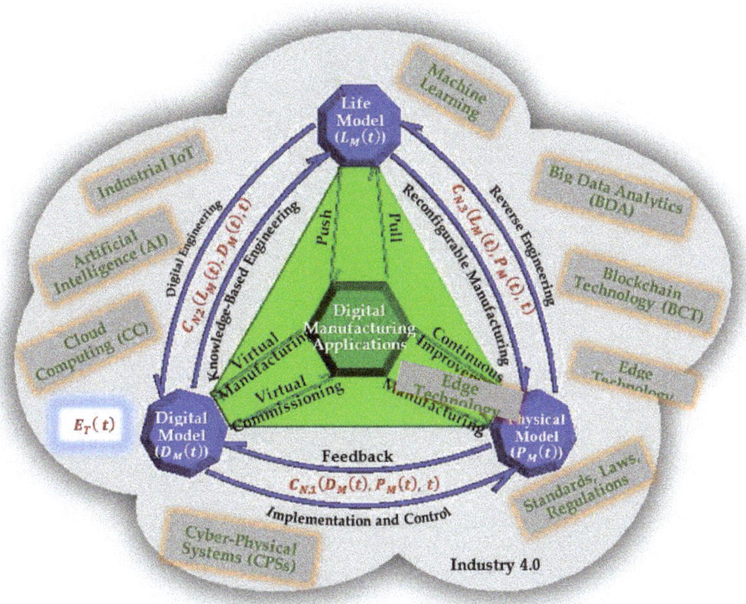

Figure 2. Generic model of digital triad *DT-II* (t).

In a *DT-II* model, a physical model $P_M(t)$ represents one, or a group of, physical thing(s), such as the parts, products, processes, systems, or other tangible things, and a digital model $D_M(t)$ is a virtual model of $P_M(t)$. One physical model $P_M(t)$ may need multiple $D_M(t)$ to represent its behaviors from different perspectives. A life model $L_M(t)$ represents the changes of physical things over time; it keeps historical data, components, templates, and knowledge that are used to design, analyze, optimize, and re-configure the physical models. The interactions of the three models, $P_M(t)$, $D_M(t)$, and $L_M(t)$, are represented by $C_N(L_M(t), D_M(t), P_M(t), t)$. There are three types of interactions $C_N(L_M(t), D_M(t), P_M(t), t)$ in *DT-II*, (i.e., $C_{N,1}(D_M(t), P_M(t))$, $C_{N,2}(L_M(t), D_M(t), t)$, and $C_{N,3}(L_M(t), P_M(t), t)$). In addition, a particular *DT-II* model can be instantiated by specifying a set of enablers $E_T(t)$ to create, operate, and sustain the models. As discussed in Section 2.3, the most commonly mentioned enablers in $E_T(t)$ are IoT, CC, AI, CPS, ML, BDA, BCT, machine learning (ML), reference models, standardizations, and edge technologies.

The proposed *DT-II* serves as a modelling solution to *FR1, FR2, FR3, FR4,* and *FR5* of a smart manufacturing system at the element level as,

$$
\{FR\}_E \qquad \{MR\}_R \qquad \{MS\}_E
$$

$$
\begin{Bmatrix} FR1 \\ FR2 \\ FR3 \\ FR4 \\ FR5 \end{Bmatrix} =
\begin{bmatrix} \times & & & & \times \\ & \times & & & \times \\ & & \times & & \times \\ & & & \times & \times \\ & & \times & \times & \times & \times \end{bmatrix}
\begin{Bmatrix} P_M(t) \\ L_M(t) \\ D_M(t) \\ C_{N,1}(t) \\ C_{N,2}(t) \\ C_{N,3}(t) \\ E_T(t) \end{Bmatrix}
\tag{2}
$$

where $\{FR\}_E$ and $\{MS\}_E$ are the set of functional requirements (FR) and modelling solutions at the element level, respectively; $\{MR\}_E$ is a matrix for mapping from $\{MS\}_E$ and $\{FR\}_E$; "$\times$" and "$\bigcirc$" represent a closely and loosely relevant mapping, respectively.

3.4. Internet of Digital Triad Things (IoDTT) as a Reference Model

A smart manufacturing system is, in fact, a system of systems (*SoS*). Despite the distributed, decentralized, and heterogeneous nature of system elements, an abstract element model based on *DT-II* can only represent the characteristics at the element level, rather than all system characteristics, including the needs for accessing virtual resources and reconfiguring systems over time.

It should be noted that the existing reference models are highly diversified and lack the generality required for users to understand, analyze, engineer, improve, optimize, manage, control, and maintain the systems in specific applications. System models are fundamental to enterprise engineering, integration, and management [63]. A reference SM model should support (1) the integration of the data, knowledge, and wisdom of all stakeholders, from suppliers to users, (2) the evaluation of system options based on high-level performance indicators (KPIs), (3) the adaptation to dynamic changes and uncertainties, (4) the seamless connection of information technology infrastructure, and (5) the affordable cost of modelling, simulation, and data analytics [25]. A reference model traditionally consists of multiple views to look at the transformation of manufacturing businesses from different views. For example, Vernadat [63] included functional, business, organizational, information, infrastructural, product, economic, and collaboration views as the main facets of a reference model. Essakly et al. [64] developed a reference framework to evaluate digitalization solutions for SMEs. The impact of technology adoption was assessed across 16 main fields of action. Moghaddam et al. [65] discussed the existing works on the development of reference SM models; popular reference SM architectures include Reference Architectural Model Industrie 4.0 (RAMI4.0), the Industrial Internet Reference Architecture (IIRA), IBM Industry 4.0, and NIST smart manufacturing. These architectures were service-oriented and the manufacturing assets were digitized and integrated as on-demand services, and SM enabled the collaboration and integration of manufacturing assets via smart plug-and-produce systems. Yang et al. [66] argued that the main objective of SM is to improve the flexibility and adaptability of manufacturing systems; therefore, SM should be data-driven and should support knowledge-based engineering. The reference SM model created by Part and Febriani [67] for welding operations was service-oriented, since virtual manufacturing assets became mandatory for dealing with the ever-growing complexity, scale, and system dynamics.

To model the interaction of system elements in SM, the concept of the Internet of Digital Triad Things (IoDTT), developed by the authors, is introduced here. Figure 3 illustrates a representation of IoDTT for SM. A smart manufacturing system is developed upon an information infrastructure $\{E_T(t)\}$ that consists of all of the accessible tools to acquire, process, mine, and utilize data regarding the system elements for decision making support. IoDTT uses a flat architecture in which all of the system elements $\{DT\text{-}II\ (t)\}$ are networked and interact. Each *DT-II* (t) is reconfigurable to sustain its lifecycle, and the lifecycle of SM is represented as a series of (*SoS* (t) at specific times. A smart manufacturing system, at a specific time t_i, is formed by one host digital triad $DT\text{-}II_h$ (t_i) with selected others. The life model $L_{m,h}$ (t_i) of the host digital triad selects a set of appropriate digital triads and constructs them as a system of systems (*SoS* (t_i)). The digital and physical models, $D_{m,h}$ (t_i) and $P_{m,h}$ (t_i), are upgraded to adapt to the changes resulting from their interactions with others.

A smart manufacturing system in its lifecycle (\{$SoS(t)$\})

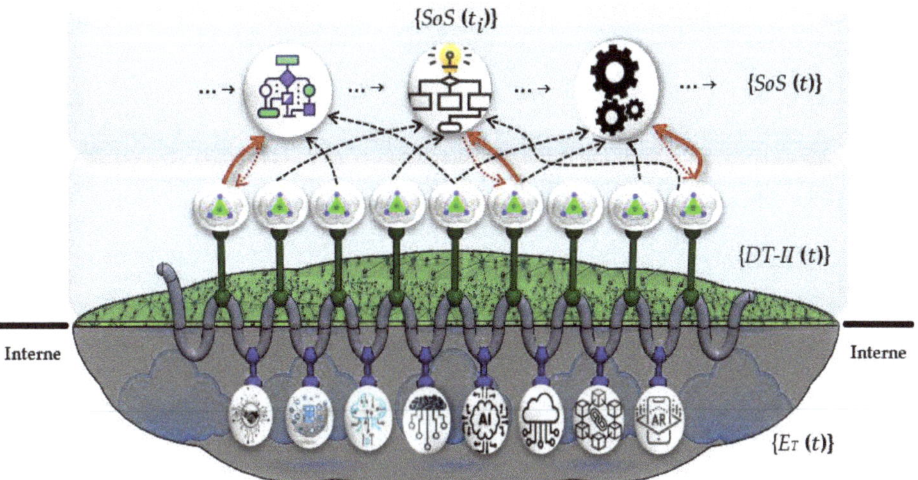

Information Infrastructure (\{$E_T(t)$\} = \{$CPS, CC, BDA, IoT, BCT, AI, ML, AR \dots.$\})

Figure 3. Internet of Digital Triad Things (IoDTT) as reference model for SM.

The proposed IoDTT serves as the modelling solution to *FR6* and *FR7* of a smart manufacturing system at the *SoS* level as,

$$
\begin{Bmatrix} \{FR\}_S \\ \{FR\}_E \\ \{FR\}_N \end{Bmatrix} = \begin{Bmatrix} \{FR\}_S \\ \{FR\}_E \\ FR6 \\ FR7 \end{Bmatrix} = \begin{bmatrix} \{MR\}_S \\ \{MR\}_E & \times \\ & \times & \times \\ & & \times \end{bmatrix} \begin{Bmatrix} \{MS\}_S \\ \{MS\}_E \\ \{SoS(t)\} \\ \{E_T(t)\} \end{Bmatrix} \tag{3}
$$

where $\{FR\}_E$, $\{MR\}_E$, and $\{MS\}_E$ are the functional requirements, mapping relations, and modelling solutions determined in Equation (1) at the element level; $\{FR\}_N = \{FR6, FR7\}^T$ are the functional requirements at the system level; $\{SoS(t)\}$ is a set of system configurations in the system lifecycle; $\{E_T(t)\}$ is the information infrastructure including all of the accessible enabling technologies. "×" and "○" represent a closely and loosely relevant mapping, respectively.

4. Discussion on Generality and Specialty in Designing and Implementing Custom Smart Manufacturing Systems

The designs involved in a modular robotic system in Figure 4 were used to illustrate the applications of an ad hoc approach to designing and implementing a smart manufacturing system, and to elaborate the need for developing a generic design methodology for the design of smart manufacturing systems.

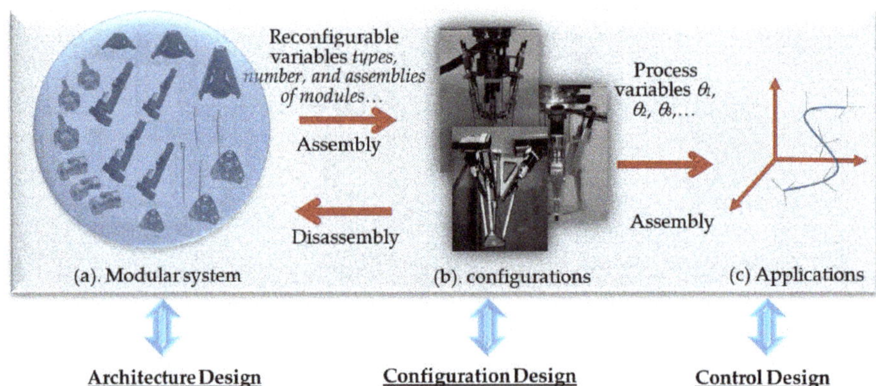

Architecture Design

FRs: Adaptability, reconfigurability, scalability, diagosability, resience

DPs: mechanical, electric, and electronic, sensing…..

Configuration Design

FRs: motion, payload, accuracy….

DPs: modular types, numbers, topology, assemblies and base selections…..

Control Design

FRs: stability, robustness, convergence, delay,…

DPs: joint displacement, velocities, accelerations, torques, uncertainty, disturbances, ….

Figure 4. Example of designing and implementing a smart manufacturing system using an ad hoc approach (Reprinted with permission from ref. [4]. 2021, Springer Nature).

Robots play increasingly critical roles in modern manufacturing [62,68]. However, conventional robots were sophisticated and applicable only to specific types of tasks. Robots in SM should be advanced to reconfigure themselves, both in hardware and software aspects, so that the same system can be applied to different types of tasks in uncertain environments. As shown in Figure 4a, the architecture of a smart robotic system is modularized; it consists of various types of functional modules, rather than an integral robotic structure. Different modules can be selected and assembled in different ways to build different configurations for given tasks over time. Similar to the designs of other complex systems, ad hoc approaches are used in designing and implementing a modular robotic system. The three main design issues are architecture design, configuration design, and control design [69,70]. Architecture design determines the available functional modules and their possible interfaces. The design of a functional module is encapsulated, and the internal change of one module does not affect its interaction with other modules. A system's architecture must offer as many configuration variants as possible, subjected to a given pool of available modules. The more configurations a system can generate, the better the system can deal with changing tasks in a dynamic environment. Architecture design is also involved in upgrading a smart system. Configuration design involves the selection and assembly of modules in a robot instance to optimally perform a given task. An active module has its local controller, an assembled robot has its system-level goals, and the control is designed to coordinate system modules to fulfill system-level tasks satisfactorily. Control design is involved during the phase of system operation.

Several research teams worldwide have contributed to the design theories and methodologies of modular robotic systems. However, every group emphasized the specialties of their proposed systems, and these systems are designed and implemented in ad hoc ways in general. Bi and Lang [69] and Bi et al. [70] discussed the limitations of ad hoc system approaches, including repetitive design efforts, inconsistency in system upgradations, the difficulty in scaling systems, and, most importantly, the lack of predictivity for future technologies to maintain the sustainability of a smart system.

It seems that any complex system, including a smart manufacturing system, should be customized to its specific application; the design of a smart manufacturing system is carried out on a case-by-case basis. However, a smart manufacturing system emphasizes its continuity from one system configuration to another to achieve adaptability and sustainability over time. The authors argue that it is critical to investigate the commonalities of different smart systems, since a general design methodology addressing the identified commonalities will help to significantly alleviate the aforementioned challenges.

It should be noted that any system design can be viewed as a transformation from the given inputs to the expected outputs, and different smart manufacturing systems share many commonalities in such a transformation. While the implementation of a specific system deals with the definitions of customers' needs and the expected outputs that are specialized, many commonalities can be identified. Taking the example of smart manufacturing systems: (1) customer requirements share the commonalities of the quantification of adaptability, resilience, and configurability; (2) the expected system designs share the commonalities of digitization, modularization, and high levels of automation and autonomy; (3) the design transformations share time-dependence, dynamics, concurrence, and continuous improvement. Therefore, the generality of a design methodology and the specialty of system implementation are not conflicted with each other. While a generic design methodology cannot be used to substitute all of the design efforts in defining quantifiable functional requirements, system measures, and modelling system behaviors to evaluate design solutions, following the generic guidance and design procedure will help users to reduce repetitive efforts, achieve consistency for system upgradation, enhance system scalability, and increase the visibility of prosperous technologies for long-term system sustainability.

5. Summary

To promote the application of digital technologies in manufacturing, particularly for small and medium sized companies, and for continuous improvement practice, we aimed to develop a generic methodology for the design of smart manufacturing systems. A concise definition of smart manufacturing (SM) was provided (based on the newly developed concepts of digital triad (DT-II) and Internet of Digital Triad Things (IoDTT), which covers the common requirements and enabling digital technologies of various manufacturing systems that are customized for specific applications. Axiomatic design theory was proposed to formulate system designs, or to reconfigure systems into mathematic models by defining the functional requirements (FRs), identifying the feasible design solutions (DSs), and evaluating system smartness based on the mapping of FRs and DSs. It should be noted that ADT is widely adopted as a systematic design approach in designing complex systems [71,72]. Part II of this paper will provide the details of ADT applications in the design of smart manufacturing, and three case studies will be introduced to demonstrate the generality and applicability of using the proposed method for designing smart manufacturing systems. The proposed concepts have theoretical and practical significance for exploring the essentials of different smart manufacturing systems, so that a systematic design methodology can be developed to guide the design of smart manufacturing systems with diversified applications. It should be noted that the proposed concepts and design methods are generic, and are used as the systematic guides in designing and analyzing a smart manufacturing system. However, every manufacturing system must be customized to its specific applications, and additional design effort is required to prioritize FRs, identify the pool of DSs, and develop performance metrics for the appropriate mapping of FRs and DSs when using the proposed concepts and design methods.

Author Contributions: Conceptualization, Z.B. and W.-J.Z.; methodology, Z.B. and C.W.; background study, C.L., L.X. and Z.B.; formal analysis, W.-J.Z., C.L., L.X.; investigation, Z.B., W.-J.Z., C.W., C.L. and L.X.; writing—original draft preparation, Z.B.; writing—review and editing, W.-J.Z., C.W., C.L. and L.X. All authors have contributed substantially to the report and they have read and agreed to the published version of the manuscript.

Funding: This research received no external funding.

Institutional Review Board Statement: Not applicable.

Informed Consent Statement: Not applicable.

Data Availability Statement: Not applicable.

Conflicts of Interest: The authors declare no conflict of interest.

List of Abbreviations

5V:	Variety, Volume, Velocity, Veracity, and Value
ADT:	Axiomatic Design Theory
AI:	Artificial Intelligence
AM:	Agile Manufacturing
AM:	Additive Manufacturing
AR:	Augmented Reality
BCT:	Blockchain Technology
BDA:	Big Data Analytics
BPM:	Business Process Management
CAD:	Computer Aided Design
CADM:	computer Aided dDesign and Manufacturing
CAE:	Computer Aided Engineering
CAM:	Computer Aided Manufacturing
CC:	Cloud Computing
CESMII	Clean Energy Smart Manufacturing Innovation Institute
CI:	Continuous Improvement
CIM:	Computer Integrated Manufacturing
CNC:	Computer Numerical Control
DM:	Digital Manufacturing
DSs:	Design Solutions
ERP:	Enterprise Resource Planning
ES:	Enterprise Systems
GT:	Group Technologies
HCPS:	Human-Cyber Physical Systems
HRI:	Human-Robot Interactions
IDM:	Intelligent Digital Mesh
IEEE:	Institute of Electrical and Electronics Engineers
IIoT:	Industrial Internet of Things
IIRA:	Industrial Internet Reference Architecture
IM:	Intelligent Manufacturing
IOS:	International Organization for Standardization
ITU;	International Telecommunication Union
KPIs:	Key Performance Indicators
LP:	Lean Production
ML:	Machine Learning
MRP:	Material Resource Planning
NC:	Numerical Controls
NIST:	National Institute of Standards and Technology
NSF:	National Science Foundation
PDM:	Product Data Management
PLM:	Product Lifecycle Management
QC:	Quality Controls
RAMI:	Reference Architectural Model Industrie
RFID:	Radio Frequency IDentification
SMEs:	Small to Midsize Enterprises
SoA:	Service-oriented Architecture
SoS:	System of Systems
TQM:	Total Quality Management

References

1. Bi, Z.M. Embracing Internet of things (IoT) and big data for industrial informatics. *Enterp. Inf. Syst.* **2017**, *11*, 949–951. [CrossRef]
2. Bi, Z.M. Revisiting system paradigms from the viewpoint of manufacturing sustainability. *Sustainability* **2011**, *3*, 1323–1340. [CrossRef]
3. Bi, Z.M.; Zhang, W.J. *Chapter 1: Human Civilization, Products, and Manufacturing. Practical Guide to Digital Manufacturing—First-Time-Right from Digital Twin to Physical Twin*; Springer International Publishing: Cham, Switzerland, 2021; ISBN 978-3-030-70303-5.
4. Bi, Z.M.; Zhang, W.J. Chapter 2: Computer-Aided Design. In *Practical Guide to Digital Manufacturing—First-Time-Right from Digital Twin to Physical Twin*; Springer International Publishing: Cham, Switzerland, 2021; ISBN 978-3-030-70303-5.
5. Blomeke, S.; Richert, J.; Mennenga, M.; Thiede, S.; Spengler, T.S.; Herrmann, C. Recycling 4.0—mapping smart manufacturing solutions to remanufacturing and recycling operations. *Procedia CIRP* **2020**, *90*, 600–605. [CrossRef]
6. Baroroh, D.K.; Chu, C.H.; Wang, L. Systematic literature review on augmented reality in smart manufacturing: Collaboration between human and computational intelligence. *J. Manuf. Syst.* **2020**. [CrossRef]
7. Bi, Z.M.; Lang, S.Y.T.; Shen, W.M. Reconfigurable manufacturing systems: The state of the art. *Int. J. Prod. Res.* **2008**, *46*, 967–992. [CrossRef]
8. Bi, Z.M.; Xu, L.D.; Wang, C. Internet of things for enterprise systems of modern manufacturing. *IEEE Trans. Industr. Inform.* **2014**, *10*, 1537–1546.
9. Guchhait, R.; Pareek, S.; Sarkar, B. How does a radio frequency identification optimize the profit in an unreliable supply chain management? *Mathematics* **2019**, *7*, 490. [CrossRef]
10. Sardar, S.K.; Sarkar, B.; Kim, B. Integrating machine learning, radio frequency identification, and consignment policy for reducing unreliability in smart supply chain management. *Processes* **2021**, *9*, 247. [CrossRef]
11. Ullah, M.; Sarkar, B. Recovery-channel selection in a hybrid manufacturing-remanufacturing production model with RFID and production quality. *Int. J. Prod. Econ.* **2020**, *219*, 360–374. [CrossRef]
12. Dey, B.K.; Pareek, S.; Tayyab, M.; Sarkar, B. Autonomation policy to control work-in-process inventory in a smart production system. *Int. J. Prod. Res.* **2020**, *59*, 1258–1280. [CrossRef]
13. Dey, B.K.; Bhuniys, S.; Sarkar, B. Involvement of controllable lead time and variable demand for a smart manufacturing under a supply chain management. *Expert Syst. Appl.* **2021**, *184*, 115464. [CrossRef]
14. Ghobakhloo, M. Industry 4.0, digitization, and opportunities for sustainability. *J. Clean. Prod.* **2020**, *252*, 119869. [CrossRef]
15. Ghobakhloo, M.; Ching, N.T. Adopting of digital technologies of smart manufacturing in SMEs. *J. Ind. Inf. Integr.* **2019**, *16*, 100107.
16. Meindl, B.; Ayala, N.F.; Mendonca, J.; Frank, A.G. The four smarts of Industry 4.0: Evolution of ten years of research future perspectives. *Technol. Forecast. Soc. Change* **2021**, *168*, 120784. [CrossRef]
17. Deloitte. The Smart Factory: Responsive, Adaptive, Connected Manufacturing. 2021. Available online: https://www2.deloitte.com/content/dam/insights/us/articles/4051_The-smart-factory/DUP_The-smart-factory.pdf (accessed on 22 September 2021).
18. Li, Q.; Tang, Q.; Chan, I.; Wei, H.; Pu, Y.; Jiang, H.; Li, J.; Zhou, J. Smart manufacturing standardization: Architectures, reference models and standards framework. *Comput. Ind.* **2018**, *101*, 91–106. [CrossRef]
19. Manufacturing Operations Management. A Brief History of Smart Manufacturing. 2021. Available online: https://www.manufacturing-operations-management.com/manufacturing/smart-manufacturing/ (accessed on 22 September 2021).
20. Lu, Y.; Morris, K.C.; Frechette, S. *Current Standards Landscape for Smart Manufacturing Systems, NIST Interagency/Internal Report (NISTIR)*; National Institute of Standards and Technology: Gaithersburg, MD, USA, 2016. [CrossRef]
21. Lu, Y.; Witherell, P.; Jones, A. Standard connections for IIoT empowered smart manufacturing. *Manuf. Lett.* **2020**, *26*, 17–20. [CrossRef] [PubMed]
22. SMLC. Implementing 21st Century Smart Manufacturing—Workshop Summary Report. 2011. Available online: https://www.controlglobal.com/assets/11wppdf/110621_smlc-smart-manufacturing.pdf (accessed on 22 September 2021).
23. dek/din. The German Standardization Roadmap Smart City. 2014. Available online: https://www.dke.de/resource/blob/778248/d2afdaf62551586a54b3270ef78d2632/the-german-standardization-roadmap-smart-city-version-1-0-data.pdf (accessed on 22 September 2021).
24. US Government. Made in China 2025 and the Future of American Industry: Hearing before the Committee on Small Business and Entrepreneurship United States Senate. 2019. Available online: https://www.govinfo.gov/content/pkg/CHRG-116shrg35699/pdf/CHRG-116shrg35699.pdf (accessed on 22 September 2021).
25. Davis, J.F.; Edgar, T.F.; Porter, J.; Bernaden, J.; Sarli, M. Smart manufacturing, manufacturing intelligence and demand-dynamic performance. *Comput. Chem. Eng.* **2012**, *47*, 145–156. [CrossRef]
26. Ahuett-Garza, H.; Kurfess, T. A brief discussion on the trends of habilitating technologies for industry 4.0 and smart manufacturing. *Manuf. Lett.* **2018**, *15*, 60–63. [CrossRef]
27. Romero, M.; Guedria, W.; Tanetto, H.; Barafort, B. Towards a characterization of smart systems: A systematic literature review. *Comput. Ind.* **2018**, *120*, 103234.
28. Lenz, J.; MacDonald, E.; Harik, R.; Wuest, T. Optimizing smart manufacturing systems by extending the smart products paradigm to beginning of life. *J. Manuf. Syst.* **2020**, *57*, 274–286. [CrossRef]
29. Filleti, R.A.P.; Silva, D.A.L.; da Silva, E.J.; Ometto, A.R. Productive and environmental performance indicators analysis by a combined LCA hybrid model and real-time manufacturing process monitoring: A grinding unit process application. *J. Clean. Prod.* **2017**, *161*, 510–523. [CrossRef]

30. Maggi, F.; Balduzzi, M.; Vosseler, R.; Rosler, M.; Quadrini, W.; Tavola, G.; Pogliani, M.; Quarta, D.; Zanero, S. Smart factory security: A case study on a modular smart manufacturing system. *Procedia Comput. Sci.* **2021**, *180*, 665–675. [CrossRef]

31. Viriyasitava, W.; Xu, L.; Bi, Z.M. Specification patterns of service-based applications using blockchain technology. *IEEE Trans. Comput. Soc. Syst.* **2020**, *7*, 886–896. [CrossRef]

32. Viriyasitava, W.; Xu, L.; Bi, Z.M. Sapsomboon, A. Blockchain-based business process management (BPM) framework for service composition in Industry 4.0. *J. Intell. Manuf.* **2020**, *31*, 1737–1748. [CrossRef]

33. Viriyasitava, W.; Xu, L.; Bi, Z.M.; Hoonsopon, D. Blockchain technology for applications in Internet of Things—mapping from system design perspective. *IEEE Internet Things J.* **2019**, *6*, 8155–8168. [CrossRef]

34. Viriyasitava, W.; Xu, L.; Bi, Z.M. Blockchain and Internet of Things for modern business process in digital economy—the state of the art. *IEEE Trans. Comput. Soc. Syst.* **2019**, *6*, 1420–1432. [CrossRef]

35. Viriyasitava, W.; Xu, L.; Bi, Z.M.; Hoonsopon, D.; Charoenruk, N. Managing QoS of Internet-of-Thing services using blockchain. *IEEE Trans. Comput. Soc. Syst.* **2019**, *6*, 1357–1368. [CrossRef]

36. Felice, F.; Petrillo, A.; Zomparelli, F. A bibliometric multicriteria model on smart manufacturing from 2011 to 2018. *IFAC PapersOnLine* **2018**, *51–11*, 1643–1648. [CrossRef]

37. Mittal, S.; Khan, M.A.; Romero, D.; Wuest, T. A critical review of smart manufacturing& Industry 4.0 maturity models: Implication for smart and medium-sized enterprises (SMEs). *J. Manuf. Syst.* **2018**, *49*, 194–214.

38. Mittal, S.; Khan, M.A.; Romero, D.; Wuest, T. Building blocks for adopting smart manufacturing. *Procedia Manuf.* **2019**, *34*, 978–985. [CrossRef]

39. Kusiak, A. Fundamentals of smart manufacturing: A multi-thread perspective. *Annu. Rev. Control* **2019**, *47*, 214–220. [CrossRef]

40. Byrne, G.; Damm, O.; Monostori, L.; Teti, R.; van Houten, F.; Wegener, K.; Wertheim, R.; Sammler, F. Towards high performance living manufacturing systems—A new convergence between biology and engineering. *CIRP J. Manuf. Sci. Technol.* **2021**. [CrossRef]

41. Wang, B.; Tao, F.; Fang, X.; Liu, C.; Liu, Y.; Friheit, T. Smart manufacturing and intelligence manufacturing: A comparative review. *Engineering* **2020**, *7*, 738–757. [CrossRef]

42. Kulvatunyou, B.; Oh, H.; Ivezic, N.; Nieman, S.T. Standards-based semantic integration of manufacturing information: Past, present and future. *J Manuf Syst.* **2020**, *52*, 184–197. [CrossRef] [PubMed]

43. Bi, Z.M. On adaptive robot system for manufacturing applications. Ph.D. Thesis, University of Saskatchewan, Saskatoon, SK, Canada, 2002.

44. Brad, S.; Murar, M. Employing smart units and servitization towards reconfigurability of manufacturing processes. *Procedia CIRP* **2015**, *30*, 498–503. [CrossRef]

45. Feldner, B.; Herber, P. A qualitative evolution of IPv6 for the industrial Internet of things. *Procedia Comput. Sci.* **2018**, *134*, 377–384. [CrossRef]

46. Kumar, A. Methods and materials for smart manufacturing: Additive manufacturing, Internet of things, flexible sensors and soft robotics. *Manuf. Lett.* **2018**, *15*, 122–125. [CrossRef]

47. Barletta, I.; Despeisse, M.; Johansson, B. The proposal of an environmental break-event point as assessment method of product-service systems for circular economy. *Procedia CIRP* **2018**, *72*, 720–725. [CrossRef]

48. Jaspert, D.; Ebel, M.; Eckardt, A.; Poeppelbuss, J. Smart retrofitting in manufacturing: A systematic review. *J. Clean. Prod.* **2021**, *312*, 127555. [CrossRef]

49. Zenisek, J.; Wild, N.; Wolffartsberger, J. Investigating the potential of smart manufacturing technologies. *Procedia Comput. Sci.* **2021**, *180*, 507–516. [CrossRef]

50. Rodger, J.M.; Bey, N.; Alting, L.; Hauschild, M.Z. Life cycle targets applied in highly automated car body manufacturing—method and algorithm. *J. Clean. Prod.* **2018**, *194*, 786–799. [CrossRef]

51. Siiskonen, M.; Malmqvist, J.; Folestad, S. Integrated product and manufacturing system platforms supporting the design of personalized medicines. *J. Manuf. Syst.* **2020**, *56*, 281–295. [CrossRef]

52. Kamble, S.S.; Gunasekaran, A.; Ghadge, A.; Raut, R. A performance measurement system for industry 4.0 enabled smart manufacturing system in SMMEs—a review and empirical investigation. *Int. J. Prod. Econ.* **2020**, *229*, 107853. [CrossRef]

53. Phuyal, S.; Bista, D.; Bista, R. Challenges, opportunities and future directions of smart manufacturing: A state of art review. *Sustain. Future* **2020**, *2*, 100023. [CrossRef]

54. Uysal, M.P.; Mergen, A.E. Smart manufacturing in intelligent digital mesh: Integration of enterprise architecture and software product line engineering. *J. Ind. Inf. Integr.* **2021**, *22*, 100202.

55. Osterrieder, P.; Budde, L.; Friedli, T. The smart factory as a key construct of industry 4.0: A systematic literature review. *Int. J. Prod. Econ.* **2020**, *221*, 107476. [CrossRef]

56. Xia, Q.; Jiang, C.; Yang, C.; Zheng, X.; Pan, X.; Shuai, Y.; Yuan, S. A method towards smart manufacturing capabilities and performance measurement. *Procedia Manuf.* **2019**, *39*, 851–858. [CrossRef]

57. Helu, M.; Morris, K.; Jung, K.; Lyons, K.; Leong, S. Identifying performance assurance challenges for smart manufacturing. *Manuf. Lett.* **2015**, *6*, 1–4. [CrossRef]

58. Ramos, A.R.; Ferreira, J.C.E.; Kumar, V.; Garza-Reves, L.A.; Cherrafi, A. A lean and cleaner production benchmarking method for sustainability assessment: A study of manufacturing companies in Brazil. *J. Clean. Prod.* **2018**, *177*, 218–231. [CrossRef]

59. Manoharan, S.; Haapala, K.R. A grey box software framework for sustainability assessment of composed manufacturing processes: A hybrid manufacturing case. *Procedia CIRP* **2019**, *80*, 440–445. [CrossRef]

60. Mahmood, K.; Otto, T.; Golova, J.; Kangru, T.; Kuts, V. An approach to analyze the performance of advanced manufacturing environment. *Procedia CIRP* **2020**, *93*, 628–633. [CrossRef]

61. Mahmoud, A.M.; Grace, J. A generic evaluation framework of smart manufacturing systems. *Procedia Comput. Sci.* **2019**, *161*, 1292–1299. [CrossRef]

62. Arden, N.S.; Fisher, A.C.; Tyner, K.; Yu, L.X.; Lee, S.I.; Kpacha, M. Industry 4.0 for pharmaceutical manufacturing: Preparing for the smart factories o the future. *Int. J. Pharm.* **2020**, *602*, 120554. [CrossRef] [PubMed]

63. Vernadat, F. Enterprise modelling: Research review and outlook. *Comput. Ind.* **2020**, *122*, 103265. [CrossRef]

64. Essakly, A.; Wichmann, M.; Spengler, T.S. A reference framework for the holistic evaluation of industry 4.0 solutions for small and medium-sized enterprises. *IFAC PaperOnLine* **2019**, *52–13*, 427–432. [CrossRef]

65. Moghaddam, M.; Cadavid, M.N.; Kenley, C.R.; Deshmukh, A.V. Reference architectures for smart manufacturing: A critical review. *J. Manuf. Syst.* **2018**, *49*, 215–225. [CrossRef]

66. Yang, S.; Navarathna, P.; Ghosh, S.; Bequette, B.W. Hybrid modeling in the era of smart manufacturing. *Comput. Chem. Eng.* **2020**, *140*, 106874. [CrossRef]

67. Part, H.S.; Febriani, R.A. Modeling a platform for smart manufacturing system. *Procedia Manuf.* **2019**, *38*, 1660–1667.

68. Gualtieri, L.; Rauch, E.; Vidoni, R. Emerging research fields in safety and ergonomics in industrial collaborative robotics: A systematic literature review. *Robot Comput. Integr. Manuf.* **2021**, *67*, 101998. [CrossRef]

69. Bi, Z.M.; Lang, S.Y.T.; Zhang, D.; Orban, P.E.; Verner, M. Integrated design toolbox for tripod based parallel kinematic machines. *ASME J. Mech. Design* **2007**, *129*, 799–807. [CrossRef]

70. Bi, A.M.; Lang, S.Y.T. Kinematic and dynamic models of a tripod system with a passive leg. *IEEE ASME Trans. Mechatron.* **2006**, *11*, 108–111. [CrossRef]

71. Bi, Z.M.; Jin, Y.; Maropoulos, P.; Zhang, W.J.; Wang, L.H. Internet of things (IOT) and big data analytics for digital manufacturing (DM). *Int. J. Prod. Res.* **2021**. [CrossRef]

72. Bi, Z.M.; Wang, G.; Thompson, J.; Ruiz, D.; Rosswurm, J.; Roof, S.; Guandique, C. System framework of adopting additive manufacturing in mass production line. *Enterp. Inf. Syst.* **2021**, 1–24. [CrossRef]

Article

Generic Design Methodology for Smart Manufacturing Systems from a Practical Perspective. Part II—Systematic Designs of Smart Manufacturing Systems

Zhuming Bi [1,*], Wen-Jun Zhang [2], Chong Wu [3,*], Chaomin Luo [4] and Lida Xu [5]

[1] Department of Civil and Mechanical Engineering, Purdue University Fort Wayne, Fort Wayne, IN 46805, USA
[2] Department of Mechanical Engineering, University of Saskatchewan, Saskatoon, SK S7N 5A9, Canada; Chris.Zhang@usask.ca
[3] School of Economics and Management, Harbin Institute of Technology, Harbin 150001, China
[4] Department of Electrical & Computer Engineering, Mississippi State University, Starkville, MS 39762, USA; chaomin.luo@ece.msstate.edu
[5] Department of Information Technology, Old Dominion University, Norfolk, VA 23529, USA; lxu@odu.edu
* Correspondence: biz@pfw.edu (Z.B.); wuchong@hit.edu.cn (C.W.)

Citation: Bi, Z.; Zhang, W.-J.; Wu, C.; Luo, C.; Xu, L. Generic Design Methodology for Smart Manufacturing Systems from a Practical Perspective. Part II—Systematic Designs of Smart Manufacturing Systems. *Machines* **2021**, *9*, 208. https://doi.org/10.3390/machines9100208

Academic Editor: Angelos P. Markopoulos

Received: 26 July 2021
Accepted: 7 September 2021
Published: 23 September 2021

Publisher's Note: MDPI stays neutral with regard to jurisdictional claims in published maps and institutional affiliations.

Abstract: In a traditional system paradigm, an enterprise reference model provides the guide for practitioners to select manufacturing elements, configure elements into a manufacturing system, and model system options for evaluation and comparison of system solutions against given performance metrics. However, a smart manufacturing system aims to reconfigure different systems in achieving high-level smartness in its system lifecycle; moreover, each smart system is customized in terms of the constraints of manufacturing resources and the prioritized performance metrics to achieve system smartness. Few works were found on the development of systematic methodologies for the design of smart manufacturing systems. The novel contributions of the presented work are at two aspects: (1) unified definitions of digital functional elements and manufacturing systems have been proposed; they are generalized to have all digitized characteristics and they are customizable to any manufacturing system with specified manufacturing resources and goals of smartness and (2) a systematic design methodology has been proposed; it can serve as the guide for designs of smart manufacturing systems in specified applications. The presented work consists of two separated parts. In the first part of paper, a simplified definition of smart manufacturing (SM) is proposed to unify the diversified expectations and a newly developed concept digital triad (DT-II) is adopted to define a generic reference model to represent essential features of smart manufacturing systems. In the second part of the paper, the axiomatic design theory (ADT) is adopted and expanded as the generic design methodology for design, analysis, and assessment of smart manufacturing systems. Three case studies are reviewed to illustrate the applications of the proposed methodology, and the future research directions towards smart manufacturing are discussed as a summary in the second part.

Keywords: smart manufacturing; information technologies (IT); system of systems (SoS); digital manufacturing (DM); digital twins (DT-I); digital triad (DT-II); cyber-physical systems; Internet of Things (IoT); Internet of Digital Triad Things (IoDTT); big data analytics (BDA); cloud computing (CC); axiomatic design theory (ADT)

1. Introduction

Smart manufacturing (SM) has been identified as one of the prioritized areas to strengthen a nation's economy in both developed and developing countries. The studies in smart manufacturing technologies have attracted a great deal of attention from researchers in multiple disciplines. On the one hand, SM is a comprehensive solution to manufacturing systems with the integration of recent information technologies (IT); on the other hand, every smart manufacturing system is customized to the needs of a specific enterprise with limited resources and its own interests of business domains, strategies, and performance metrics. This leads to a high diversity for developers and users to understand the true meanings of SM, and accordingly, to select and integrate existing technologies adequately in the context of given circumstances of manufacturing businesses. To fill this gap, this paper attempts to unify the definition of SM with a newly proposed concept called digital triad (DT-II) to cover all common features of digital technologies towards system smartness; the contents of system smartness can be tailored to the needs of specific companies especially in terms of flexibility, scalability, adaptability, and resilience. Moreover, the concept of Internet of Digital Triad Things (IoDTT) is proposed as a system reference model to deal with the integrations of digital solutions at upper levels. The rationales of DT-II and IoDTT have been elaborated in the first part of the paper [1]. In the second part here, these concepts will be used to develop a systematic methodology for designs of smart manufacturing systems. To this end, the rest of the paper is organized as follow. In Section 2, the design of a smart manufacturing system is formulated to define a design space for the discussed functional requirements (FRs) of SM [1]; the design space consists of commonly adopted digital technologies including DT-I, CPS, IoT, CC, AI, VM, BDA, SoA, and BCT. In Section 3, the methods for the evaluation and comparison of design solutions (DS) are discussed, and the focus is put on the quantification as well as system performance indicators in dealing with the changes and uncertainties in dynamic business environments. In Section 4, axiomatic design theory (ADT) is adopted as a systematic methodology in designing a smart manufacturing system; a general design procedure is proposed with the detailed discussions of customizing FRs and DSs to specific applications. In Section 5, three case studies are introduced to illustrate how ADT can be applied in customizing a smart manufacturing systems when the performance indicators for system smartness are given. In Section 6, the innovation of the presented work is summarized for its theoretical and practical significance and the authors' future works in advancing the concepts of DT-II, IoDDT, and the ADT-based design methodology for smart manufacturing systems are outlined.

2. Design of Smart Manufacturing Systems

Since a smart manufacturing system aims for system adaptability in dealing with changes and uncertainties, and such a capability is achieved by either the flexibility of system elements or the configurability at system level, design of a smart manufacturing system involves three iterative phases in its system lifecycle, i.e., system design phase, system operation phase, and system reconfiguration phase, as shown in Figure 1. At the system design phase, a set of functional requirements (FRs) is defined, available physical assets and accessible virtual assets are considered to define a design space with all feasible design solutions (DSs), and system performances are ranked to define a set of prioritized performance metrics. At the system operation phase, design analysis and synthesis are performed to optimize a system, and thus implement it in application; all smart things in the system are monitored to determine if system elements or the whole system has to be reconfigured to meet the identified changes. At the system reconfiguration phase, either system elements or the whole system is reconfigured to make the smart manufacturing system sustainable.

Many researchers have investigated the design methodologies of intelligent manufacturing systems. Unglert et al. [2] proposed a computational design synthesis to analyze reconfigurable manufacturing cells where functional modules were reorganized to balance the capability and capacity of system. It was used in the concurrent design of system configurations. Kurgan et al. [3] proposed an integrated design methodology and considered manufacturing requirements at the phase of system design; it led to the cost saving of 18% and time reduction of 17% in the case study. SM is a type of reconfigurable systems that is sustainable over time. A system configuration of a reconfigurable system consists of a set of functional building blocks that are selected and assembled from available physical and virtual manufacturing assets for specific tasks [4–7]. A reconfigurable system allows the additions, removals, and modifications of functional modules without affecting the functions of other modules, and this helps scale the capacity of productions [8]. A reconfigurable system is characterized by its modularity, integrability, customization, convertibility, scalability, diagnosability, mobility, and adaptability. The high-level building blocks of SM were classified by Mittal et al. [9], and the most commonly used ones were intelligent controls, data-driven production managements, data analytics, smart products, smart materials, interoperability, data sharing, and standards.

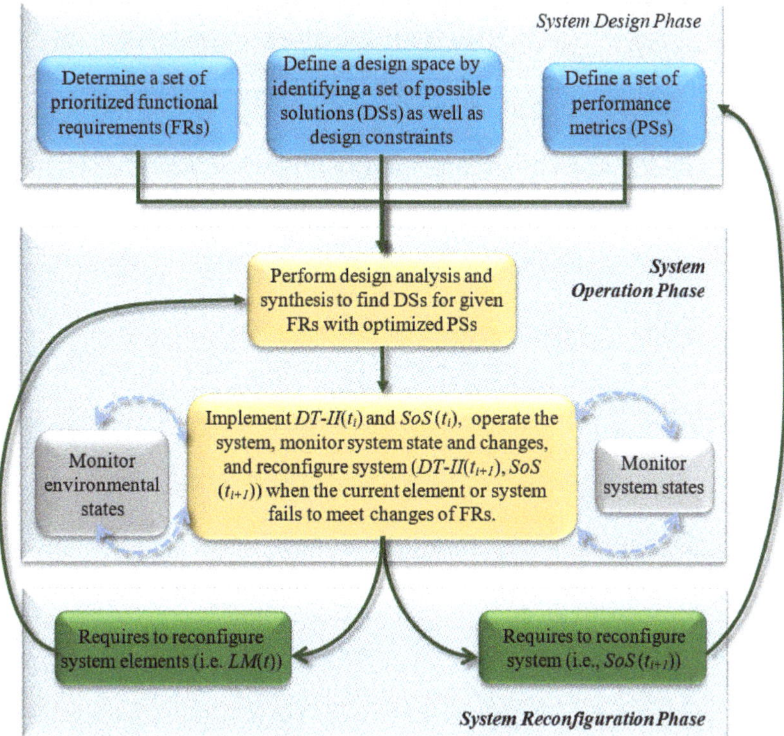

Figure 1. Three phases in design of a smart manufacturing system.

Here, the axiomatic design theory (ADT) is used to describe the design procedure of smart manufacturing systems. However, the authors' main interests are (1) the determinations of functional requirements (FRs) and design spaces by the solutions, and (2) the evaluation of the mappings from design solutions (DSs) to functional requirements (FRs). How to decompose FRs to meet the independence principle and minimized information principle is of less interest in this section. As the matter of fact, for a complex and multidisciplinary system, every system element is conceptually coupled with others, which makes it impractical to achieve the independence of functional requirements in decomposition. Since FRs have been discussed in part I, we discuss common digital technologies as design solutions (DSs) and performance evaluations (PSs) for system smartness as follows.

A smart manufacturing system has been modelled as an instance of the Internet of Digital Triad Things (IoDTT), and the system consists of a set of networked digital triads. Therefore, various digital technologies are the essential enablers to satisfy the functional requirements of smart manufacturing systems. Here, commonly used digital technologies as well as the roles in SM are discussed.

2.1. Digital Technologies for Smart Materials and Processes

Similar to traditional manufacturing, SM aims to make and deliver products to users However, products from SM are generally more advanced in terms of their intelligence level, functionalities, and more importantly, their sustainability from the perspectives of economy, environment, and human society. Akrivos et al. [10] argued that future products should be sustainable, and products from SM should be made to enhance their sustainability, especially the self-healing properties of materials, products, and systems A smart product could sense occurred damage and heal the damage autonomously to sustain the product life, so that the product life is extended, even lasting permanently. The most advanced digital technologies such as extreme ultraviolet lithography (EUV) [11] and 4D or 5D printing [12] became commercially available to prepare raw materials for smart products.

2.2. Digital Twins (DT-I)

To practice the first-time right in decision-making processes, decisions should be verified and validated before they are executed in the physical world. Modelling and simulation tools used to be standalone system that were widely applied to predict system behaviors based on assumed conditions, changes, disturbances, and randomness [13] DT-I has been advanced from traditional modelling and simulation approaches in the sense that physical and virtual models are connected and interacted directly. The concept of digital twins (DT-I) has enabled the interactions and integrations of information and physical worlds in real-time manners. For example, augmented reality was adopted specifically for the interaction interfaces in smart manufacturing [14]. DT-I could be multidimensional: for example, geometric, physical capability, rule, and behavior models in the five-dimensional model [15]. DT-I could be expanded to represent reconfigurable systems at different granularities and for corresponding missions. In a recent literature review by Semeraro et al. [16], DI-I itself was treated as a system paradigm where virtual models were embedded as indispensable components to model the behaviors of physical components and make smart decisions for system operations.

2.3. Cyber-Physical Systems (CPS)

CPS is related closely to DT-II; while the former focuses on the interactions of digital and physical twins in the cyber and physical worlds, respectively. CPS supports real-time communication and interaction of cyber and physical systems to close control loops, monitor system states in real-time, and adjust system behaviors promptly when the changes are detected [17,18]. The CPS-based system elements are applicable to many tasks such as communication, controls, scalability, validation and verifications, and system managements [19,20]. CPS can be data-driven to optimize system controls using real-time data collected from the physical world; CPS makes a manufacturing system smart by improving the responsiveness, adaptability, and predictivity of system elements and facilitating the collaboration of stakeholders in the entire process of mass customization of products [21].

2.4. Human Cyber-Physical Systems (HCPS)

Due to a high-level of uncertainty and complexity, many fully automated solutions involve in a high cost, and such solutions become impractical due to limited cost-effectiveness. Therefore, humans still play indispensable roles in making manufacturing systems flexible and adaptable. D'Addona et al. [22] analyzed the needs of human operations for cognition in practicing adaptive automation through case studies. Cognition was utilized to respond out-of-the-loop conditions such as abnormal transitions, and skill loss, automation-induced errors, adaptable behaviors, and inappropriate trusts in collaborative manufacturing processes. Humans must be in the loop to achieve the desired performances of production. In such a way, the safety and comfortableness of workers were well balanced in uncertain environments. Humans are integrated with CPS and human cyber-physical systems (HCPS). Enabling technologies such as augmented reality (AR) are critical to support human-machine collaborations and interactions. Baroroh et al. [23] showed that AR were used in SM in implementing interactions, manufacturing processes, machine functions, and knowledge explorations.

2.5. Internet of Things (IoT)

IoT seems essential to SM due to two main reasons: (1) SM is data-driven, and real-time data about smart things, stakeholders, and business environments must be available to support the decision-making processes at any level and scope of manufacturing businesses and (2) SM requires the access to virtual resources over the Internet to deal with manufacturing businesses over product lifecycles. SM adopts IoT as the information infrastructure to network manufacturing assets such as products, parts, machine tools, sensors, and decision-making units in the heterogeneous environment. IoT allows a smart manufacturing system to sense system elements and environment, access virtual services over the Internet, and provide abundant data to drive decision-making processes at all levels and scopes of manufacturing businesses in system lifecycles [24–26]. IoT consists of countless smart things that are built over the Internet with wired and wireless communications [27,28].

2.6. Cloud Computing (CC)

Modern manufacturing systems tend to be decentralized and distributed, while decision-making activities for smart manufacturing are facing a number of challenges, such as (1) sharing and accessing data in distributed environments, (2) storing and maintaining an ever-increasing amount of data in the network, (3) the high demand of computing to optimize decisions with the limited local computing resources, and (4) the complexity of coordination, collaboration, and interoperation of manufacturing resources over the Internet. Cloud computing (CC) is built upon service-oriented architecture (SoA); every task that manipulates data, accesses virtual resource, or run a model for decision-making can be treated as a service (XaaS), and any system element with a limited computing capability can utilize CC to support its decision-making activities [29] To select services to meet manufacturing needs in cloud manufacturing, Huang and Wu [30] developed a two-layer trust fuzzy model which consisted of time, cost, availability, reliability, and safety at the first-layer and 13 other indexes at the second-layer.

2.7. Artificial Intelligence (AI)

In making a decision for a manufacturing business, it is an ideal scenario that (1) an explicit mathematic model is available to represent the relations of inputs, outputs, and system parameters, (2) all of the required data of the decision model are available and accurate when a decision is made, and (3) the computation is manageable by the system to reach decisions in time. Unfortunately, most of decision-making processes are too complex to develop explicit mathematic models, and the required data in decision-making are often incomplete, ambiguous, and not free of error. Artificial intelligence (AI) is a cognitive science for data mining and decision-making support in the fields of data analytics, image processes, robotics, natural language processes, and machine learning. AI became the frontier of manufacturing technologies in future industrial systems, and was integrated with industrial IoT (IIoT), BDA, CC, and CPS to support industrial operations in an efficient, flexible, and sustainable way [31].

2.8. Virtual Manufacturing (VM)

Utilizing virtual assets over the Internet transferred certain manufacturing businesses into services, and, accordingly, the manufacturing system became a product-service system. This leveraged the flexibility and capability of a smart manufacturing system to deal with the complexity and changes [32]. To access virtual assets over the Internet, manufacturing assets must support interoperations. Adamczyk et al. [33] introduced a knowledge-based expert system to support the semantic interoperations in SM; note that the semantic interoperations must address the divergence and misinterpretation of heterogenic data from various sources. Landolfi et al. [34] referred to the use of virtual assets as manufacturing as a service (MaaS). A MaaS based platform was developed to connect service vendors, suppliers, and customers to enterprises directly for system-level optimization.

2.9. Big Data Analytics (BDA)

SM is built upon IoT. To achieve a high system diagnosability, predictivity, and responsiveness, SM must rely on big data connected over the Internet to support its decision-making systems at all levels and scopes. SM applications involved the challenges to assure integrity, quality, privacy, availability, scalability, transformation, legitimacy, surveillance, and governance of data [19]. BDA and SM used to be investigated in the fields of information technologies and intelligent manufacturing, respectively. These concepts were recently bridged due to close correspondences [35]. Ren et al. [36] analyzed emerging issues and potential solutions when BDA is integrated as one of critical technologies to deal with big data and data-driven decision-makings from the perspective of product lifecycles. Tao et al. [37] discussed the importance of big data analytics from the historical perspective, identified the bottlenecks of BDA in utilizing abundant data in developing SM, and proposed a conceptual framework to integrate BDA tools for effective data-driven SM.

SM fully utilized advanced data analytics tools to support decision making activities at various domains and levels of manufacturing businesses. Accordingly, with the increase of volume, velocity, and variety of data acquired from business environment, it posed the challenge of processing data efficiently [38]. On the other hand, BDA also relies on reliable and trustworthy data from IoT to reduce knowledge and information for decision makings; therefore, BDA relates to numerous activities in an information flow including data collection, sharing, processing, and fusion. Wang and Luo [39] proposed a reference framework to take advantage of digital-twin models to fuse the data from virtual and physical models seamlessly.

2.10. Blockchain Theologies (BCT)

SM applications involved the challenge of assuring integrity, quality, privacy, availability, scalability, transformation, legitimacy, surveillance and governance of data, value transfer, and manufacturing services [19]. In contrast to traditional standalone information systems, SM is networked and its system boundaries are open; SM is more vulnerable in terms of security, privacy, and safety. Tuptuk and Hailes [40] discussed the challenges in securing smart manufacturing systems in terms of existing vulnerability, potential cyberattacks, awareness and preparations for security loopholes, and security measures. The blockchain technology (BCT) has been explored to embed trustworthiness and visibility in SM. Viriyasitavat et al. [41–43] developed a few algorithms to (1) select partners and compose them as virtual enterprises over the Internet and (2) assure the privacy, trustiness, and security in value transfer and interoperations of business partners.

3. Performance Metrics (PMs) for System Smartness

The expectations for SM can be classified into functional requirements (FRs) and performance metrics (PMs). On the one hand, FRs are a set of hard goals that a manufacturing system must achieve; FRs are treated as design constraints in developing DSs. On the other hand, PMs are a set of soft goals for which DSs of a smart manufacturing system should be optimized. Since a system solution is specific to given applications, the classification of system expectations for FRs and PMs is different from one system to another. The common FRs of smart manufacturing systems have been discussed in Section 3.2; in this section, the commonly used PMs will be discussed.

Researchers have proposed many evaluation models and performance metrics for manufacturing systems from different perspectives. However, performance metrics can conflict with each other. Jiang et al. [44] discussed the contradictions of metrics in multi-objective optimizations (MOO). Based on their relevance, the metrics were classified into capacity, convergence, diversity, and convergence-diversity, and one performance evaluation model was developed to achieve the consistencies of Pareto fronts in (MOO). In evaluating system sustainability, Moldavska and Welo [45] emphasized the importance of sustainable development goals and proposed incorporating these goals in evaluating system sustainability. Auer et al. [46] assessed manufacturing systems from the perspective of products, and life-cycle assessment and life-cycle costing was used to determine the impact of manufacturing systems on the eco-environment. Jung et al. [47] tried to map strategy-level performance metrics to the structure of SM, and the identified challenges were the selection of performance metrics, the correspondences of performance metrics and manufacturing activities, and the representation of system models for comparisons. Some researchers investigated the impact of certain information technologies on the performances of a smart manufacturing system. For example, Kiesel et al. [48] discussed the roles of 5G in reducing the latency of critical applications since 5G was able to accelerate digital transformation in the sense that the latency could be below 1 milliseconds (ms) in monitoring and controlling complex production processes. The potential economic benefits were quantified. Barletta et al. [49] valued SM for its contribution to environmental sustainability, and they presented the assessment model and tools to evaluate sustainability readiness of smart manufacturing systems. Ante et al. [50] proposed a hierarchical structure of key performance indicators (KPIs) to measure the performance of smart manufacturing systems; the performances were evaluated at strategic, tactical, and operational levels, and the dependences of system elements were taken into consideration in quantifying KPIs. Zhang et al. [51] emphasized the impact of disposed products on economy and environment, and they suggested combining the gray correlation decision-making trials to evaluate products in layout designs of manufacturing plants. Both centralized and decentralized production systems were modelled by Moutzis et al. [52] and system performances were evaluated from the perspectives of lead time, cost, flexibility, throughout, and environmental impact relevant to transportation. It is a difficult challenge for a small and medium-sized enterprise (SME) to choose an appropriate digital solution for specific decision-making. Martin et al. [53] argued that the core value of smart manufacturing was to utilize data to predict behaviors of cyber physical production systems, and they adopted the value stream mapping method (VSM) to analyze and compare smart manufacturing solutions.

Quantitative evaluations are critical in the analysis and synthesis of system designs. Georgoulias et al. [54] indicated that existing empirical evaluation models were only applicable to specific applications, and they argued that an evaluation model or algorithm should be generic, holistic, and quantitative. Georgoulias et al. [55] further developed an evaluation model to quantify the system flexibility (i.e., product flexibility, capacity flexibility, and operation flexibility) in dealing with the changes of management processes in manufacturing organization; the developed model was used to optimize the system for better effectiveness and competitiveness. Youssef et al. [56] used the universal generation function to evaluate the availability of manufacturing assets; it considered the changes of production rates and demands in assessing system configurations. Cagno et al. [57] proposed a framework to measure the sustainability of enterprises and the sustainability was assessed based on economic, social, and environmental indicators. In the framework by Farias et al. [58], the green performance and leanness were emphasized in determining the assessment criteria and metrics of manufacturing systems. Junior et al. [59] proposed a balanced scorecard method to evaluate system sustainability based on the correspondences of economic, environmental, and social lines to the learning and growing, process, and market and financial perspectives. Cai and Lai [60] evaluated the system sustainability from the perspective of energy flow within manufacturing plants. Unfortunately, the information for the assessment model would not be available until the physical system was built and in operation and the statistical data were collected and available for use. Brennan [61] discussed the need for holonic manufacturing systems to develop corresponding performance metrics. A holonic system was required to handle disturbance, support human integration, and provide reliability, robustness, and flexibility in coping with changes, and a system design was evaluated based on reliability, responsiveness, flexibility, cost, and assets. Mahmood et al. [62] used modeling and simulation to assess the performance of applied technologies in integrated production lines while only the performances at shop-floor level were evaluated without the consideration of external partners and end-users. Burggra et al. [63] compared the performances of artificial intelligence (AI) and human beings in job-shop scheduling of a cyber-production management system using a reinforcement learning algorithm. Ottesjo et al. [64] proposed an assessment tool to measure the level of digitization of SMEs, and it aimed to analyze administrative awareness and technical capabilities and identify digitalization gaps for SMEs to advance their manufacturing systems from the system lifecycle perspective.

The following section will focus on the smartness of system that is exhibited in its system lifecycle.

3.1. Visibility, Diagnosability, and Predictivity

A smart manufacturing system must be a closed system that is responsive to internal and external changes; the primary condition is that the system possesses an ability to understand the past, present, and future of the system. Visibility, diagnosability, and predictivity reflect the levels of system smartness in detecting changes and disturbances, diagnosing and troubleshooting problems, and predicting the trends of changes based on data collected from various sources over manufacturing systems. System visibility relies on the sensors and instrumentations installed on smart things, and diagnosability and predictivity rely on the capabilities of advanced information technologies such as AI, CC, and BDA [1].

3.2. Upgradability

Upgradability measures how easily a system or system element can be upgraded to newly developed technologies. Manufacturing technologies are essential tools to run manufacturing businesses, and manufacturing technologies are continuously evolving with the advance of fundamental science and technologies, especially digital technologies. To prolong the lifecycle, manufacturing systems should be modularized so that individual functional modules can be maintained and upgraded with a minimized impact on systems [65,66]. The smartness of a system can be measured by the upgradability of adopting technologies, enterprise systems, and decision-making units at various levels and domains.

3.3. Adaptability

Adaptability measures the capability of a system to deal with changes and uncertainties [67]. Adaptability is measured from system outputs; correspondingly, adaptability is achieved by the flexibility of system elements and the reconfigurability among system elements. Internal or external changes and uncertainties can only be tackled by the changes that can be possibly implemented on system elements or system configurations; therefore, system adaptability can alternatively be measured on (1) internal adjustable components, (2) modular system architecture, and (3) a combination of adjustable and modular components and use of external assets [68,69]. In particular, a modularized architecture makes a system reconfigurable to meet new manufacturing needs by reconfiguring its physical and logical structures. Note that aiming at high-level adaptability involves in an increased cost and complexity in general. The challenges in developing a reconfigurable system are high initial investments, long-term of investment returns, limited system performances at reconfiguration and ramp up phases, and the complexity of task-oriented configuration designs [66,70–72].

3.4. Resilience

The wish-list of future manufacturing systems provided by O'Connell et al. [73] emphasized system resilience. Resilience refers to the system ability to achieve high-level objectives (i.e., adaptation, sustainability, and reliability) in the presence of unpredicted changes and disturbances [74–77]. In particular, adaptation referred to the enhanced ability to achieve desired goals in a dynamic environment, including the ability to reduce vulnerability to threats and adverse disturbances. To improve system resilience, Zhang et al. [78] developed a dynamic model to control a reconfigurable electronic assembly line that was subjected to spatio-temporal disruptions. Resilience dynamics were analyzed by max-plus algebra, the analyzed results were used to generate digital twins, and the control of the assembly line was implemented over an open reconfigurable architecture.

3.5. Flexibility

Flexibility is similar to adaptability, but it is measured on system elements rather than system outputs. Lafou et al. [79] defined flexibility as the ability of a manufacturing system to deal with variations of products, and system flexibility was quantified based on the mappings of products and manufacturing assets. They commented that modularity and standardization of manufacturing resources and products generally minimized the introduction costs of new variations. System flexibility can be achieved by software, hardware, or a combination of both. The flexibility of a hardware system must be supported by the corresponding software system. Keddis et al. [80] discussed the flexibility of data-driven communication to match the flexibility of adaptable hardware system.

All of the performance indicators are driven from manufacturing systems; therefore, performance indicators are associated with each other in certain ways. It is important to understand their correspondences. For example, Lufi and Besenfelder [81] investigated the dependence of robustness on system flexibility of manufacturing systems since system flexibility tackled with volatile and unpredictable environments and a manufacturing system should make a trade-off between optimization and robustness for the best interest of system performance. Mass personalization needs high-level flexibility and responsiveness of a manufacturing system to make personalized products in small batch sizes cost-effectively. Traditional manufacturing systems have their limits in reconfiguring systems to accommodate changes, and SM should be capable of self-reconfiguring and optimizing to achieve flexible, autonomous, and error-tolerant productions in turbulent business environments. System elements in a self-organizing system are distributed, adaptable, self-autonomic, and supportive to bottom-up reconfiguration [82].

3.6. Sustainability

Alike to humankind, a smart manufacturing system aims ultimately at a long system lifespan; manufacturing enterprises are facing an increasing pressure to optimize system sustainability in addition to traditional performance measures such as reliability, cost, and productivity. Sustainability becomes necessary to consider in decision-making processes over system lifecycles. A manufacturing process is a type of mechanical, chemical, electrical, or biological transformation that can be modelled by energy generation, transfer, storage, or consumption. Hoang et al. [83] developed a mathematic model with thermodynamic, physical-thermodynamic, and economic-thermodynamic indicators to estimate energy efficiency of manufacturing systems. Huang and Badurdeen [84] investigated the impacts of products and processes respectively in evaluating system sustainability; the framework for sustainability evaluation included the metrics involved at five stages. In the assessment by Jiang et al. [85], system sustainability was quantified for decomposing system-level mission into device-level manufacturing processes and integrating data from device-level to enterprise-level executions. Sustainability was evaluated comprehensively from economic, environmental, and social perspectives. SMEs have limited resources to pursue system sustainability as prioritized business objectives; the required sustainability is treated as the constraint of business rather than the performance to be optimized. Singh et al. [20] introduced an expert system to quantify system sustainability for SMEs. Sustainability becomes mandatory simply because the ecosystem of the earth and desired quality of humankinds could not be maintained without sustainable manufacturing [86].

Zhang et al. [87] developed a business case to make system-level decisions in SMEs; it considered system dynamics in assessing sustainability and lifecycle costing of products.

4. Systematic Methodology for Design of Smart Manufacturing Systems

The main functional requirements enabling digital technologies and a complete list of expectations of a smart manufacturing system have been discussed in Sections 3 and 4; however, it is extremely rare that an enterprise has the access to any digital technologies when the enterprise needs, and a manufacturing system can be designed and implemented from scratch. It is impractical to design an ideal smart manufacturing system without physical constraints. In practice, the smartness of a manufacturing system will be iteratively improved by upgrading and incorporating more digital methodologies in continuous improvement (CI).

To make system complexity manageable at each iteration, the axiomatic design theory (ADT) is adopted in Figure 1 to narrow down a set of FRs, DSs, and PMs that are most critical to given applications, and the rest of FRs, DSs, and PMs should be formulated as design constraints based on available manufacturing assets and current system states. In other words, design of a smart manufacturing system at each iteration only involves (1) one or a few metrics relevant to system smartness (i.e., flexibility, visibility, sustainability, resilience, and even some traditional metrics such as efficiency and agility) and (2) one or a few corresponding digital triads (DT-II) or the configuration in IoDTT. Available manufacturing assets and given marketing conditions are formulated as design constraints.

5. Case Studies

Three examples of manufacturing system designs by the authors and their collaborators are introduced here to illustrate how the proposed methodology was applied in the system development to increase system smartness in continuous improvement (CI). Note that the application scenarios were specified, the design solutions (DSs) were limited to certain digital technologies, and system smartness was associated with the system performance of interests in achieving specified functional requirements (FRs) in the given applications. In all of these three cases, FRs and system smartness were interpreted and defined based on customers' needs. The design space of DSs were for digital technologies and determined by system developers based on accessible manufacturing resources, and the following discussions were limited to using the proposed methodology to formulate a smart system design problem. Interested readers might find the details and rad data of these design examples in the corresponding publications [5,7,88–90].

5.1. Case Study 1: BDA for Visibility and Diagnosability in Continuous Improvement (CI)

The purpose of the first case study was to show that the definition of system smartness in a smart manufacturing system can be customized to the prioritized key performance indicators (KPIs). In other words, pursuing a smart manufacturing system is a long-term effort of continuous improvement, and targeted system smartness should be as specific as possible to be measured quantitatively. The process was applicable to system design in any sectors. With real-time data collected from the things in the physical world and simulation models in the digital world, the decision-making processes at any level and domain could be data driven to improve system responsiveness, since big data helps to improve system smartness in terms of the visibility of system states and changes and the diagnosability of defects and malfunctions.

Figure 2 shows a case where system smartness was defined for visibility and diagnosability, and digital technologies for data collection and analysis were identified as the design solutions of interest. In the developed solution, BDA was incorporated in an enterprise system, heterogeneous and data were analyzed and processed to make the scale of the datasets manageable, and the decisions for the actions in continuous improvement could be made promptly. Note that the data of past, present, and prediction could be maintained in a life model together with digital and physical things as DT-II in system implementation [88–90].

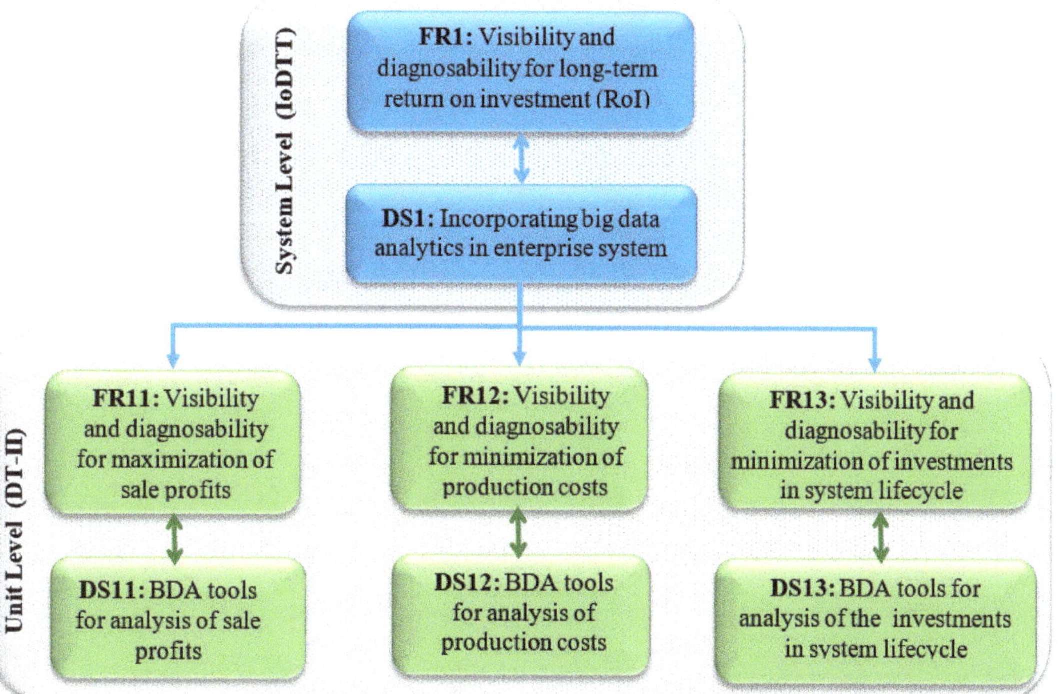

Figure 2. Case study I–Improve visibility and diagnosability by big data analytics (BDA).

5.2. Case Study 2: Incorporating Additive Manufacturing for Flexibility and Adaptability

The purpose of the second case study was to show that the proposed design methodology was ready to be applied at the phase of system operations when some system performances were found unsatisfactory, and the solution to critical processes must be obtained in enhancing system smartness. In such a case, DSs were for certain manufacturing processes, and system smartness was related to unsatisfactory performances of interests. The process was applicable to design problems in system operations in any continuous improvement practice. In general, a manufacturing system transfers raw materials into final products through a series of manufacturing processes. When an enterprise aims at system smartness for dealing with the changes and disturbance in its material flow, the hardware systems must have flexibility and capabilities to accommodate these changes in manufacturing processes. From this perspective, incorporating more and more advanced digital technologies in production systems helps to improve system smartness in terms of the adaptability and robustness.

Figure 3 showed a case where system smartness was defined for high-level flexibility and adaptability in dealing with unavoidable defects occurring to production lines; system flexibility and adaptability was directly measured by the direct run rate (DRR) of products, i.e., the percentage of products that meet the requirements of quality at the first try. A product might be damaged due to numerous potential interactions of tooling and products over production lines. System flexibility and adaptability was measured by a set of decomposed FRs shown in Figure 3 and Table 1. Additive manufacturing (AM) was introduced as the design solutions (DSs) to enhance the capabilities of the manufacturing system in producing protective tools when they are needed. In the developed solution, 3D printers were introduced to make protective parts for problematic tools where product defects occurred. A number of DT-II units were developed to implement the whole process from monitoring production lines to detecting defects on products, identifying problematic tools, generating and verifying digital models for protective parts, producing parts, and finally to mounting parts on assembling tools in the production lines. According to ADT, the design solutions (DSs) in Figure 4 were developed to fulfill the identified FRs [7].

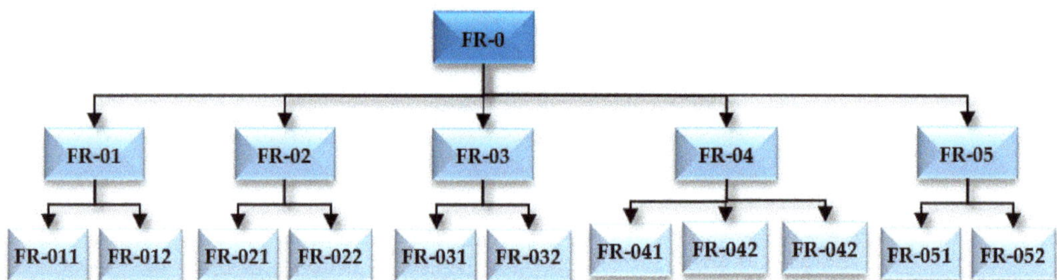

Figure 3. Decomposition of system smartness (flexibility and adaptability) in case study 2 (Reprinted with permission from ref. [7]. 2021 Taylor & Francis).

Table 1. The meanings of decomposed *FRs* in case study 2 (Reprinted with permission from ref. [7]. 2021 Taylor & Francis).

FRs	Description
FR-0:	Develop the solution to improve DRR of truck assembly line by integrating AM processes
FR-01:	Utilize the data of truck quality inspection for surface defects, identify the sources (workstations and assistive tools) of defects. *FR-011:*　Detect surface defects. *FR-012:*　Identify problematic assembling processes and assistive tools.
FR-02:	Develop and model parts as the protective solutions to identified defects. *FR-021:*　Utilize information of assistive tools. *FR-022:*　Optimize design for strength, fabrication time, and cost.
FR-03:	Provide the tested physical solutions to assembly workstations in less than 24 h. *FR-031:*　Perform tests on physical parts for material strength. *FR-032:*　Perform simulation for functional validation and process optimization.
FR-04:	Standardize the procedure and practice of AM processes. *FR-041:*　Maintain normal operations of AM machines. *FR-042:*　Provide guides and training manuals for operators and procedure. *FR-043:*　Standardize the interactions of functional modules.
FR-05:	Routinize the operations of AM machines with the aid of inventory, design library, planning and scheduling of printing jobs for cost reduction. *FR-051:*　Build and maintain design libraries for knowledge-based engineering *FR-052:*　Manage the inventory of protective parts.

5.3. Case Study 3: Using IoT for Automation

The purpose of the third case study was to show that the proposed design methodology can be extended to design any systems or products as long as FRs, DSs, and performance metrics (PMs) could be tailored to the specified applications. The proposed design methodology is generally applicable to designs of any smart systems or products, since the systems are tailored to given applications by defining system smartness and feasible design solutions of most interests. As mentioned before, system smartness can be defined for high-level adaptability and sustainability in dynamic environment; system smartness can also be defined for some traditional performance metrics such as efficiency, agility, robustness, cost-effectiveness, and degree of automation. Such design types have a significant advantage in practice since adopting digital technologies has a direct impact on these system metrics.

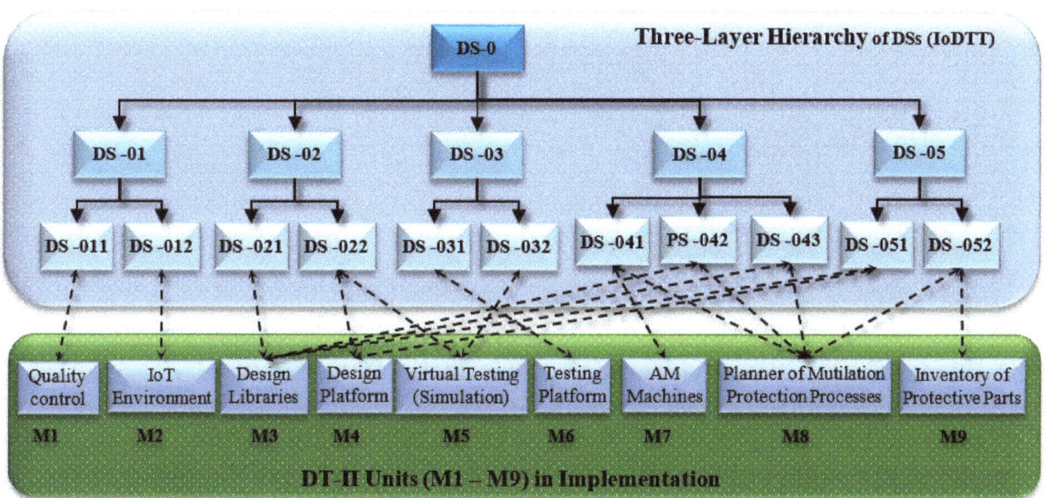

Figure 4. Proposed digital solutions in case study 2 (Reprinted with permission from ref. [7]. 2021 Taylor & Francis).

Figure 5 showed a case of designing an automatic fuel-recharging system. System smartness was defined for the degree of automation in performing relevant tasks for drivers to refill gas on vehicles. System smartness was measured by the minimized cost for fully automated refueling services at gas stations. The IoT-based technologies were explored as the design solutions to minimize users' interferences cost-effectively. The system-level goal was decomposed into four FRs, i.e., FR11 for 'data collection and processing', FR12 for 'controls for abnormal events', FR13 for 'refueling operation', and FR14 for 'controls for normal events'. To meet FR11, various sensors and instrumentations such as vision, laser scanning, bar-code scanning, compliant sensors, and controllable platform were used as DSs to detect incoming vehicles and determine the relative position and orientation of fuel spout; embedded chips and apps on phones can be integrated with the Internet of Things (IoT) database to obtain customers' intent and payment information and collect information about vehicle and fuel. To meet FR13, gantry systems, robots, and sophisticated mechanisms were integrated with multi-functional tools to access a fuel port, open a fuel-filler cover, retrieve a refueling tool, close a fuel-filler cover, and reset the refueling tool. To meet FR12 and FR14, the system-level controls for normal and abnormal events were implemented as stand-alone systems or IoT-enabled apps; in addition, all processing parameters could be specified manually through the interfaces of programmable logic controllers (PLC) or IoT-based apps.

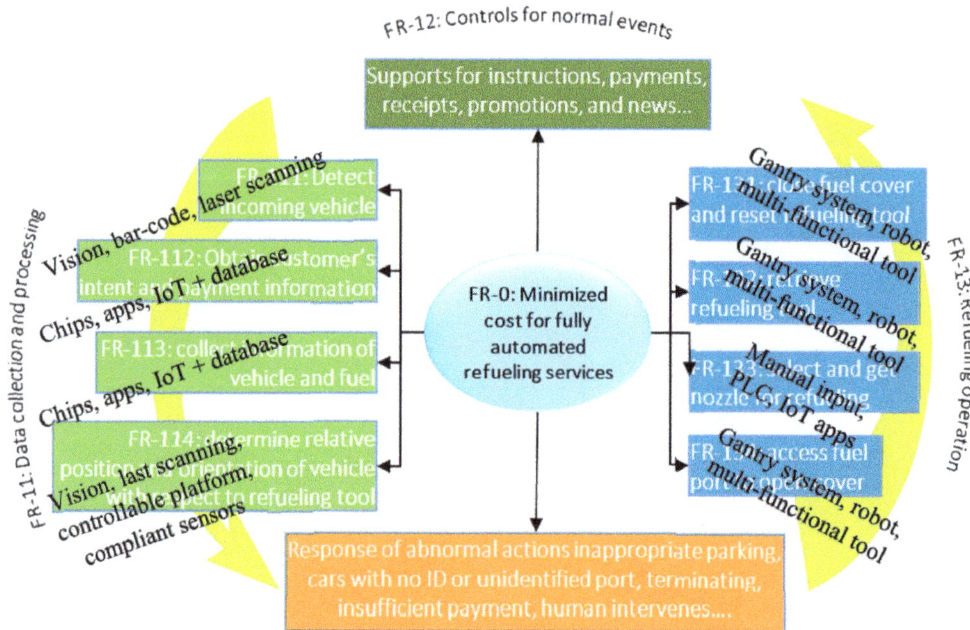

Figure 5. Design solutions (DSs) to fully automated refueling (Reprinted with permission from ref. [5]. 2021 Emerald Publishing Limited).

6. Conclusions and Future Directions

To the authors' knowledge, limited works are available on the development of systematic methodologies for designs of smart manufacturing systems. The novel contributions of the presented work were made at two aspects: (1) unified definitions of digital elements and manufacturing systems have been proposed; they are generalized to have all of the digitized characteristics and they are customizable to be applied in any manufacturing system with specified manufacturing resources and goals of smartness and (2) a systematic design methodology has been proposed; it can serve as the guide for designs of smart manufacturing systems in certain applications from a practical perspective. Note that 'practical perspective' here refers to the views of specific enterprises with given resources, technology accesses, and the interests of business domains, strategies, and performance indicators including costs.

The proposed design methodology deals with the high diversified systems by some customizing efforts in defining prioritized goals of system smartness and affordable digital technologies in achieving system goals; in addition, the performance metrics for system smartness must be quantifiable so that different design solutions can be analyzed, evaluated, and compared to optimize system solutions. Future research efforts will be needed in many areas such as (1) developing quantifiable performance metrics for system smartness of interest and synergizing multiple performance metrics when they are considered simultaneously; (2) establishing design libraries which include commonly design solutions (DSs); (3) developing some design templates which correspond to design solutions (DSs) and functional requirements (FRs) with consideration of the sustainability at both of component and system levels; (4) using BDA and AI in dealing with the combinatory complexity in tailoring digital solutions to specific manufacturing applications' effective computing; (5) making a trade-off between system reconfigurations and the utilization of virtual assets in system lifecycle; (6) developing some systematic approaches to verify and validate a system before it will be actually implemented in physical world; (7) developing the standardized procedures for design of smart manufacturing systems based on the proposed methodology; (8) reshaping the proposed design methodology as a standardized procedure in designing smart manufacturing systems; (9) adapting the proposed design methodology to supply network systems since they are a model of the social-technological-economic system—a reality of mankind.

Author Contributions: Conceptualization, Z.B. and W.-J.Z.; methodology, Z.B. and C.W.; implementation and writing, Z.B., W.-J.Z., C.W., C.L. and L.X. All authors have read and agreed to the published version of the manuscript.

Funding: This research received no external funding.

Institutional Review Board Statement: Not applicable.

Informed Consent Statement: Not applicable.

Data Availability Statement: Not applicable.

Conflicts of Interest: The authors declare no conflict of interest.

Abbreviations

ADT:	Axiomatic Design Theory
AI:	Artificial Intelligence
AM:	Additive Manufacturing
AR:	Augmented Reality
BCT:	Blockchain Technology
BDA:	Big Data Analytics
BPM:	Business Process Management
CC:	Cloud Computing
CI:	Continuous Improvement
DM:	Digital Manufacturing
DSs:	Design Solutions
DT-I:	Digital Twins
DT-II:	Digital Traid
ERP:	Enterprise Resource Planning
FRs:	Functional Requirements
HCPS:	Human-Cyber Physical Systems
HRI:	Human-Robot Interactions
IoDTT:	Internet of Digital Triad Things
IoT:	Internet of Things
IIoT:	Industrial Internet of Things
IT:	Information Technologies
KPIs:	Key Performance Indicators
MaaS:	Manufacturing as a service
ML:	Machine Learning
MOO:	Multi-Objective Optimizations
PLC:	Programmable Logic Controller
PMs:	Performance Metrics
SM:	Smart Manufacturing
SME:	Small to Midsize Enterprises
SoA:	Service-oriented Architecture
SoS:	System of Systems
VSM:	Value stream mapping
SaaS:	Any decision making as a service

References

1. Bi, Z.M.; Zhang, C.W.J.; Wu, C.; Luo, C.; Xu, L. Generic design methodology for smart manufacturing systems from practical perspective, part I—digital triad concept and its application as new definition of system reference model. *Mach. Simultaneously under review*.
2. Unglert, J.; Jauregui-Becker, J.; Hoekstra, S. Computational design synthesis of reconfigurable cellular manufacturing systems: A design engineering model. *Procedia CIRP* **2016**, *57*, 374–379. [CrossRef]
3. Kurgan, A.; Maggiore, P.; Golkar, A. Integrated design methodology for improved system manufacturability. In Proceedings of the 2020 IEEE International Symposium on Systems Engineering (ISSE), Vienna, Austria, 12 October–12 November 2020 [CrossRef]
4. Bi, Z.M.; Miao, Z.H.; Zhang, B.; Zhang, W.J. The state of the art of testing standards for integrated robotic systems. *Robot. Comput. Integr. Manuf.* **2020**, *63*, 101893. [CrossRef]
5. Bi, Z.M.; Luo, C.; Miao, Z.; Zhang, B.; Zhang, C.W.J. Automatic robotic recharging systems –development and challenges. *Ind. Robot* **2021**, *48*, 95–109. [CrossRef]
6. Bi, Z.M.; Miao, Z.H.; Zhang, B.; Zhang, W.J. A framework for performance assessment of heterogeneous robotic systems. *IEEE Syst. J.* **2021**, *15*, 1191–1201. [CrossRef]
7. Bi, Z.M.; Wang, G.; Thompson, J.; Ruiz, D.; Rosswurm, J.; Roof, S.; Guandique, C. System framework of adopting additive manufacturing in mass production line. *Enterp. Inf. Syst.* **2021**. [CrossRef]
8. Morgan, J.; Halton, M.; Qiao, Y.; Breslin, J.G. Industry 4.0 smart reconfigurable machines. *J. Manuf. Syst.* **2021**, *59*, 481–506. [CrossRef]
9. Mittal, S.; Khan, M.A.; Romero, D.; Wuest, T. Building blocks for adopting smart manufacturing. *Procedia Manuf.* **2019**, *34*, 978–985. [CrossRef]
10. Akrivos, V.; Haines-Gadd, M.; Mativenga, P.; Charnlet, F. Improved metrics for assessment of immortal materials and products. *Procedia CIRP* **2019**, *80*, 596–601. [CrossRef]
11. ASML. EUV Lithography Systems Twinscan Nxe: 3400 B. 2021. Available online: https://www.asml.com/en/products/euv-lithography-systems/twinscan-nxe3400b (accessed on 7 September 2021).
12. Haleem, A.; Javaid, M.; Singh, R.P.; Suman, R. Significant roles of 4D printing using smart materials in the field of manufacturing. *Adv. Ind. Eng. Polym. Res.* **2021**. [CrossRef]
13. Barbosa, C.; Azevedo, A. Towards a hybrid multi-dimensional simulation approach for performance assessment of MTO and ETO manufacturing environments. *Procedia Manuf.* **2018**, *17*, 852–859. [CrossRef]
14. Zhang, Y.; Kwok, T.-H. Design and interaction interface using augmented reality for smart manufacturing. *Procedia Manuf.* **2018**, *26*, 1278–1286. [CrossRef]
15. Zhang, C.; Xu, W.; Liu, J.; Liu, Z.; Zhou, Z.; Pham, D.T. A reconfigurable model approach for digital twin-based manufacturing system. *Procedia CIRP* **2019**, *83*, 118–125. [CrossRef]
16. Semeraro, C.; Lezoche, M.; Panetto, H.; Dassisti, M. Digital twin paradigm: A systematic literature review. *Comput. Ind.* **2021**, *130*, 103469. [CrossRef]
17. Leng, J.; Wang, D.; Shen, W.; Li, X.; Liu, Q.; Chen, X. Digital twins-based smart manufacturing system design in Industry 4.0: A review. *J. Manuf. Syst.* **2021**, *60*, 119–137. [CrossRef]
18. Liu, Q.; Leng, J.; Yan, D.; Zhang, D.; Wei, L.; Yu, A.; Zhao, R.; Zhang, H.; Chen, X. Digital twin-based designing of the configuration, motion, control and optimization model of a flow-type smart manufacturing system. *J. Manuf. Syst.* **2021**, *58*, 52–64. [CrossRef]
19. Singh, H. Big data, industry 4.0 and cyber-physical systems integration: A smart industry context. *Mater. Today Proc.* **2021**, *46*, 157–162. [CrossRef]
20. Singh, S.; Olugu, E.U.; Musa, S.N. Development of sustainable manufacturing performance evaluation expert system for small and medium enterprises. *Procedia CIRP* **2016**, *40*, 608–613. [CrossRef]
21. Wang, X.; Wang, Y.; Tao, F.; Liu, A. New paradigm of data-driven smart customization through digital twin. *J. Manuf. Syst.* **2021**, *58*, 270–280. [CrossRef]
22. D'Addona, D.M.; Bracco, F.; Bettoni, A.; Nishino, N.; Carpanzano, E.; Bruzzone, A.A. Adaptive automation and human factors in manufacturing: An experimental assessment for a cognitive approach. *CIRP Ann. Manuf. Technol.* **2018**, *67*, 455–458. [CrossRef]
23. Baroroh, D.K.; Chu, C.H.; Wang, L. Systematic literature review on augmented reality in smart manufacturing: Collaboration between human and computational intelligence. *J. Manuf. Syst.* **2020**. [CrossRef]
24. Bi, Z.M.; Wang, G.; Xu, L. A visualization platform for internet of things in manufacturing applications. *Internet Res.* **2016**, *26*, 377–401. [CrossRef]
25. Bi, Z.M.; Wang, G.; Xu, L.; Thompson, M.; Mir, R.; Nyikos, J.; Mane, A.; Witte, C.; Sidwell, C. IoT-based system for communication and coordination of football robot team. *Internet Res.* **2017**, *27*, 162–181. [CrossRef]
26. Bi, Z.M.; Liu, Y.F.; Krider, J.; Buckland, J.; Whiteman, A.; Beachy, D.; Smitch, J. Real-time force monitoring of smart grippers for Internet of things (IoT) applications. *J. Ind. Inf. Integr.* **2018**, *11*, 19–28. [CrossRef]
27. Dey, B.K.; Pareek, S.; Tayyab, M.; Sarkar, B. Autonomation policy to control work-in-process inventory in a smart production system. *Int. J. Prod. Res.* **2020**, *59*, 1258–1280. [CrossRef]
28. Ullah, M.; Sarkar, B. Recovery-channel selection in a hybrid manufacturing-remanufacturing production model with RFID and production quality. *Int. J. Prod. Econ.* **2020**, *219*, 360–374. [CrossRef]

29. Henzel, R.; Herzwurm, G. Cloud manufacturing: A state-of-the-art survey of current issues. *Procedia CIRP* **2018**, *72*, 947–952. [CrossRef]
30. Huang, L.; Wu, C. Selection approach of cloud manufacturing resource for manufacturing enterprises based on trust evaluation. In Proceedings of the 2020 Prognostics and Health Management Conference (PHM-Besançon), Besancon, France, 4–7 May 2020; pp. 309–313.
31. Lee, J.; Davari, H.; Singh, J.; Pandhare, V. Industrial artificial intelligence for industry 4.0-based manufacturing systems. *Manuf. Lett.* **2018**, *18*, 20–23. [CrossRef]
32. Shihundla, T.B.; Mpofu, K.; Adenuga, O.T. Integrating product-service systems into the manufacturing industry: Industry 4.0 perspectives. *Procedia CIRP* **2019**, *83*, 8–13. [CrossRef]
33. Adamczyk, B.S.; Szejka, A.L.; Junior, O.C. Knowledge-based expert system to support the semantic interoperability in smart manufacturing. *Comput. Ind.* **2020**, *115*, 103161. [CrossRef]
34. Landolfi, G.; Barni, A.; Izzo, G.; Fontana, A.; Bettoni, A. A MaaS platform architecture support data sovereignty in sustainability assessment of manufacturing systems. *Procedia Manuf.* **2019**, *38*, 548–555. [CrossRef]
35. Bi, Z.M.; Cochran, D. Big data analytics with applications. *J. Manag. Anal.* **2015**, *1*, 249–265. [CrossRef]
36. Ren, S.; Zhang, Y.; Liu, Y.; Sakao, T.; Huisingh, D.; Almeida, C.M.V.B. A comprehensive review of big data analytics throughout product lifecycle to support sustainale smart manufacturing: A framework, challenges and future research directions. *J. Clean. Prod.* **2019**, *210*, 1343–1365. [CrossRef]
37. Tao, F.; Qi, Q.; Liu, A.; Kusiak, A. Data-driven smart manufacturing. *J. Manuf. Syst.* **2018**, *48*, 157–169. [CrossRef]
38. Wang, J.; Ma, Y.; Zhang, L.; Gao, R.X.; Wu, D. Deep learning for smart manufacturing: Methods and applications. *J. Manuf. Syst.* **2018**, *48*, 144–156. [CrossRef]
39. Wang, P.; Luo, M. A digital twin based big data virtual and real fusion learning reference framework supported by industrial internet towards smart manufacturing. *J. Manuf. Syst.* **2021**, *58*, 16–32. [CrossRef]
40. Tuptuk, N.; Hailes, S. Security of smart manufacturing systems. *J. Manuf. Syst.* **2018**, *47*, 93–106. [CrossRef]
41. Viriyasitava, W.; Xu, L.; Bi, Z.M.; Hoonsopon, D. Blockchain technology for applications in Internet of Things-mapping from system design perspective. *IEEE Internet Things J.* **2019**, *6*, 8155–8168. [CrossRef]
42. Viriyasitava, W.; Xu, L.; Bi, Z.M. Blockchain and Internet of Things for modern business process in digital economy-the state of the art. *IEEE Trans. Comput. Soc. Syst.* **2019**, *6*, 1420–1432. [CrossRef]
43. Viriyasitava, W.; Xu, L.; Bi, Z.M.; Hoonsopon, D.; Charoenruk, N. Managing QoS of Internet-ofThing services using blockchain. *IEEE Trans. Comput. Soc. Syst.* **2019**, *6*, 1357–1368. [CrossRef]
44. Jiang, S.; Ong, Y.S.; Zhang, J.; Feng, L. Consistencies and contradictions of performance metrics in multi-objective optimization. *IEEE Trans. Cybern.* **2014**, *44*, 2391–2404. [CrossRef] [PubMed]
45. Moldavska, A.; Welo, T. A holistic approach to corporate sustainability assessment: Incorporating sustainable development goals into sustainable manufacturing performance evaluation. *J. Manuf. Syst.* **2019**, *50*, 53–68. [CrossRef]
46. Auer, J.; Bey, N.; Schafer, J.M. Combined lie cycle assessment and life cycle costing in the eco-care-matrix: A case study on the performance of a modernized manufacturing system for glass containers. *J. Clean. Prod.* **2017**, *141*, 99–109. [CrossRef]
47. Jung, K.; Morris, K.C.; Lyons, K.W.; Leong, S.; Cho, H. Mapping strategic goals and operational performance metrics for smart manufacturing systems. *Procedia Comput. Sci.* **2015**, *44*, 184–193. [CrossRef]
48. Kiesel, R.; van Roessel, J.; Schmitt, R.H. Quantification of economic potential of 5G for latency critical applications in production. *Procedia Manuf.* **2020**, *52*, 113–120. [CrossRef]
49. Barletta, I.; Despeisse, M.; Hoffenson, S.; Johansson, B. Organisational sustainability readiness: A model and assessment tool for manufacturing companies. *J. Clean. Prod.* **2021**, *284*, 125404. [CrossRef]
50. Ante, G.; Facchini, F.; Mossa, G.; Digiesi, S. Developing a key performance indicators tree for lean and smart production systems. *IFAC PapersOnLine* **2018**, *51*, 13–18. [CrossRef]
51. Zhang, Z.; Li, H.; Du, Y. End of life vehicle disassembly plant layout evaluation integrating gray correlation and decision making trial and evaluation laboratory. *IEEE Access* **2020**, *8*, 141446–141455. [CrossRef]
52. Moutzis, D.; Doukas, M.; Psarommatis, F. A multi-criteria evaluation of centralized and decentralized production networks in a highly customer-driven environment. *CIRP Ann. Manuf. Technol.* **2012**, *61*, 427–430. [CrossRef]
53. Martin, N.L.; Der, A.; Herrmann, C.; Thiede, S. Assessment of smart manufacturing solutions based on extended value stream mapping. *Procedia CIRP* **2020**, *93*, 371–376. [CrossRef]
54. Georgoulias, K.; Papakostas, N.; Makris, S.; Chryssolouris, G. A toolbox approach for flexibility measurements in diverse environments. *Ann. CIRP* **2007**, *56*, 423–426. [CrossRef]
55. Georgoulias, K.; Papakostas, N.; Chryssolouris, G.; Stanev, S.; Krappe, H.; Ovtcharova, J. Evaluation of flexibility for the effective change management of manufacturing organizations. *Robot. Comput. Integr. Manuf.* **2009**, *25*, 888–893. [CrossRef]
56. Youssef, A.M.A.; Mohib, A.; ElMaraghy, H.A. Availability assessment of multi-state manufacturing systems using universal generating function. *CIRP Ann.* **2006**, *55*, 445–448. [CrossRef]
57. Cagno, E.; Neri, A.; Howard, M.; Brenna, G.; Trianni, A. Industrial sustainability performance measurement systems: A novel framework. *J. Clean. Prod.* **2019**, *230*, 1354–1375. [CrossRef]
58. Farias, L.M.S.; Santos, L.C.; Gohr, C.F.; de Oliveira, L.C.; da Silva, A.M.H. Criteria and practices for lean and green performance assessment: Systematic review and conceptual framework. *J. Clean. Prod.* **2019**, *218*, 746–762. [CrossRef]

59. Junior, A.N.; de Oliveira, M.C.; Helleno, A.L. Sustainability evaluation model for manufacturing systems based on the correction between triple bottom line dimensions and balanced scorecard perspectives. *J. Clean. Prod.* **2018**, *190*, 84–93. [CrossRef]

60. Cai, W.; Lai, K.H. Sustainability assessment of mechanical manufacturing systems in the industrial sector. *Renew. Sustain. Energy Rev.* **2021**, *146*, 110169. [CrossRef]

61. Brennan, R.W. Towards the assessment of holonic manufacturing systems. *IFAC Proc.* **2006**, *35*, 399–404. [CrossRef]

62. Mahmood, K.; Otto, T.; Golova, J.; Kangru, T.; Kuts, V. An approach to analyze the performance of advanced manufacturing environment. *Procedia CIRP* **2020**, *93*, 628–633. [CrossRef]

63. Burggra, P.; Wagner, J.; Koke, B.; Bamberg, M. Performance assessment methodology for AI-supported decision-making in production management. *Procedia CIRP* **2020**, *93*, 891–896. [CrossRef]

64. Ottesjo, B.; Nystrom, S.; Nafors, D.; Berglund, J.; Johansson, B.; Gullander, P. A tool for holistic assessment of digitalization capabilities in manufacturing SMEs. *Procedia CIRP* **2020**, *93*, 676–681. [CrossRef]

65. Bi, Z.M.; Zhang, W.J. Modularity technology in manufacturing: Taxonomy and issues. *Int. J. Adv. Manuf. Technol.* **2001**, *18*, 381–390. [CrossRef]

66. Bi, Z.M. On Adaptive Robot System for Manufacturing Applications. Ph.D. Thesis, Department of Mechanical Engineering, University of Saskatchewan, Saskatoon, SK, Canada, 2002.

67. Goncalves, G.; Reis, J.; Pinto, R.; Peschl, M. Adaptability in smart manufacturing systems. In Proceedings of the INTELLI 2018, the Seventh International Conference on Intelligent Systems and Applications, Venice, Italy, 24–28 July 2018; pp. 36–42.

68. Bi, Z.M.; Wang, L.; Lan, S.Y.T. Current status of reconfigurable assembly systems. *Int. J. Manuf. Res.* **2007**, *2*, 303–327. [CrossRef]

69. Bi, Z.M.; Zhang, W.J.; Chen, I.M.; Lang, S.Y.T. Automated generation of the D-H parameters for configuration design of modular manipulators. *Robot. Comput. Integr. Manuf.* **2007**, *23*, 553–562. [CrossRef]

70. Bortolini, M.; Ferrari, E.; Galizia, F.G.; Mora, C. Implementation of reconfigurable manufacturing in the Italian context: State-of-the-art and trends. *Procedia Manuf.* **2018**, *39*, 591–598. [CrossRef]

71. Viriyasitava, W.; Xu, L.; Bi, Z.M. Specification patterns of service-based applications using blockchain technology. *IEEE Trans. Comput. Soc. Syst.* **2020**, *7*, 886–896. [CrossRef]

72. Viriyasitava, W.; Xu, L.; Bi, Z.M.; Sapsomboon, A. Blockchain-based business process management (BPM) framework for service composition in Industry 4.0. *J. Intell. Manuf.* **2020**, *31*, 1737–1748. [CrossRef]

73. O'Connell, D.; Walker, B.; Abel, N.; Grigg, N. The Resilience, Adaptation, and Transformation Assessment Framework: From Theory to Application. 2015. Available online: https://www.stapgef.org/sites/default/files/documents/CSIRO-STAP-Resilience-Adaptation-Transformation-Assessment-Framework-Report.pdf (accessed on 7 September 2021).

74. Zhang, W.J.; Lin, Y. Principles of design of resilient systems and its application to enterprise information systems. *Enterp. Inf. Syst.* **2010**, *4*, 99–110. [CrossRef]

75. Zhang, T.; Zhang, W.J.; Gupta, M.M. Resilient robots: Concept, review and future directions. *Robotics* **2017**, *6*, 22. [CrossRef]

76. Zhang, T.; Zhang, W.J.; Gupta, M.M. An under-actuated self-reconfigurable robot and the reconfiguration evolution. *Mech. Mach. Theory* **2018**, *124*, 248–258. [CrossRef]

77. Zhang, W.J.; van Luttervel, C.A. Towards a resilient manufacturing system. *Ann. CIRP* **2011**, *60*, 469–472. [CrossRef]

78. Zhang, D.; Xie, M.; Yan, H.; Liu, Q. Resilience dynamic modeling and control for a reconfigurable electronic assembly line under spatio-temporal disruptions. *J. Manuf. Syst.* **2021**, *60*, 852–863. [CrossRef]

79. Lafou, M.; Mathieu, L.; Pois, S.; Alocket, M. Manufacturing system flexibility: Product flexibility assessment. *Procedia CIRP* **2016**, *41*, 99–104. [CrossRef]

80. Keddis, N.; Burdalo, J.; Kainz, G.; Ziid, A. Increasing the adaptability of manufacturing systems by using data-centric communication. In Proceedings of the 2014 IEEE Emerging Technology and Factory Automation (ETFA), Barcelona, Spain, 16–19 September 2014.

81. Lufi, N.; Besenfelder, C. Flexibility based assessment of production system robustness. *Procedia CIRP* **2014**, *19*, 81–86.

82. Qin, Z.; Lu, Y. Self-organizing manufacturing network: A paradigm towards smart manufacturing in mass personalization. *J. Manuf. Syst.* **2021**, *60*, 35–47. [CrossRef]

83. Hoang, A.; Do, P.; Lung, B. Energy efficiency performance-based prognostics for aided maintenance decision-making: Application to manufacturing platform. *J. Clean. Prod.* **2017**, *142*, 2838–3857. [CrossRef]

84. Huang, A.; Badurdeen, F. Sustainable manufacturing performance evaluation: Integrating product and process metrics for systems level assessment. *Procedia Manuf.* **2016**, *8*, 563–570. [CrossRef]

85. Jiang, Q.; Liu, Z.; Liu, W.; Li, T.; Cong, W.; Zhang, H.; Shi, J. A principal component analysis based three-dimensional sustainability assessment model to evaluate corporate sustainable performance. *J. Clean. Prod.* **2018**, *187*, 625–637. [CrossRef]

86. Ghobakhloo, M. Industry 4.0, digitization, and opportunities for sustainability. *J. Clean. Prod.* **2020**, *252*, 119869. [CrossRef]

87. Zhang, H.; Veltri, A.; Calvo-Amodio, J.; Haapala, K.R. Making the business case for sustainable manufacturing in small and medium-sized manufacturing enterprise: A systems decision making approach. *J. Clean. Prod.* **2021**, *287*, 125038. [CrossRef]

88. Cochran, D.; Jafri, M.U.; Chu, A.K.; Bi, Z.M. Incorporating design improvement with effective evaluation using manufacturing system design decomposition (MSDD). *J. Ind. Inf. Integr.* **2016**, *2*, 65–74. [CrossRef]

89. Cochran, D.; Kim, Y.S.; Foley, J.; Bi, Z.M. Use of the manufacturing system design decomposition for comparative analysis and effective design of production systems. *Int. J. Prod. Res.* **2017**, *55*, 870–890. [CrossRef]
90. Cochran, D.; Kinard, D.; Bi, Z.M. Manufacturing system design meets big data analytics for continuous improvement. *Procedia CIRP* **2016**, *50*, 647–652. [CrossRef]

43

machines

MDPI

Article

Optimization of Material Supply in Smart Manufacturing Environment: A Metaheuristic Approach for Matrix Production

Tamás Bányai

Institute of Logistics, University of Miskolc, 3515 Miskolc, Hungary; alttamas@uni-miskolc.hu

Abstract: In the context of Industry 4.0, the matrix production developed by KUKA robotics represents a revolutionary solution for flexible manufacturing systems. Because of the adaptable and flexible manufacturing and material handling solutions, the design and control of these processes require new models and methods, especially from a real-time control point of view. Within the frame of this article, a new real-time optimization algorithm for in-plant material supply of smart manufacturing is proposed. After a systematic literature review, this paper describes a possible structure of the in-plant supply in matrix production environment. The mathematical model of the mentioned matrix production system is defined. The optimization problem of the described model is an integrated routing and scheduling problem, which is an NP-hard problem. The integrated routing and scheduling problem are solved with a hybrid multi-phase black hole and flower pollination-based metaheuristic algorithm. The computational results focusing on clustering and routing problems validate the model and evaluate its performance. The case studies show that matrix production is a suitable solution for smart manufacturing.

Keywords: cyber-physical system; heuristics; logistics; matrix production; optimization; smart manufacturing

Citation: Bányai, T. Optimization of Material Supply in Smart Manufacturing Environment: A Metaheuristic Approach for Matrix Production. *Machines* **2021**, *9*, 220. https://doi.org/10.3390/machines9100220

Academic Editors: Zhuming Bi, Lida Xu, Puren Ouyang and Angelos P. Markopoulos

Received: 25 August 2021
Accepted: 22 September 2021
Published: 29 September 2021

Publisher's Note: MDPI stays neutral with regard to jurisdictional claims in published maps and institutional affiliations.

1. Introduction

Thanks to digitization and Industry 4.0 technologies and solutions, today's economy is in the middle of significant transformation processes regarding the fulfilment of customers' demands. Production companies must apply the solutions of the fourth industrial revolution to improve their efficiency. The ever-changing production and service sector requires the improvement of these attributes. Logistics and material handling operations have more and more importance related to the purchasing, production, distribution, and reverse processes, and they have a significant impact on the strategic, tactical, and operative level of enterprise systems.

As Figure 1 shows, Industry 4.0 technologies offer new innovation accelerators, like augmented and virtual reality, cloud and fog computing related to big data problems, additive manufacturing, Internet of Thing (IoT), autonomous standardized production and material handling resources, smart tools, gentelligent products, simulation and digital twin solutions, cyber security, and system integration. These Industry 4.0 technologies are important influencing factors for manufacturing processes [1,2] and they lead to the appearance of dynamic manufacturing networks [3].

Augmented and virtual reality is a key technology for smart manufacturing because it makes it possible to realize an interactive human–machine interaction in a real-world environment while the components of the physical world are extended by perceptual information. Augmented and virtual reality can be used in training, design, manufacturing, operation, services, sales, and marketing. In the field of manufacturing, the most important applications are quality control and total quality management; maintenance operations, especially in a dangerous environment; assembly work instructions; and performance monitoring [4].

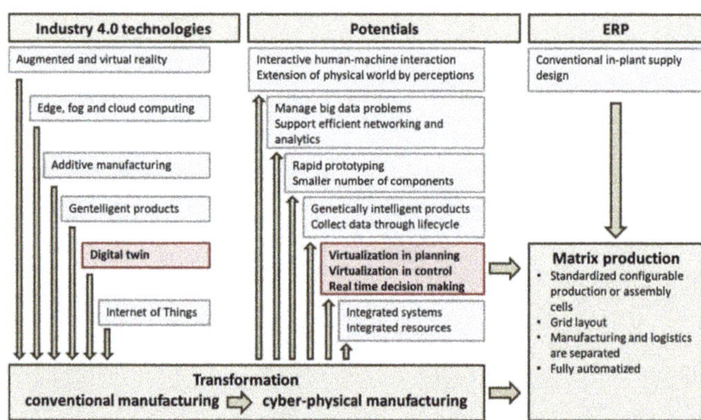

Figure 1. Industry 4.0 technologies as new innovation accelerators and their impact on matrix production.

Complex manufacturing systems generate unprecedented amounts of data that are difficult to handle with traditional computing methods. Cloud, edge, and fog computing make it possible to manage big data problems. Big data is coming from a wide range of sensors from manufacturing systems. Cloud and fog computing integrate servers, storages, databases to support efficient networking, analytics, and intelligence solutions [5].

The introduction of additive manufacturing will have a great impact on the supply chain processes and logistics solutions, because both external and in-plant material flow solutions will change dramatically. It is caused by the fact that this technology is based on the building of 3D objects by adding layer-upon-layer of various materials, like plastic, metal, or organic materials [6].

The new concept of gentelligent products aims to develop genetically intelligent products and components, which collect data through their lifecycle and bequeath them to the next generation in various time spans. The appearance of gentelligent products has a great impact on big data problems [7].

The application of digitalization-based technologies enables the virtualization of product and process planning and control [8]. Digital twins represent an integrated probabilistic simulation of complex products or processes using physical models, sensor updates, and cloud-based information to mirror the product or process of its corresponding twin [9,10]. Digital twin technology makes it possible to convert conventional manufacturing systems into cyber-physical systems, and this transformation can lead to the improvement of the design process of in-plant material supply, adding a real-time phase to the conventional in-plant supply process. In conventional manufacturing systems, the real time optimization is almost impossible, because real time optimization is based on real time data and status information. Using digital twin technology and smart sensor networks, real time data and status information can be collected from the physical system, and a real time model for discrete event simulation can be generated to perform scenario analysis for real time decision making.

The Internet of Things describes an integrated system of computers and mechanical machines provided with unique identifiers. The IoT in manufacturing systems makes it possible to transfer data through a network among manufacturing equipment (standardized production cells and assembly cells), materials handling machines (autonomous mobile robots and automated guided vehicles), intelligent tools, gentelligent products, and ERP systems [11].

The Industry 4.0 technologies make it possible to transform conventional manufacturing processes to cyber-physical manufacturing processes to aim for higher flexibility, productivity, availability, cost-efficiency, energy-efficiency, and sustainability. The fulfilment of more and more diverse customers' demands requires more and more sophisticated, flexible, and intelligent solutions based on these technologies both inside and outside of the production plants in all fields of industry including automotive industry as a flagship.

The in-plant material supply solutions are commonly based on milk-run material supply, especially in the field of automotive industry. KUKA AG (one of the world's leading specialists in automation) offered a new, revolutionary solution for flexible manufacturing, transforming conventional manufacturing into cyber-physical manufacturing with the application of Industry 4.0 technologies. This new solution is the matrix production. With its new demonstration plant opened on March 2018 in Augsburg, KUKA demonstrates the advantages of this matrix production under real conditions. In a matrix production system, standardized configurable production or assembly cells are arranged in a grid layout. Manufacturing and logistics are separated and fully automatized. The matrix production system uses various Industry 4.0 technologies, like robots and turntables in the production and assembly cells, autonomous guided vehicles, digital twin support for real time control, prediction, and performance analysis. However, as a journalist wrote [12], "However, all theory is gray." There is a huge number of open questions focusing on manufacturing and logistics.

Manufacturing systems of increased complexity face a number of new design and operation problems that can be addressed by the opportunities provided by the Fourth Industrial Revolution. In the case of matrix production, the material supply of standardized configurable production or assembly cells is one of the most important tasks of logistics, because the separated manufacturing and logistics and the increased flexibility require new models and methods. This article focuses on the optimization of in-plant supply in matrix production. The highlights of the article are the following: (1) integrated model to solve the in-plant material supply problem in matrix production system, which enables both the conventional and real time planning of in-plant material supply; (2) integrated solution of assignment and routing problems based on heuristic optimization algorithms.

The article is organized as follows. Section 2 presents a systematic literature review, which summarizes the research background of in-plant supply optimization in manufacturing systems. Section 3 is the problem description including the mathematical model of integrated assignment and routing problem in matrix production systems. Section 4 presents a metaheuristic optimization algorithm to solve the integrated assignment and routing problem, based on flower pollination and black hole heuristics. Section 5 demonstrates the numerical results. Conclusions, managerial impacts, and future research directions are discussed in the remaining part of the article.

2. Literature Review

Within the frame of the systematic literature, the main scientific results, scientific gaps, and bottlenecks are identified and described [13]. The optimization of logistics and supply chain design and control of manufacturing systems has been researched in the past 30 years. The first articles in this field were published before 2000, focusing on heuristic optimization of rough-mill yield with production priorities [14], optimum allocation of jobs on machine-tools [15], and facility location problem for large-scale logistics [16]. The number of published research papers has increased; it shows the importance of the optimization of manufacturing-related supply chain solutions.

The literature introduces a wide range of design methods used to solve problems of manufacturing-related processes, like unified decomposition, decision-making methods, queuing theory, data-driven modelling, fuzzy description, and heuristic and metaheuristic algorithms and simulation.

Researchers solved a simultaneous planning task of an integrated production, inventory, and inbound transportation problem as a mixed-integer linear program and proposed a three-phase unified decomposition heuristic [17]. A bi-objective nonlinear programming model was proposed as a decision-making tool to select the carriers between supply chain levels with emphasis on the environmental factors [18], and the problem was solved with a multi-objective meta-heuristic imperialist competitive algorithm. For the solution of coordination problems of production planning and transportation planning, a mixed-integer linear programming model and a non-linear programming model were supposed, with a decomposition-based heuristic and a Lagrangian relaxation method [19]. Service load balancing, task scheduling, and transportation optimization problem were formulated as a new queuing network for parallel scheduling of multiple processes and orders from customers to be supplied [20]. Data-driven decision-making models are more and more important in manufacturing, especially in the field of cyber-physical manufacturing and logistics. The design and operation of manufacturing-related logistics and supply problems can be managed using data-driven models and methods [21]. Simulation models can be used both for the design of machines [22] and for the optimization of systems and processes. Simulation techniques can be used as a decision support method for process improvement of intermittent production systems [23]. A hybrid approach of discrete event simulation integrated with location search algorithm was used to solve a cells assignment problem in an assembly facility [24]. An ontology-driven, component-based framework shows the application of Jellyfish-type simulation models [25]. The suggested integration of simulation and encompassing mathematical optimization reduced the complexity of the assembly facility and generated alternative assignments in two phases.

Various heuristic and metaheuristic algorithm make it possible to solve NP-hard optimization problems in manufacturing systems. Service load balancing, scheduling, and logistics optimization in cloud manufacturing are solved with a genetic algorithm [26]. A supply chain configuration problem of manufacturing plants, distributors, and retailers is formulated as an integer-programming model and solved with an ant colony optimization-based heuristic [27]. A new mathematical model for multi-product economic order quantity model with imperfect supply batches was supposed by researchers. They developed three robust possibilistic programming approaches and solved the problems with two novel meta-heuristic algorithms named water cycle and whale optimization algorithms [28]. The whale optimization algorithm was also used to solve a production-distribution network problem [29]. A novel integrated bacteria foraging algorithm embedding a five-phase based heuristic was supposed to solve an integrated model of facility transfer and production planning in dynamic cellular manufacturing-based supply chain [30]. The design problems of closed-loop supply chains represent a special form of manufacturing-related supply problems, where disassembly operations are performed instead of manufacturing. An optimized disassembly process is required for efficient remanufacturing and recycling of returned products. The dynamic lot-sizing and vehicle routing problem of this integrated process was solved with a two-phase iterative heuristic [31]. Time- and capacity-related constraints of manufacturing-related logistics are usually taken into consideration as hard constraints, but they are in truth soft constraints, because they are influenced by more external factors and their stochastic environment. Soft constraints can be taken into consideration using biased-randomized algorithms as an effective methodology to cope with NP-hard and non-smooth optimization problems in many practical applications [32]. One optimization approach uses set partitioning and another approach employs the concept of seed routes to determine the solution of an integrated production, inventory, and distribution model for supplying retail demand locations from a production facility [33]. Iterated greedy algorithm solved the optimization problem of makespan for the distributed no-wait flow shop scheduling problem [34]. Other interesting solutions are represented by hybrid algorithms, like a hybrid genetic algorithm for multi-product competitive supply chain network design with price-dependent demand [35], a hybrid firefly-chaotic simulated annealing approach for facility layout problem [36], or a prioritized K-mean clustering hybrid genetic algorithm for discounted fixed charge transportation problems [37]. Manufacturing and in-plant supply processes are typical uncertain environments, where fuzzy modelling and fuzzy optimization offer suitable tools, and fuzzy approach can easily integrate with other analytical or heuristic algorithms [38].

Several scenarios and case studies related to the research field were assessed and evaluated in various articles. The case studies of manufacturing-related logistics and supply chain problems are generally focusing on traditional manufacturing, cloud manufacturing [26], or dynamic cellular manufacturing [30], and only a few of them are discussing the logistics and in-plant supply problems of cyber-physical manufacturing systems, especially the matrix production concept. The most important fields of case studies are from the automotive industry, but valuable case studies have been published in the fields of perishable inventory systems [39], biofuel supply [40], fast moving parts [41], garment manufacturing [42], rice supply chain [43], luxury watches [44], or winery [45].

In this article, black hole and flower pollination heuristic is used. Albert Einstein was the first scientist who predicted the existence of black holes in 1916. American astronomer John Wheeler was the denominator of black holes. When a star burns out, it may collapse, or fall into itself. In the case of smaller stars, they become a neutron star or a white dwarf, while in the case of larger stars they will create a stellar black hole. Black holes are invisible, but the environment outside of the Schwarzschild radius can be analyzed. The black holes have a great impact on particles near them. If the distance between the particle and the core of the black hole is smaller than the Schwarzschild radius, the particle can move in any direction, but in the other case, the space-time is deformed and the particle will be absorbed by the black hole. The black hole heuristic is based on this phenomenon of black holes in the outer space [46]. There are various applications of the black hole heuristic, like discrete sizing optimization of planar structures [47], feature selection and classification on biological data [48], and optimization of consignment-store-based supply chain [49] or for urban traffic network control [50].

Flower pollination-based heuristic belongs to the bio-inspired algorithms [51]. This algorithm is used in various fields, like identifying essential proteins [52], multi-level image thresholding [53], visual tracking [54], EEG-based person identification [55], or double-floor corridor allocation problem [56]. The solutions of the mathematical problems are represented by pollen grains, and the optimization process is based on the moving of these grains in the search space modelled by biotic, probiotic, and self-pollination. The algorithm can be described in four important steps.

As the above-mentioned content analysis shows, existing studies focus on the analytical and heuristic optimization of both conventional and cyber-physical manufacturing systems, while only a few of them consider the energy efficiency aspects of in-plant material supply in cyber-physical systems.

More than 50% of the articles were published in the last 5 years. This result indicates the scientific potential of the design of in-plant supply solution of cyber-physical manufacturing environment. The articles that addressed the design and control problems of the manufacturing system and their material supply problems are focusing on conventional manufacturing, and only a few of them describe the logistic problems of cyber-physical manufacturing. Therefore, this research topic still needs more attention and research. According to that, the focus of this research is the modelling and optimization of in-plant supply of the matrix production system, focusing on cell assignment and routing problems.

Table 1 summarizes the main contributions of the related research works from the main contribution and the focus on manufacturing, optimization method, and sustainability point of view. As the analysis shows, a wide range of research works focus on the optimization of conventional manufacturing systems from technology and in-plant supply point of view, and these works are using both analytical methods and heuristics. There are some research works related to the in-plant supply optimization in cyber-physical systems, but these researches are focusing on KPIs (Key Performance Indicators). The table identifies a research gap, because the in-plant material supply of cyber-physical systems has not been extensively published until now. As a consequence related to the analysis shown in Table 1, the main contributions of this article are the followings: (1) model framework of autonomous guided vehicles-based supply of matrix production; (2) mathematical description of cell assignment and routing problem in matrix production; (3) computational method based on flower pollination algorithm to solve the assignment and routing problem in matrix production; and (4) computational results of the described model to validate the models and the methods.

Table 1. Authors' contributions related to the optimization of cyber-physical production systems including I4.0 and heuristic optimization approaches.

Research	Contribution	Optimization		Manufacturing		Sustainability
		Analytical	Heuristics	Conventional	Cyber-physical	
Rosin et al., 2020 [1]	Application of principles and tools of I4.0 in lean management				✓	
Skapinyecz et al., 2018 [2]	Optimal selection of logistics service providers in Industry 4.0	✓			✓	
Tchoffa et al., 2019 [3]	Extension of federated interoperability framework in I4.0	✓			✓	
Alcácer et al., 2019 [4]	Information and communication technologies in I4.0				✓	
Dastjerdi et al., 2016 [5]	Impact of fog computing on IoT solutions				✓	
Huang et al., 2013 [6]	Additive manufacturing and sustainability				✓	✓
Wu et al., 2010 [7]	Magnetic magnesium for data storage in gentelligent products			✓	✓	✓
Guo et al., 2019 [8]	Modular based flexible digital twin for factory design				✓	
Tao et al., 2018 [9]	Digital twin-enabled product design, manufacturing, and service				✓	
Ding et al., 2019 [10]	Digital twin-based cyber-physical production system				✓	
Cui et al., 2020 [11]	Big data applications			✓	✓	
Schahinian, 2020 [12]	Concept of matrix production				✓	
Bányai et al., 2019 [13]	Real time optimization of matrix production systems		✓		✓	
Azarm et al., 1991 [14]	Production priorities in the heuristic optimization of rough-mill yield		✓	✓		
Kops et al., 1994 [15]	Optimum allocation of jobs on machine tools	✓		✓		
Hidaka et al., 1997 [16]	Facility location for large-scale logistics using heuristics		✓	✓		
Chitsaz et al., 2019 [17]	Joint optimization of production and distribution		✓	✓		
Eydi et al., 2020 [18]	Decision making for supplier and carrier selection		✓	✓		✓
Feng et al., 2018 [19]	Integrated production and transportation planning		✓	✓		✓
Ghomi et al., 2019 [20]	Optimization in cloud manufacturing		✓		✓	
Sadati et al., 2018 [21]	Identification of significant control variables in manufacturing		✓	✓	✓	
Haberer et al., 2016 [22]	Optimization of a crawler track unit	✓		✓		
Tamás, 2017 [23]	Simulation-enabled decision making in manufacturing processes	✓		✓		
Saez-Mas et al., 2020 [24]	Hybrid approach for cell assignment problems		✓	✓		
Bohács et al., 2017 [25]	Ontology-driven framework for Jellyfish-type simulation	✓	✓	✓		
Ghomi et al., 2019 [26]	Optimization of queueing problems in cloud manufacturing		✓		✓	
Hong et al., 2018 [27]	Multi-stage supply chain optimization		✓	✓		✓
Khalilpourazari et al., 2019 [28]	Analysis of impact of defective supply batches		✓	✓		
Mehranfar et al., 2019 [29]	Sustainability oriented product distribution		✓	✓		✓
Liu et al., 2017 [30]	Impact of facility transfer on cellular manufacturing		✓	✓	✓	
Habibi et al., 2017 [31]	Integrated optimization f collection and disassembly		✓	✓		
Juan et al., 2020 [32]	Soft constraints in production optimization		✓	✓	✓	
Russel, 2017 [33]	Optimization in production routing		✓	✓		
Shao et al., 2017 [34]	No wait flow shop scheduling optimization		✓	✓		
Saghaeeian et al., 2018 [35]	Multi-product competitive supply chain network design		✓	✓	✓	
Tayal et al., 2018 [36]	Facility layout optimization from big data point of view		✓	✓	✓	
Tari et al., 2018 [37]	Discounted fixed charge transportation problems		✓	✓		✓
Sakalli et al., 2018 [38]	Integrated stochastic production and distribution planning		✓	✓		
Abouee-Mehrizi et al., 2019 [39]	Design of perishable inventory systems with Markov decision process	✓		✓		
Aboytees et al., 2020 [40]	Optimization of hub-and-spoke network problems		✓	✓		✓
Behfard et al., 2018 [41]	Optimization of last time buy problem for fast moving parts		✓	✓		
Ma et al., 2018 [42]	Resource sharing optimization		✓	✓		
Cheraghalipour et al., 2019 [43]	Agricultural supply chain optimization for wide geographic range		✓	✓		✓
Respen et al., 2017 [44]	Perturbations in production plan, demand, and dispatching		✓	✓		
Varas et al., 2018 [45]	Lot sizing for uncertain demands		✓	✓		
Hatamlou, 2013 [46]	Heuristic data clustering		✓			
Gholizadeh et al., 2019 [47]	Discrete sizing optimization with heuristics		✓			
Pashaei et al., 2017 [48]	Binary black hole heuristics		✓			
Bányai et al., 2017 [49]	Consignment-store-based supply chain optimization		✓	✓		✓
Khooban et al., 2017 [50]	Fuzzy logic-based urban traffic network control		✓			
Lei et al., 2019 [51]	Flower pollination heuristics		✓			
Lei et al., 2018 [52]	Application of flower pollination heuristics		✓			
Shen et al., 2018 [53]	Multi-level image thresholding with flower pollination heuristics		✓			

Table 1. *Cont.*

Research	Contribution	Optimization		Manufacturing		Sustain-ability
		Analyti-cal	Heuris-tics	Conven-tional	Cyber-physical	
Gao et al., 2018 [54]	Visual tracking with flower pollination heuristics		✓			
Rodrigues et al., 2016 [55]	Binary flower pollination algorithm		✓			
Guan et al., 2019 [56]	Double-floor corridor allocation		✓			
Kherabadi et al., 2017 [57]	Gravitational search algorithm in Fuzzy controllers		✓			
Szentesi et al., 2021 [58]	Process optimization for distribution logistics	✓		✓		
Bányai et al., 2017 [59]	Optimization of blending technologies	✓		✓		
Hardai et al., 2021 [60]	Logistics aspects of I4.0			✓		
Kundrák et al., 2019 [61]	Efficiency improvement in manufacturing technologies			✓		
This proposal	Optimization of in-plant supply for matrix production		✓		✓	✓

3. Materials and Methods

The optimization problem of the matrix production-based in-plant supply has two stages. Within the frame of the first stage, the various production orders must be assigned to the available standardized production cells, while the second phase focuses on the optimal routing of automated guided vehicles. The structure of the integrated assignment and routing model can be seen in Figure 2.

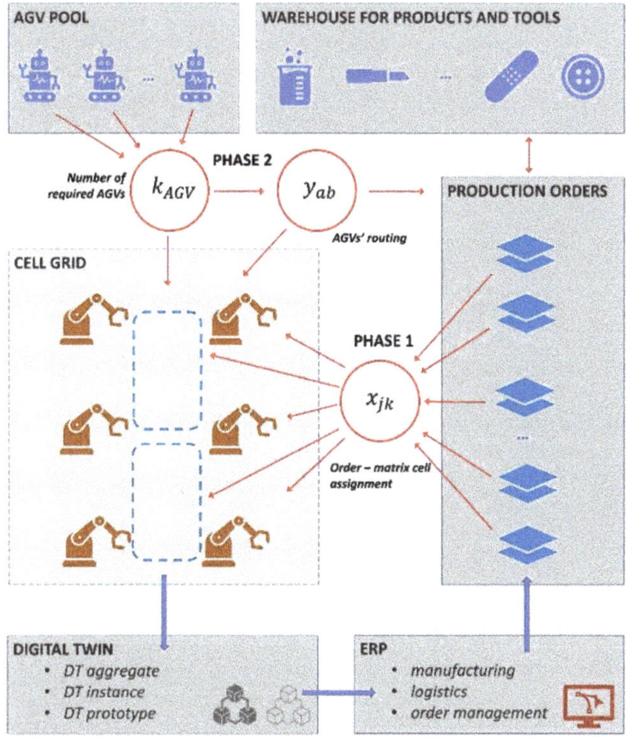

Figure 2. Integrated model of assignment and routing problem in a cyber-physical manufacturing environment.

Phase 1 includes the assignment of production orders to the grid cells. Production orders are generated by the Enterprise Resource Planning (ERP) using the results of Material Requirement Planning (MRP-I) and Manufacturing Resource Planning (MRP-II). The ERP is connected to the sensors and data collection units of cyber-physical environment through a digital twin solution, which makes it possible to make real time analysis, controlling and forecasting. The size of the AGV pool defines the number of available AGVs, which has a great impact on the in-plant supply process from an availability and efficiency point of view. The more available AGV in the AGV pool, the higher the flexibility and availability, which can influence the utilization of technological resources caused by the changeover time. The second part of the matrix production system includes the storages for tools and components required for the manufacturing. The more the available tool set for required changeover operation, the higher the flexibility and resource utilization for technological resources.

Phase 2 includes the routing of AGVs available in the AGV pool. A typical route of an AGV includes the following tracks: (1) from AGV pool to the warehouse, (2) from the warehouse to the first cell grid of the scheduled route, (3) tracks among cells grids, and (4) from the last cell grid back to the AGV pool. The objective function is either resource- or sustainability-based. Resource-based objective function means the minimization of numbers of required AGVs, while sustainability-based objective means the minimization of energy-consumption of material supply operations. The input parameters of the integrated assignment and routing problem are the followings:

- τ_{ij}^{p} is the production lead time of production order i at production cell j, where $i = 1 \ldots m$ and $j = 1 \ldots n$;

- τ_{ikj}^{c} is the changeover time among production orders between production order i and production order k at production cell j, where $k = 1 \ldots m$, and $\tau_{ikj}^{c} \geq 0$ if it is possible to perform a change between production order i and k at production cell j, otherwise $\tau_{ikj}^{c} = -1$;

- a_{ij} is the availability matrix, which takes a value of 1 if the production order i can be assigned to matrix cell j, otherwise 0.

- a_{ikj}^{c} is the changeover availability matrix, which takes a value of 1 if it is possible to change from production order i to production order k at matrix cell j, otherwise 0;

- τ_{i}^{lower1} and τ_{i}^{upper} are the lower and upper time limits of finishing operation i in the first phase (assignment) of the optimization;

- τ_{i}^{lower2} and τ_{i}^{upper2} are the lower and upper time limits of finishing operation i in the second phase (routing) of the optimization;

- s_{j}^{upper1} is the upper limit of operations at production cell j;

- z_{ij} is the required toolset for production order i at matrix cell j; and

- r_{g}^{max} is the available number of required toolset g.

3.1. Assignment of Production Operations to Matrix Cells

Within the frame of this phase, the assignment problem of required production operations (production orders) to available standardized flexible production cells is described. The decision variable of the assignment problem is the assignment matrix x_{jk}, which defines that operation x_{jk} production order is assigned to the matrix cell j as k^{th} operation.

The objective function of the first phase of the optimization problem is the minimization of the total operation time within a predefined timeframe, which can be calculated as a sum of the production operations and changeover times:

$$\tau = \tau^{p} + \tau^{c}, \tag{1}$$

where τ^p is the production lead time, and τ^c is the changeover time among the various production operations of the standardized production cells. The first part of the objective function represents the total operation time, which can be calculated as follows:

$$\tau^p = \sum_{j=1}^n \sum_{k=1}^\omega \tau^p_{x_{jkj}}, \tag{2}$$

where ω_j is the number of assigned production orders to production cell j.

The second part of the objective function describes the changeover time among the scheduled operation of matrix cells depending on the assignment:

$$\tau^c = \sum_{j=1}^n \sum_{k=1}^{\omega_j - 1} \tau^c_{x_{jk}x_{jk+1}j}. \tag{3}$$

As an alternative objective function, it is also possible to take into consideration the minimization of the required time spans to fulfil all production orders:

$$\tau^a = \max_j \left(\sum_{k=1}^\omega \tau^p_{x_{jkj}} + \sum_{k=1}^{\omega_j - 1} \tau^c_{x_{jk}x_{jk+1}j} \right) \rightarrow min. \tag{4}$$

Within the frame of the assignment problem, various constraints must be taken into consideration, like time- and capacity-related constraints. The solution of the assignment problems is limited by these constraints.

Constraint 1 defines that production orders can be assigned to suitable production cells:

$$\forall j, k : a_{x_{jkj}} = 1 \rightarrow x_{jk} > 0, \tag{5}$$

Constraint 2 describes that there are production operation pairs and matrix cells, where it is not possible to perform a changeover:

$$a^c_{ikj} = 1 \rightarrow \tau^c_{ikj} \geq 0 \text{ and } a^c_{ikj} = 0 \rightarrow \tau^c_{ikj} = -1. \tag{6}$$

Constraint 3 describes that the operation of production orders must be finished between the lower and upper limit of end time, so it is not allowed to exceed these time-related constraints:

$$\forall i = x_{jk} : \tau^{lower1}_i \leq \sum_{l=1}^k \tau^p_{x_{jlj}} + \sum_{l=1}^{k-1} \tau^c_{x_{jl}x_{jl+1}j} \leq \tau^{upper1}_i. \tag{7}$$

Constraint 4 describes that the number of operations is limited at each production cell, so it is not allowed to exceed the upper limit of operations at a chosen production cell:

$$\forall j : \max_k \left(x_{jk} > 0 \right) \leq s^{upper}_j. \tag{8}$$

Constraint 5 describes that one production order can be assigned exactly to one production cell:

$$\forall j \neq j^* \wedge k \neq k^* : x_{jk} \neq x_{j^*k^*}. \tag{9}$$

Constraint 6 describes that it is not allowed to exceed the available number of toolsets within a time frame:

$$\forall t : \sum_{j=1}^n z_{x_{jkj}}(t) \leq r^{max}_g. \tag{10}$$

3.2. Routing of AGVs in Cell Grid

Within the frame of this phase, the assignment of production orders to matrix cells is given (the production plan is defined) and the optimal routing of available automated guided vehicles must be solved based on the results of the assignment problem. The decision variable of this routing problem is a matrix including permutation arrays, where one permutation array represents the optimal route of an automated guided vehicle. The y_{ab} routing matrix defines that the bth station of AGV a is the matrix cell assigned to production order y_{ab}.

The objective function of the second phase routing problem is the minimization of vehicle fleet size and the minimization of energy consumption of in-plant supply:

$$k_{AGV} \to min. \text{ and } c \to min. \tag{11}$$

where k_{AGV} is the required number of AGVs and c is the calculated energy consumption.

The minimization of the fleet size can be described as the maximum size of fleet within the frame of the time frame:

$$k_{AGV} = \max_b(y_{ab} > 0) \to min. \tag{12}$$

The minimization of the energy consumption cannot be defined as the minimization of the routes, because energy consumption depends on the weight of the load:

$$c = c^I + c^{II} + c^{III} \tag{13}$$

where c^I is the energy consumption of the AGVs from the warehouse to the first station (matrix cell) of the in-plant supply route, c^{II} is the energy consumption of the AGVs among the stations (matrix cells), while c^{III} is the energy consumption of the AGVs from the last station (matrix cell) to the warehouse.

The energy consumption of the AGV from the warehouse to the first station (matrix cell) of the in-plant supply route can be defined as a function of length of the route and the weight of the load:

$$c^I = \sum_{a=1}^{k_{AGV}} \left(l_{0j(y_{a1})} \sum_{b=1}^{b_a^{max}} q_{y_{ab}} \right), \tag{14}$$

where b_a^{max} is the number of stations of in-plant supply route a, $q_{y_{ab}}$ is the weight of the load for production order scheduled as station b of route a, and $l_{0j(y_{a1})}$ is the length of the transportation between the warehouse and the first matrix cell of the route.

The energy consumption of the AGV among matrix cells can be defined as follows:

$$c^{II} = \sum_{a=1}^{k_{AGV}} \left(\sum_{b=1}^{b_a^{max}-1} \left(l_{j(y_{ab})j(y_{ab+1})} \sum_{d=b}^{b_a^{max}} q_{y_{ad}} \right) \right), \tag{15}$$

where $j(y_{ab})$ is the matrix cell ID assigned to the production order, which is scheduled to route a as station b.

The energy consumption of the AGV from the last matrix cell of the in-plant supply route and the warehouse can be defined as follows:

$$c^{III} = \sum_{a=1}^{k_{AGV}} \left(l_{j(y_{ab_a^{max}})0} q_{y_{ab_a^{max}}} \right), \tag{16}$$

where $q_{y_{ab_{max(a)}}}$ is the weight of the load for production order scheduled to the last station of in-plant supply route a, and $l_{j(y_{ab_a^{max}})0}$ is the length of the transportation between the last matrix cell of route a and the warehouse.

Within the frame of this routing problem, various constraints must be taken into consideration, like time-, capacity- and energy consumption-related constraints. The solution of the routing problem is limited by these constraints.

Constraint 1 defines that it is not allowed to exceed the maximum number of stations within one supply route:

$$\forall a : b_a^{max} = \max_b(y_{ab} >_a^{max}) \leq v_a^{max}, \tag{17}$$

where v_a^{max} is the upper limit of the number of stations assigned to route a.

In the case of electric AGVs and heavy loadings, it is important to take into consideration the impact of weight and route length on the energy consumption, because in the case of heavy loadings the transportation route can be limited. Energy consumption constraints can be transformed to material flow intensity constraints, because we can define a proportion of energy consumption and material flow intensity (product of length and weight).

Constraint 2 defines that it is not allowed (and not possible) to exceed the material flow intensity, which depends on the weight of loading and length of route:

$$\forall a : q_a^I + q_a^{II} + q_a^{III} \leq q_a^{max}, \tag{18}$$

where

$$q_a^I = l_{0j(y_{a1})} \sum_{b=1}^{b_a^{max}} q_{y_{ab}} \tag{19}$$

$$q_a^{II} = \sum_{b=1}^{b_a^{max}} \left(l_{j(y_{ab})j(y_{ab+1})} \sum_{d=b}^{b_a^{max}} q_{y_{ab}} \right) \tag{20}$$

$$q_a^{III} = l_{j(y_{ab_a^{max}})0} q_{y_{ab_a^{max}}} \leq q_a^{max} \tag{21}$$

Constraint 3 defines that it is not allowed to exceed the upper and lower limit of arrival time at the matrix cells:

$$\forall y_{ab} : \tau_i^{lower2} \leq \sum_{d=0}^{b-1} \tau_{j(y_{ad})j(y_{ad+1})}^t + \tau_{j(y_{ad+1})}^t \leq \tau_i^{upper2} \tag{22}$$

where $\tau_{j(y_{ad})j(y_{ad+1})}^t$ is the transportation time between matrix cells assigned to the station b of route a, and $\tau_{j(y_{ad+1})}^h$ is the material handling time (loading and unloading) at matrix cell assigned to the station $d + 1$ of route a. The lower and upper limit for arrival time depends on the assignment matrix.

Constraint 4 defines that it is not allowed to exceed the upper limit of capacity (weight or volume) of automated guided vehicles:

$$\forall a : \sum_{b=0}^{b_a^{max}} q_{y_{ab}} \leq q_a^{max} \tag{23}$$

where q_a^{max} is the upper limit of capacity of route (or vehicle) a.

Constraint 5 defines that supply demands can be transported only with appropriate vehicles:

$$\forall y_{ab} : a(y_{ab}) \in \Xi_{y_{ab}} \tag{24}$$

where $\Xi_{y_{ab}}$ is the set of vehicles appropriate for transportation of required materials and tools of production order y_{ab} from the warehouse to the assigned matrix cell. The description of nomenclatures used in the mathematical model can be seen in Appendix A.

To solve the above-described integrated assignment and routing problem, a multiphase optimization algorithm will be described.

4. Results

The multiphase solution algorithm includes the optimization of assignment of production orders to matrix cells and the routing of autonomous guided vehicles among AGV pool, warehouse, and matrix cells. The solution of the assignment problem is based on a black-hole heuristic, while the routing (which also includes a virtual scheduling of production orders) is solved with a flower pollination-based heuristic.

4.1. Black-Hole Heuristic for the Assignment Problem

This population-based heuristic can be summarized in five major steps. The first step is the generation of an initial population of stars representing the initial solutions of the real problem. The coordinates of the generated stars describe the decision variables of the optimization problem. The decision variable of the above-described assignment problem is the assignment matrix, which defines the assignment of production orders to matrix cells, so the initial solutions of the black hole algorithm can be defined as follows:

$$X^{0\alpha} = \left[x_{jk}^{0\alpha} \right] \tag{25}$$

where $x_{jk}^{0\alpha}$ is the ID of the production order assigned to the matrix cell j as kth operation of the initial solution α. The initial solution matrix has m numbers, where $x_{jk}^{0\alpha} \geq 1$. $\alpha = 1...\lambda$, and λ is the number of initial solutions.

The second step is the evaluation of the initial solutions with the objective function and calculate the gravity force of the star.

$$e_{jk}^{\mu\alpha} = \max_{j} \left(\sum_{k=1}^{\omega} \tau_{x_{jk}j}^{p} + \sum_{k=1}^{\omega_j - 1} \tau_{x_{jk}x_{jk+1}j}^{c} \right) \tag{26}$$

where μ is the iteration step and $\mu = 0$ directly after the initialization of the solution matrix. We can write that

$$x_{jk}^{\mu\alpha} \geq 1 \rightarrow e_{jk}^{\mu\alpha} > 0 \tag{27}$$

The third phase is to find the best solution in this iteration step. This best solution is dedicated as the black hole of the search space and all other stars representing worst solutions will move toward this solution. We can also define more black holes, but in this case the algorithm is like gravity force algorithm [57].

$$e_{BH}^{\mu} = \max_{\alpha} \left(e_{jk}^{\mu\alpha} \right) = \max_{\alpha} \left(\max_{j} \left(\sum_{k=1}^{\omega} \tau_{x_{jk}j}^{p} + \sum_{k=1}^{\omega_j - 1} \tau_{x_{jk}x_{jk+1}j}^{c} \right) \right) \tag{28}$$

The fourth phase of the black hole heuristic is to move the stars towards the black holes. The speed and distance of moving depends on the value of objective function, which is represented by the gravity force of the star.

$$x_{jk}^{\mu\alpha} = x_{jk}^{\mu-1,\alpha} + round \left(rnd \left| x_{BH}^{\mu-1} - x_{jk}^{\mu-1,\alpha} \right| \right) \tag{29}$$

Stars reaching the event horizon described by the value of Schwarzschild radius will be absorbed and a new star will be initialized. After this step, various termination criteria can be taken into consideration, like computational time or the measure of convergence.

Within the frame of a scenario including 16 production orders and 9 matrix cells, this paper will demonstrate the described model and the results of the black hole heuristic-based assignment optimization. We can define both the availability matrix of matrix cells and the operation time of matrix cells for each production order. Table 2 shows the operation time of production orders. It is not necessary to describe both matrices, because the operation time can be defined as a ∞ value if the production order cannot be fulfilled in the matrix cell.

We can define the changeover time of matrix cells between production orders. This changeover time is caused by the various required tool sets of production orders. If the production orders are changed at a matrix cell, the following operations are required: (1) take down the used tool set of the matrix cell, (2) collect remaining components of previous production order, (3) transport the old tool set to the tool storage and the remaining components to the warehouse, (4) transport the new required tool set to the next production order from the tool store to the matrix cell, (5) transport the required components from the warehouse to the matrix cell, and (6) set up the new tool set of the production order. These changeover times for this scenario are summarized as a total changeover time in Table 3.

Table 2. Operation time of production orders [min].

Production Order ID	Matrix Cell ID								
	1	2	3	4	5	6	7	8	9
1	2.5	3.3	4.0	5.6	7.9	1.2	9.5	4.8	1.5
2	5.4	4.8	7.4	8.6	6.1	1.3	9.8	9.7	3.6
3	8.0	1.0	9.6	9.9	9.9	1.6	8.8	6.5	5.7
4	9.1	1.1	2.7	2.1	6.9	1.5	9.2	3.9	9.9
5	7.6	1.8	2.0	4.8	5.4	1.3	8.5	4.9	5.5
6	6.8	1.6	7.2	3.8	4.3	1.6	9.3	2.3	5.3
7	9.2	8.0	6.9	8.0	7.5	.17	8.6	1.5	5.6
8	4.5	7.1	8.0	1.7	2.1	1.3	9.0	8.7	8.0
9	6.8	5.3	9.5	4.2	2.6	1.2	8.5	7.2	9.0
10	4.6	6.1	4.1	6.6	2.4	1.2	9.0	6.1	5.0
11	8.3	5.6	5.0	3.9	8.7	1.1	9.6	8.3	5.2
12	7.6	9.1	8.2	8.8	5.6	1.1	9.6	2.5	5.9
13	2.0	4.5	6.3	7.7	3.1	1.1	9.6	3.4	5.8
14	4.4	5.9	3.2	2.7	1.0	1.5	8.8	6.5	2.6
15	7.9	6.9	6.7	8.3	1.8	1.4	9.1	4.6	6.5
16	8.1	4.5	9.0	8.4	8.3	1.4	9.2	8.2	4.9

Table 3. Changeover time between production operations (OID = Production order ID) [min].

OID	OID															
	1	2	3	4	5	6	7	8	9	10	11	12	13	14	15	16
1	0	1.8	3.0	3.5	2.9	8.3	2.0	7.5	7.0	1.1	1.0	2.1	4.3	6.5	7.7	7.8
2	7.3	0	4.1	7.2	5.0	8.3	8.9	8.7	7.5	4.0	9.5	3.4	8.2	9.7	9.7	2.6
3	1.1	9.6	0	2.6	1.1	6.8	7.9	5.9	4.1	5.8	8.2	7.4	8.1	6.5	1.7	7.2
4	5.5	1.1	7.1	0	7.1	6.1	9.1	8.4	7.6	8.3	7.1	5.4	2.7	4.3	9.3	2.1
5	8.3	4.4	1.2	7.9	0	4.9	7.8	1.2	8.7	4.2	9.1	8.5	8.2	7.1	9.3	9.9
6	5.6	2.5	3.4	8.2	4.4	0	1.9	2.9	6.5	6.8	4.8	9.2	9.0	7.9	5.5	4.0
7	5.4	3.9	9.4	9.6	5.4	5.2	0	1.5	9.7	5.6	2.5	2.8	6.2	5.4	2.1	9.9
8	5.3	8.1	2.8	5.6	5.4	9.8	4.5	0	1.5	2.0	4.0	3.1	2.6	8.8	8.2	3.6
9	9.4	6.1	9.0	4.2	6.0	2.4	7.6	1.2	0	8.6	9.1	2.5	8.4	2.7	1.0	5.1
10	1.5	4.7	7.7	8.8	1.6	8.2	1.5	9.0	3.1	0	3.5	6.2	1.8	3.9	2.6	5.4
11	4.4	5.1	1.1	8.6	3.1	8.6	5.7	6.2	5.5	3.3	0	3.8	2.2	5.7	1.9	3.9
12	6.4	4.0	2.2	5.6	1.6	8.0	5.0	5.2	7.8	3.2	5.0	0	1.8	5.8	8.2	3.6
13	3.0	3.6	1.0	8.8	3.6	8.6	9.2	2.0	9.5	7.0	8.1	6.8	0	3.0	9.0	8.3
14	1.2	1.5	4.4	8.0	6.3	6.9	4.4	6.9	5.4	5.2	1.9	5.6	4.2	0	2.8	4.6
15	7.4	1.2	4.6	1.7	9.2	7.2	8.3	9.9	2.3	9.6	9.4	9.0	4.2	5.9	0	5.5
16	2.1	6.7	5.8	6.8	2.2	7.9	5.3	9.2	1.0	8.9	8.4	3.7	4.4	6.0	1.3	0

The time constraints can be defined as the lower and upper limits of the beginning and finishing of production order-related operations. Table 4 shows the time-based constraints of the scenario. The ∞ value of upper time limit defines that there is no time limit for this production order.

Table 4. Time constraints of production time (PTC = production time constraints. OID = Production order ID. BMIN = beginning time lower limit. BMAX = beginning time upper limit. FMIN = finishing time lower limit. FMAX = finishing time upper limit) [min].

PTC	OID															
	1	2	3	4	5	6	7	8	9	10	11	12	13	14	15	16
Bmin	0	2.6	6.3	0	0	0	0	2.2	0	2.2	0	4.4	7.1	0	0	0
Bmax	2.3	5.3	∞	∞	∞	9.4	2.2	7.2	8.7	3.9	5.5	7.3	9.1	∞	∞	2.1
Fmin	2.1	5.6	3.4	2.2	8.1	3.2	1.1	5.5	4.1	7.6	4.2	6.1	6.6	2.4	8.1	1.1
Fmax	3.2	∞	∞	∞	9,8	∞	5.5	8.8	8.9	9.1	6.3	7.9	∞	∞	∞	3.8

Figure 3 shows the result of the black hole heuristic-based solution algorithm. The value of the objective function is 9.1 min, which means that the last production order will be finished in 9.1 min, which is the cycle time of the 16 production orders. This numerical result shows that the described optimization algorithm can take the time-related constraints into consideration and the algorithm makes it possible to find an optimal solution for the in-plant supply optimization problem. As Figure 3 shows, in the case of the first scenario the algorithm takes a wide range of the predefined constraints into consideration, including the production time (or lead time) constraint, and the upper and lower limit of beginning and ending time for the production process. At first glance, it may seem that the changeover time in matrix cell 6 could be relocated to the matrix cells 3 or 5, thereby reducing the total manufacturing time, but this is not the case as changeover operations and idle times are not freely moveable due to technological limitations.

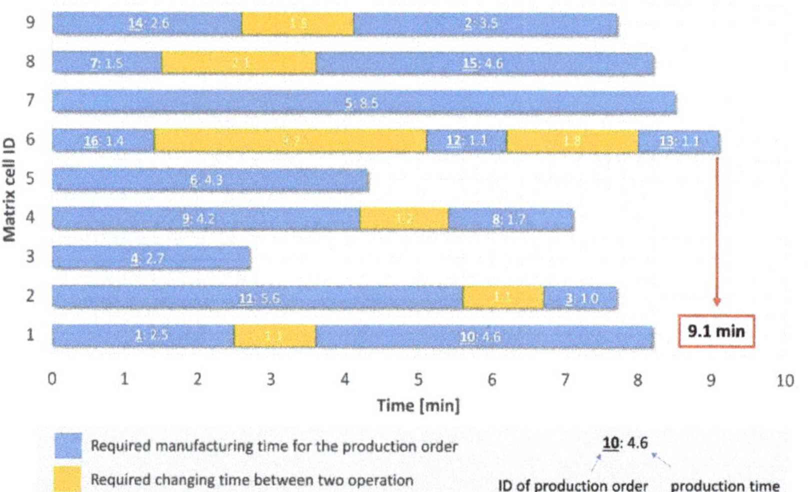

Figure 3. Gantt chart of the working process resulting from the optimal assignment of production orders to matrix cells in scenario A.

The total idle time of the matrix cells within the time window of the fulfilment of the 16 production demands is 18.4 min. The distribution of the idle time among the matrix cells is shown in Figure 4.

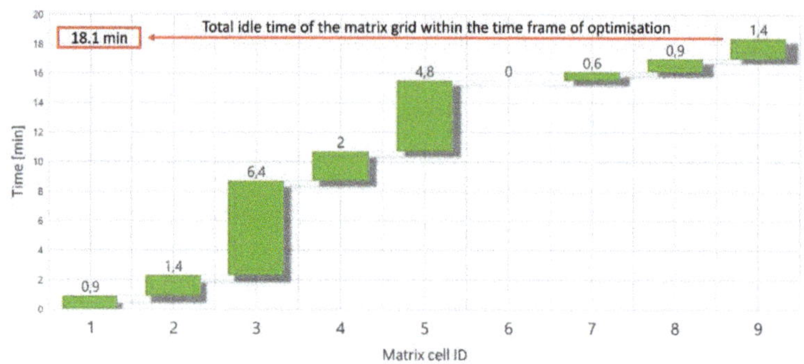

Figure 4. Idle time distribution among matrix cells within the cycle time of the 16 production orders in scenario A.

The technological and logistics resources of the matrix production system are usually state-of-the-art technologies and have expensive operation costs; therefore, it is important to optimize their idle time in order to increase their utilization. In the case of an even distribution of idle time, the production time could be reduced in this case as well, however, as in the case of the changeover time, the time-related constraints and the availability of technological resources do not allow this. The distribution of idle time and the changeover time depends on the flexibility and availability of matrix cells. Higher availability and flexibility makes it possible to produce a wider range of products, which can lead to increased changeover time.

Figure 5 shows the results of a second scenario, where the same operation and changeover times were used, but the number of available matrix cells was reduced to six and the solution was not limited by the time constraints of the previous scenario. The value of the objective function is 13.8 min, which means that the last production order will be finished in 13.8 min. As Figure 5 shows, the number of available standardized configurable productions or assembly cells has a great impact on the results of in-plant supply processes from a time and capacity point of view. However, in the matrix production system, the processes of technology and logistics are separated, but the decreased number of available technological resources influences the required logistics resources and the computation result shows a higher time span for the working process. In this case, the technological resources must have an increased flexibility and availability for the same manufacturing time. If the availability and flexibility of matrix cells does not increase, the decreased number of technological resources will result in a longer time period being required to complete production, even with a better distribution of idle times.

The total idle time of the matrix cells within the time window of the fulfilment of the 16 production demands is 9.6 min. The distribution of the idle time among the matrix cells is shown in Figure 6. The result shows that the decreased available technological resource influences also the idle time. In this case, the distribution of idle time is more even, but this change in the distribution of idle time has no positive impact on the required manufacturing time because of the decreased number of technological resources.

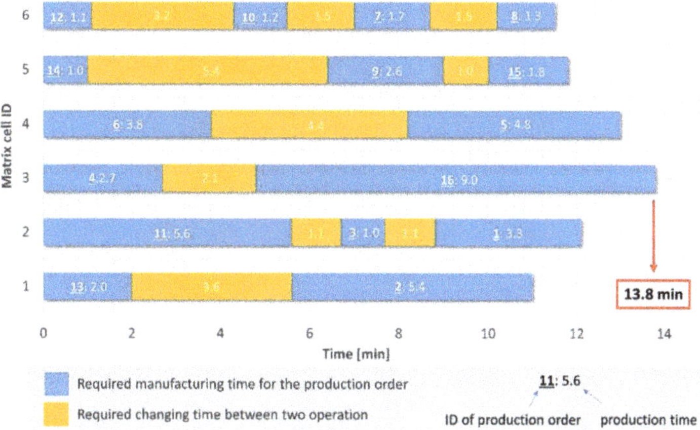

Figure 5. Gantt chart of the working process resulting from the optimal assignment of production orders to matrix cells in scenario B.

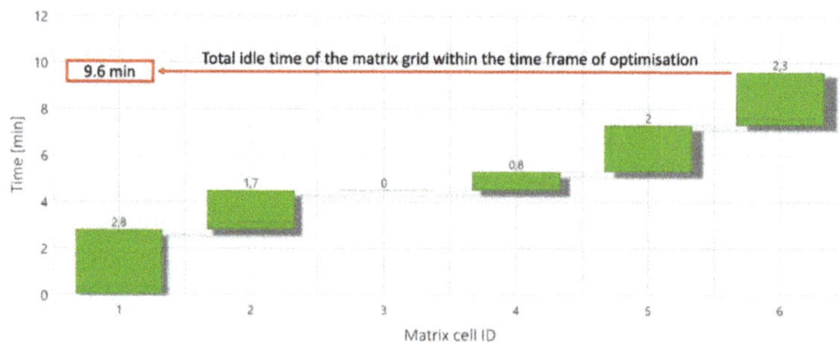

Figure 6. Idle time distribution among matrix cells within the cycle time of the 16 production orders in scenario B.

4.2. Flower Pollination Heuristic for Routing Problem

The first step of the optimization algorithm is the initialization steps, where the basic parameters of the algorithm regarding the real problem and the process of optimization will be defined. The parameters of the real problem are the size and dimension of the search space, as well as the impact of constraints on the search space. The parameters of the algorithm are the followings: switching process between global and local search (biotic and probiotic pollination), termination criteria (computation time, iteration steps, or convergence), and the number of initial solutions (pollen grains).

The second step is the initialization of the solutions, which means the definition of the pollen grains in the search space (pastureland).

$$Y^{0\alpha} = \left[y_{ab}^{0\alpha} \right] \tag{30}$$

where $y_{ab}^{0\alpha}$ is the ID of the matrix cell assigned to route a as bth station as the initial solution α. The initial solution matrix has m numbers, where $y_{ab}^{0\alpha} \geq 1$. $\alpha = 1...\lambda$, and λ is the number of initial solutions.

The next step is the evaluation of the pollen grains, which is based on the objective function of the routing problem describing the minimization of the energy consumption of the routes defined by solution α in iteration step μ:

$$e_{ab}^{\mu\alpha} = \sum_{a=1}^{k_{AGV}} \left(\sum_{b=1}^{b_a^{max}-1} \left(l_{j(y_{ab})j(y_{ab+1})} \sum_{d=b}^{b_a^{max}} q_{y_{ad}} \right) \right) + \\ + \sum_{a=1}^{k_{AGV}} \sum \left(l_{0j(y_{a1})} \sum_{b=1}^{b_a^{max}} q_{y_{ab}} \right) + \sum_{a=1}^{k_{AGV}} \left(l_{j(y_{ab_a^{max}})0} q_{y_{ab_a^{max}}} \right) \tag{31}$$

where μ is the iteration step and $\mu = 0$ directly after the initialization of the solution matrix. We can write that

$$y_{ab}^{\mu\alpha} \geq 1 \rightarrow e_{ab}^{\mu\alpha} > 0 \tag{32}$$

The third phase is the initialization of a decision number that defines the switch-possibility between biotic and probiotic pollination. The fourth step is the pollination depending on the type of search. In the case of global search, a biotic pollination is performed:

$$y_{ab}^{\mu+1,\alpha} = y_{ab}^{\mu,\alpha} + L(\lambda) \left(y_{ab}^{best,\alpha} - y_{ab}^{\mu,\alpha} \right) \tag{33}$$

where $L(\lambda)$ is the Levy-distribution.

In the case of local search, an abiotic pollination is performed:

$$y_{ab}^{\mu+1,\alpha} = y_{ab}^{\mu,\alpha} + \vartheta \left(y_{r_1 r_2}^{\mu,\alpha} - y_{r_3 r_4}^{\mu,\alpha} \right) \tag{34}$$

where $y_{r_1 r_2}^{\mu,\alpha}$ and $y_{r_3 r_4}^{\mu,\alpha}$ are random solutions in the iteration step μ, and ϑ is a random number between 0 and 1. To transform the continuous representation to a discrete permutation representation, the smallest position value rule was used.

Within the frame of a scenario including 15 production orders and 9 matrix cells, this paper demonstrates the described routing model of the matrix production and the results of the flower pollination-based routing optimization. The optimal assignment of production orders is given, so we can define the lower and upper time limits of production orders, as shown in Table 5.

Table 5. Optimal assignment of production orders to matrix cells and their lower and upper time limits as input parameters if there is a routing problem in the matrix grid (OID = production order ID. AMC = assigned matrix cell ID. MHT = material handling time of the production order at the matrix cell. BMin = beginning time lower limit. BMax = beginning time upper limit).

| | OID | | | | | | | | | | | | | | |
---	1	2	3	4	5	6	7	8	9	10	11	12	13	14	15
AMC	3	2	7	4	6	5	8	1	6	9	1	9	5	6	2
MHT	0.8	1.3	0.85	0.9	0.45	0.4	1.2	1.1	0.45	0.3	1.1	0.3	0.4	0.45	1.3
Bmin	1.1	0.3	1.1	4.2	5.1	3.2	2.2	0.4	1.1	2.2	6.5	4.5	5.1	3.7	7.8
Bmax	3.4	3.4	5.4	8.5	6.9	8.4	6.2	3.3	4.4	5.1	10.3	6.7	7.5	6.5	9.9

The distances among matrix cells, warehouses, and storages are shown in Table 6.

Table 6. Distances in the matrix grid [10 m].

Production Order ID	Matrix Cell ID									
	WH/ST	1	2	3	4	5	6	7	8	9
WH/ST	0	0.65	0.65	0.65	1.3	1.3	1.3	1.95	1.95	1.95
1	0.65	0	0.65	1.3	0.65	1.3	1.95	1.3	1.95	2.6
2	0.65	0.65	0	0.65	1.3	0.65	1.3	1,95	1,3	1,95
3	0.65	1.3	0.65	0	1.95	1.3	0.65	2.6	1.95	1.3
4	1.3	0.65	1.3	1.95	0	0.65	1.3	0.65	1.3	1.95
5	1.3	1.3	0.65	1.3	0.65	0	0.65	1.3	0.65	1.3
6	1.3	1.95	1.3	0.65	1.3	0.65	0	1.95	1.3	0.65
7	1.95	1.3	1.95	2.6	0.65	1.3	1.95	0	0.65	1.3
8	1.95	1.95	1.3	1.95	1.3	0.65	1.3	0.65	0	0.65
9	1.95	2.6	1.95	1.3	1.95	1.3	0.65	1.3	0.65	0

Figure 7 shows the optimal routing in the matrix grid. There are three routes in the matrix cell within the time span of routing. Six production orders are assigned to route 1 (blue), five production order are assigned to route 2 (red), and three production orders are assigned to route 3 (green). This computational result shows that more AGVs are required in the matrix production system. As presented in the chapter discussing the optimization algorithm, clusters must be formed from the manufacturing tasks. It can be seen in Figure 7 that the clustering algorithm, when designing the clusters of the production task forming each route, try to form clusters with an even number of production tasks, taking into account the time- and capacity-related constraints. The increased number of available AGVs can lead to decreased cluster, which influences the required manufacturing time and lead time.

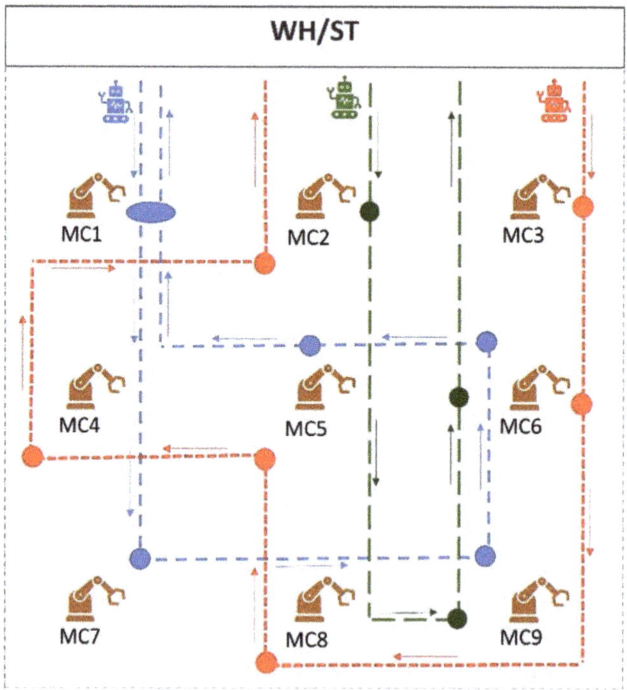

Figure 7. Three optimized in-plant supply routes in the matrix grid optimized with flower pollination-based heuristics.

Figure 8 shows the efficiency of the flower pollination-based heuristics. The scheduled production orders are between the predefined lower and upper time limits. The results show that time-related constraints also can be taken into consideration. It is especially important from the production orders point of view, because the predefined time limits, which are based on ERP data, are assumptions of the high service level in matrix production system. The time frame defined by the lower and upper limits influences the solution. In the case of a narrow time frame defined for the manufacturing of production orders, both the availability of technological resources and the availability of logistics resources must be increased to minimize the total required time frame for manufacturing all production orders.

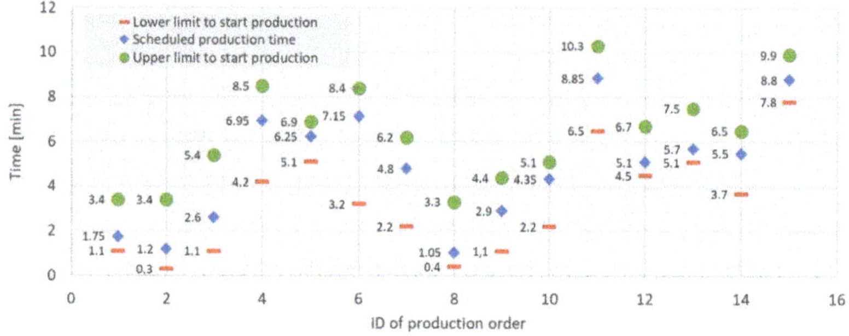

Figure 8. The distribution of scheduled production order between the related lower and upper time limits.

The assignment of production orders to matrix cells and the routing of production orders led to an optimal in-plant supply of matrix cells. The energy consumption, as the objective function of the design problem, can be calculated based on Equation (13) and is shown in Figure 9. As the computed energy consumption rates show, the energy consumptions of the in-plant supply routes are quasi-uniform, because the state-space of the heuristic optimization model representing the potential solutions of the real problem makes it possible. In the case of a decreased number of AGVs, this uniform distribution is not possible. The energy consumption has a great impact on both the operation cost and on the environmental impact. Depending on the energy generation source (oil, wind, photovoltaic, water, nuclear, biomass, etc.), we can define the emission, and this emission can be taken into consideration as a virtual emission of the manufacturing process. The energy consumption of AGVs influences the required loading of batteries, therefore, the even distribution of energy consumption makes it possible to make a more transparent loading process for the AGVs.

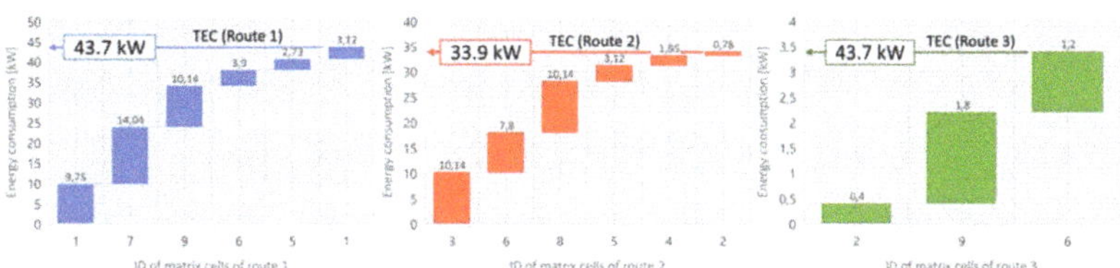

Figure 9. The distribution of energy consumption in each in-plant supply route (TEC = total energy consumption).

4.3. Challenges and Applicability in Real Industrial Environment

The above-described methodology is applicable in a real industrial environment, but there are challenges that may be faced while applying this proposed model in reality. The application is based on data from the ERP and from the digital twin. The conventional ERP data sets and real-time digital twin-enabled information for simulation-based scenario analysis and forecasting are available using standard interfaces, because standard-based interoperability is an important challenge for large, complex manufacturing systems. The optimization module for in-plant supply design can be implemented either as a part of the ERP or MES, or as an add-on software using standardized channels for information sharing. The implementation cost of these solutions can vary, add-on solutions are cheaper, but ERP-integrated optimization can lead to a more robust and stable solution. The validation of input data for digital twin is also a challenge, because the smart sensor network must have stringent dependability, especially from a reliability and availability point of view, as sensor failures can cause bad data, which influences the results of digital twin-enabled simulation and influences real time decisions. In the case of a conventional manufacturing system, the development of digital twin solutions requires new business models considering expected costs and profit as well as the design, operation, and maintenance requirements. These aspects are summarized in Figure 10.

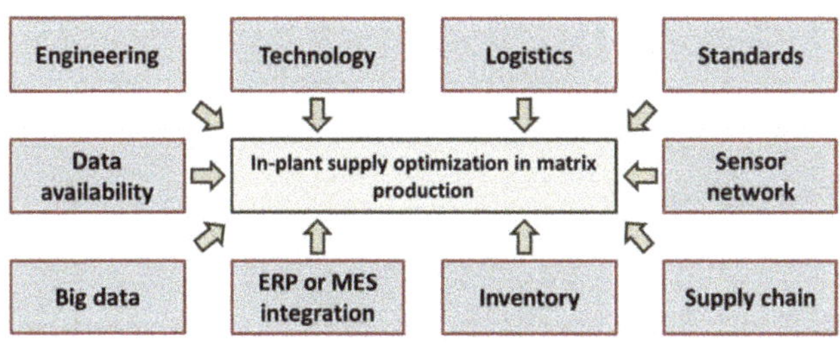

Figure 10. Challenges regarding the proposed model and method.

The practical application of the above-described methodology can be performed in many ways, depending on the available IT environment (ERP, sensor networks, simulation software). As an example, Figure 11 shows a possible solution focusing on the integration of SAP and Technomatix Plant Simulation. Technomatix Plant Simulation is a discrete event simulation software, which makes it possible to use SAP data and integrate real time data from the digital twin of the physical processes in the manufacturing system [62]. The SAP can generate a data file using Advanced Business Application Programming (ABAB) and this data file can be used by the Technomatix Plant Simulation for scenario analysis, especially in the field of production planning. The transformation of a conventional manufacturing system into a cyber-physical manufacturing system using IoT technologies makes it possible to mirror the physical manufacturing system, and the real time data including failure data and status information from the smart sensor network makes it possible to create a digital twin, which is available for the Technomatix Plant Simulation using ODBC or SQL for Oracle. The SAP data is also available as an Excel file export using Dynamic Data Exchange (DDE), Visual Basic Script (VBS) or Component Object Model (COM). The Technomatix Plant Simulation provides a built-in optimization library (BiOL) for stochastic optimization problems, and it is possible to use this heuristics-enabled solver to perform the proposed optimization tasks.

The above-described scenarios validated the presented in-plant supply model in a cyber-physical production environment and justify the fact that the matrix production, as a new production concept, is suitable for the efficient production of diversified customers' demands; not only the technology but also the logistic processes must be optimized. In this relation, efficiency means that the matrix production system makes it possible to fulfil diversified customers' demands near to the efficiency of mass production. KUKA defines this efficiency in the following context: "It (matrix production) will thus become possible to implement the manufacture of customized series as an integral part of Industry 4.0 without limitations in the context of industrial mass production [12]". The validation includes the following aspects: (1) the proposed functional model is suitable to support the in-plant supply optimization in a matrix production system; (2) the mathematical model includes time-, capacity-, and energy-related objective functions and constraints, and these objectives have a great impact on the cost-efficiency, availability, performance, energy consumption, and sustainability of the matrix production system; the computational results shows that the optimization algorithm resulted in valid solutions in the matrix production system, where time-, capacity- and energy-related constraints are taken into consideration.

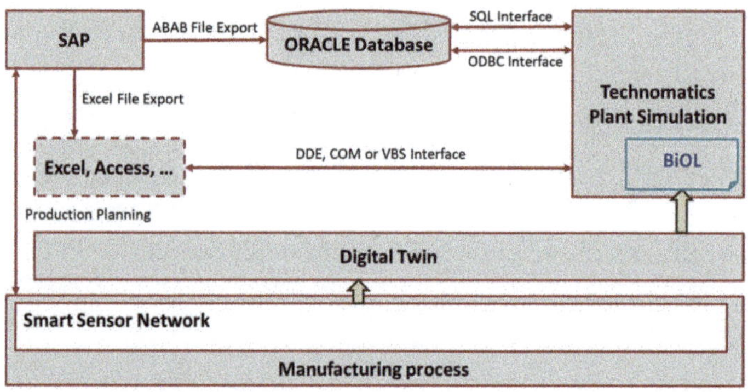

Figure 11. Practical applicability of the proposed methodology for integrating SAP and Technomatix Plant Simulation.

To summarize, the proposed model based on assignment and routing problems of matrix production makes it possible to analyze the impact of assignment of production order to matrix cells and the routing of automated guided vehicles in the matrix grid on the energy efficiency, availability, and required resources of material handling.

As the findings of the literature review show, the articles that addressed the analysis of in-plant supply are focusing on a conventional production environment, but only a few of them aimed to identify the optimization aspects of in-plant supply solutions in matrix production.

The comparison of the results with those from other studies shows that the optimization of material handling processes in cyber-physical systems still need more attention and research. The reason for this is that, in the case of cyber-physical systems, where Industry 4.0 technologies make possible the realization of flexible and efficient operation, the improvement of in-plant supply solutions and the optimization of their processes must be taken into consideration.

5. Discussion and Conclusions

The efficiency of manufacturing systems influences the efficiency of value chains, including purchasing and distribution processes; therefore, it is important to analyze the influencing factors of manufacturing systems and transform them into smart manufacturing systems using IoT technologies [58–60]. Within the frame of this research work, the authors developed an integrated model of in-plant supply based on the matrix production concept of KUKA. This model makes it possible to optimize the assignment and routing tasks of this new cyber-physical solution in the era of Industry 4.0. More generally, this paper focused on the mathematical description of the in-plant supply solutions in matrix production, including the assignment of technology and logistics (matrix cells as production resources and production order) and routing of autonomous guided vehicles. Why is so much effort being put into this research? Conventional production environments have been transformed into cyber-physical production, and this new production environment needs more attention both from a technology [61] and logistics point of view. A comparative table contrasted the proposed methodology in front of related analyzed research works, where the relationship between this solution and past literature was discussed. The existing studies include the optimization of both conventional and cyber-physical manufacturing systems, while only a few of them consider the sustainability-related aspects in matrix production and other cyber-physical manufacturing environments.

The added value of the paper is in the description of the autonomous guided vehicles-based in-plant supply in a cyber-physical environment, where production is based on standardized flexible manufacturing resources. The scientific contribution of this paper for researchers in this field is the mathematical modelling of in-plant supply in cyber-physical production including assignment, routing, and virtually scheduling. The results can be generalized because the model can be applied for different production environments. Managerial decisions can be influenced by the results of this research, because the described method makes it possible to analyze various supply strategies and make decisions regarding the size of AGV pool or strategy of warehousing of components or storage of tools and toolsets for the standardized flexible production cells. This managerial impact results from the fact that the above-mentioned algorithm takes different values of the size of the AGV pool as well as available tools required for changeovers into consideration, and the optimization results show whether or not the in-plant supply process can be performed with the given parameters.

However, there are also limitations of the study and the described model, which provides direction for further research. Within the frame of this model, stochastic parameters were not taken into consideration. In further studies, the model can be extended to a more complex model including Fuzzy sets to describe stochastic processes.

Funding: This research received no external funding.

Institutional Review Board Statement: Not applicable.

Informed Consent Statement: Not applicable.

Data Availability Statement: Not applicable.

Conflicts of Interest: The author declare no conflict of interest.

Appendix A. Description of Nomenclatures

Table A1. Description of nomenclatures used in the mathematical model.

Nomenclature	Description	Dimension
i	Production order, $i = 1 \ldots m$	[-]
j	Production cell in the matrix grid, $j = 1 \ldots n$	[-]
k	Production order, which can be defined either as a unique order, or as a lot, depending on the customers' demand. The customers' demand is available from the ERP. $k = 1 \ldots m$	[-]
τ_{ij}^{p}	Production lead time of production order i at production cell j	[min]
τ_{ikj}^{c}	Changeover time, which is the required time for the process of converting a matrix cell from the initial production process generated by the production order i to another generated by production order k at production cell j.	[min]
a_{ij}	Availability matrix, which takes value 1 if the production order i can be assigned to matrix cell j, otherwise 0. The availability depends on technological and logistic conditions and parameters.	[-]
a_{ikj}^{c}	Changeover availability matrix, which takes value 1 if it is possible to converting matrix cell from production order i to production order k at matrix cell j, otherwise 0	[-]
τ_{i}^{lower1}	Lower time limit of finishing operation i in the first phase (assignment) of the optimization.	[min]
τ_{i}^{upper1}	Upper time limit of finishing operation i in the first phase (assignment) of the optimization.	[min]
τ_{i}^{lower2}	Lower time limit of finishing operation i in the second phase (routing) of the optimization.	[min]
τ_{i}^{upper2}	Upper time limit of finishing operation i in the second phase (routing) of the optimization.	[min]
s_{j}^{upper1}	Upper limit of operations at production cell j.	[-]
z_{ij}	Required toolset for production order i at matrix cell j. The toolset is available from the tool storage and it includes tools and equipment for production and related measuring.	[-]
r_{g}^{max}	Available number of required toolset g.	[pcs]
τ^{p}	Production lead time.	[min]
τ^{c}	Changeover time among the various production operations of the standardized production cells.	[min]
ω_{j}	Number of assigned production orders to production cell j.	[pcs]
k_{AGV}	Required number of AGVs.	[pcs]
c	Calculated energy consumption.	

Table A1. *Cont.*

Nomenclature	Description	Dimension
c^{I}	Energy consumption of the AGVs from the warehouse to the first station (matrix cell) of the in-plant supply route.	[kWh]
c^{II}	Energy consumption of the AGVs among the stations (matrix cells).	[kWh]
c^{III}	Energy consumption of the AGVs from the last station (matrix cell) to the warehouse.	[kWh]
b_a^{max}	Number of stations of in-plant supply route a.	[pcs]
$q_{y_{ab}}$	Weight of the load for production order scheduled as station b of route a.	[kg]
$l_{0j(y_{a1})}$	Length of the transportation between the warehouse and the first matrix cell of the route.	[m]
$j(y_{ab})$	Matrix cell ID assigned to the production order, which is scheduled to the route a as station b.	[-]
$q_{y_{ab}_{max(a)}}$	Weight of the load for production order scheduled to the last station of in-plant supply route a.	[kg]
$l_{j(y_{ab_a^{max}})0}$	Length of the transportation between the last matrix cell of route a and the warehouse.	[m]
v_a^{max}	Upper limit of the number of stations assigned to route a.	[pcs]
$\tau^t_{j(y_{ad})j(y_{ad+1})}$	Transportation time between matrix cells assigned to the station b of route a.	[min]
$\tau^h_{j(y_{ad+1})}$	Material handling time (loading and unloading) at matrix cell assigned to the station $d+1$ of route a.	[min]
q_a^{max}	Upper limit of capacity of route (or vehicle) a.	[kg]
$\Xi_{y_{ab}}$	Set of vehicles appropriate for transportation of required materials and tools of production order y_{ab} from the warehouse to the assigned matrix cell.	[-]

References

1. Rosin, F.; Forget, P.; Lamouri, S.; Pellerin, R. Impacts of Industry 4.0 technologies on Lean principles. *Int. J. Prod. Res.* **2020**, *58*, 1644–1661. [CrossRef]
2. Skapinyecz, R.; Illés, B.; Bányai, Á. Logistic aspects of Industry 4.0. *IOP Conf. Ser. Mater. Sci. Eng.* **2018**, *448*, 12014. [CrossRef]
3. Tchoffa, D.; Figaly, N.; Ghodous, P.; Exposito, E.; Apedome, K.S.; El Mhamedi, A. Dynamic manufacturing network-from flat semantic graphs to composite models. *Int. J. Prod. Res.* **2019**, *270*, 6569–6578. [CrossRef]
4. Alcácer, V.; Cruz-Machado, V. Scanning the Industry 4.0: A Literature Review on Technologies for Manufacturing Systems. *Eng. Sci. Technol. Int. J. Jestech.* **2019**, *22*, 899–919. [CrossRef]
5. Dastjerdi, A.V.; Buyya, R. Fog Computing: Helping the Internet of Things Realize Its Potential. *Computer* **2016**, *49*, 112–116. [CrossRef]
6. Huang, S.H.; Liu, P.; Mokasdar, A.; Hou, L. Additive manufacturing and its societal impact: A literature review. *Int. J. Adv. Manuf. Technol.* **2013**, *67*, 1191–1203. [CrossRef]
7. Wu, K.H.; Gastan, E.; Rodman, M.; Behrens, B.A.; Bach, F.W.; Gatzen, H.H. Development and application of magnetic magnesium for data storage in gentelligent products. *J. Magn. Magn. Mater.* **2010**, *322*, 1134–1136. [CrossRef]
8. Guo, J.; Zhao, N.; Sun, L.; Zhang, S. Modular based flexible digital twin for factory design. *J. Ambient. Intell. Humaniz. Comput.* **2019**, *10*, 1189–1200. [CrossRef]
9. Tao, F.; Cheng, J.; Qi, Q.; Zhang, M.; Zhang, H.; Sui, F. Digital twin-driven product design, manufacturing, and service with big data. *Int. J. Adv. Manuf. Technol.* **2018**, *94*, 3563–3576. [CrossRef]
10. Ding, K.; Chan, F.T.S.; Zhang, X.D.; Zhou, G.H.; Zhang, F.Q. Defining a Digital Twin-based Cyber-Physical Production System for autonomous manufacturing in smart shop floors. *Int. J. Prod. Res.* **2019**, *57*, 6315–6334. [CrossRef]
11. Cui, Y.S.; Kara, S.; Chan, K.C. Manufacturing big data ecosystem: A systematic literature review. *Robot. Comput. Integr. Manuf.* **2020**, *62*, 101861. [CrossRef]
12. Schahinian, D. Digital Ecosystems-KUKA Launches a Pilot Plant for Matrix Production. Homepage of Hannover Messe. Available online: https://www.hannovermesse.de (accessed on 22 April 2020).
13. Bányai, Á.; Illés, B.; Glistau, E.; Machado, N.I.C.; Tamás, P.; Manzoor, F.; Bányai, T. Smart Cyber-Physical Manufacturing: Extended and Real-Time Optimization of Logistics Resources in Matrix Production. *Appl. Sci.* **2019**, *9*, 1287. [CrossRef]
14. Azarm, S.; Harhalakis, G.; Srinivasan, M.; Statton, P. Heuristic optimization of rough-mill yield with production priorities. *J. Eng. Ind. Trans. ASME* **1991**, *113*, 108–116. [CrossRef]
15. Kops, L.; Natarajan, S. Time partitioning based on the job-flow schedule—A new approach to optimum allocation of jobs on machine-tools. *Int. J. Adv. Manuf. Technol.* **1994**, *9*, 204–210. [CrossRef]

16. Hidaka, K.; Okano, H. Practical approach to a facility location problem for large-scale logistics. *J. Algorithm. Comput. Technol.* **1997**, *1350*, 12–21. [CrossRef]

17. Chitsaz, M.; Cordeau, J.F.; Jans, R. A unified decomposition matheuristic for assembly, production, and inventory routing. *Inf. J. Comput.* **2019**, *31*, 134–152. [CrossRef]

18. Eydi, A.; Fathi, A. An integrated decision making model for supplier and carrier selection with emphasis on the environmental factors. *Soft Comput.* **2020**, *24*, 4243–4258. [CrossRef]

19. Feng, P.P.; Liu, Y.; Wu, F.; Chu, C.B. Two heuristics for coordinating production planning and transportation planning. *Int. J. Prod. Res.* **2018**, *56*, 6872–6889. [CrossRef]

20. Ghomi, E.J.; Rahmani, A.M.; Qader, N.N. Service load balancing, scheduling, and logistics optimization in cloud manufacturing by using genetic algorithm. *Concurr. Comput. Pract. Exp.* **2019**, *31*, e5329. [CrossRef]

21. Sadati, N.; Chinnam, R.B.; Nezhad, M.Z. Observational data-driven modeling and optimization of manufacturing processes. *Expert Syst. Appl.* **2018**, *93*, 456–464. [CrossRef]

22. Haberer, C.; Wolfschluckner, A.; Landschützer, C. Analysis and optimization of a crawler track unit. *Konstruktion* **2016**, *68*, 76–82. [CrossRef]

23. Tamás, P. Decision Support Simulation Method for Process Improvement of Intermittent Production Systems. *Appl. Sci.* **2017**, *7*, 950. [CrossRef]

24. Saez-Mas, A.; Garcia-Sabater, J.J.; Garcia-Sabater, J.P.; Maheut, J. Hybrid approach of discrete event simulation integrated with location search algorithm in a cells assignment problem: A case study. *Cent. Eur. J. Oper. Res.* **2020**, *28*, 125–142. [CrossRef]

25. Bohács, G.; Rinkács, A. Development of an ontology-driven, component based framework for the implementation of adaptiveness in a Jellyfish-type simulation model. *J. Ambient. Intell. Smart Environ.* **2017**, *9*, 361–374. [CrossRef]

26. Ghomi, E.J.; Rahmani, A.M.; Qader, N.N. Service load balancing, task scheduling and transportation optimisation in cloud manufacturing by applying queuing system. *Enterp. Inf. Syst.* **2019**, *13*, 865–894. [CrossRef]

27. Hong, J.T.; Diabat, A.; Panicker, V.V.; Rajagopalan, S. A two-stage supply chain problem with fixed costs: An ant colony optimization approach. *Int. J. Prod. Econ.* **2018**, *204*, 214–226. [CrossRef]

28. Khalilpourazari, S.; Pasandideh, S.H.R.; Ghodratnama, A. Robust possibilistic programming for multi-item EOQ model with defective supply batches: Whale Optimization and Water Cycle Algorithms. *Neural Comput. Appl.* **2019**, *31*, 6587–6614. [CrossRef]

29. Mehranfar, N.; Hajiaghaei-Keshteli, M.; Fathollahi-Fard, M. A Novel Hybrid Whale Optimization Algorithm to Solve a Production-Distribution Network Problem Considering Carbon Emissions. *Int. J. Eng.* **2019**, *32*, 1781–1789. [CrossRef]

30. Liu, C.F.; Wang, J.F.; Leung, J.Y.T. Integrated bacteria foraging algorithm for cellular manufacturing in supply chain considering facility transfer and production planning. *Appl. Soft Comput.* **2017**, *62*, 602–618. [CrossRef]

31. Habibi, M.K.K.; Battaia, O.; Cung, V.D.; Dolgui, A. An efficient two-phase iterative heuristic for collection-disassembly problem. *Comput. Ind. Eng.* **2017**, *110*, 505–514. [CrossRef]

32. Juan, A.A.; Corlu, C.G.; Tordecilla, R.D.; de la Torre, R.; Ferrer, A. On the Use of Biased-Randomized Algorithms for Solving Non-Smooth Optimization Problems. *Algorithms* **2020**, *13*, 8. [CrossRef]

33. Russell, R.A. Mathematical programming heuristics for the production routing problem. *Int. J. Prod. Econ.* **2017**, *193*, 40–49. [CrossRef]

34. Shao, W.S.; Pi, D.C.; Shao, Z.S. Optimization of makespan for the distributed no-wait flow shop scheduling problem with iterated greedy algorithms. *Knowl. Based Syst.* **2017**, *137*, 163–181. [CrossRef]

35. Saghaeeian, A.; Ramezanian, R. An efficient hybrid genetic algorithm for multi-product competitive supply chain network design with price-dependent demand. *Appl. Soft Comput.* **2018**, *71*, 872–893. [CrossRef]

36. Tayal, A.; Singh, S.P. "Integrating big data analytic and hybrid firefly-chaotic simulated annealing approach for facility layout problem. *Ann. Oper. Res.* **2018**, *270*, 489–514. [CrossRef]

37. Tari, F.G.; Hashemi, Z. Prioritized K-mean clustering hybrid GA for discounted fixed charge transportation problems. *Comput. Ind. Eng.* **2018**, *126*, 63–74. [CrossRef]

38. Sakalli, U.S.; Atabas, I. Ant colony optimization and genetic algorithm for fuzzy stochastic production-distribution planning. *Appl. Sci.* **2018**, *8*, 2042. [CrossRef]

39. Abouee-Mehrizi, H.; Baron, O.; Berman, O.; Chen, D. Managing perishable inventory systems with multiple priority classes. *Prod. Oper. Manag.* **2019**, *28*, 2305–2322. [CrossRef]

40. Aboytes-Ojeda, M.; Castillo-Villar, K.K.; Roni, M.S. A decomposition approach based on meta-heuristics and exact methods for solving a two-stage stochastic biofuel hub-and-spoke network problem. *J. Clean. Prod.* **2020**, *247*, 119176. [CrossRef]

41. Behfard, S.; Al Hanbali, A.; van der Heijden, M.C.; Zijm, W.H.M. Last Time Buy and repair decisions for fast moving parts. *International J. Prod. Econ.* **2018**, *197*, 158–173. [CrossRef]

42. Ma, K.; Thomassey, S.; Zeng, X.Y.; Wang, L.C.; Chen, Y. A resource sharing solution optimized by simulation-based heuristic for garment manufacturing. *Int. J. Adv. Manuf. Technol.* **2018**, *99*, 2803–2818. [CrossRef]

43. Cheraghalipour, A.; Paydar, M.M.; Hajiaghaei-Keshteli, M. Designing and solving a bi-level model for rice supply chain using the evolutionary algorithms. *Comput. Electron. Agric.* **2019**, *162*, 651–668. [CrossRef]

44. Respen, J.; Zufferey, N.; Wieser, P. Three-level inventory deployment for a luxury watch company facing various perturbations. *J. Oper. Res. Soc.* **2017**, *38*, 1195–1210. [CrossRef]

45. Varas, M.; Maturana, S.; Cholette, S.; Mac Cawley, A.; Basso, F. Assessing the benefits of labelling postponement in an export-focused winery. *Int. J. Prod. Res.* **2018**, *56*, 4132–4151. [CrossRef]
46. Hatamlou, A. Black hole: A new heuristic optimization approach for data clustering. *Inf. Sci.* **2013**, *222*, 175–184. [CrossRef]
47. Gholizadeh, S.; Razavi, N.; Shojaei, E. Improved black hole and multiverse algorithms for discrete sizing optimization of planar structures. *Eng. Optim.* **2019**, *51*, 1645–1667. [CrossRef]
48. Pashaei, E.; Aydin, N. Binary black hole algorithm for feature selection and classification on biological data. *Appl. Soft Comput.* **2017**, *56*, 94–106. [CrossRef]
49. Bányai, Á.; Bányai, T.; Illés, B. Optimization of consignment-store-based supply chain with black hole algorithm. *Complexity* **2017**, *2017*, 6038973. [CrossRef]
50. Khooban, M.H.; Liaghat, A. A time-varying strategy for urban traffic network control: A fuzzy logic control based on an improved black hole algorithm. *Int. J. Biol. Inspir. Comput.* **2017**, *10*, 33–42. [CrossRef]
51. Lei, M.Y.; Zhou, Y.Q.; Luo, Q.F. Enhanced Metaheuristic Optimization: Wind-Driven Flower Pollination Algorithm. *IEEE Access* **2019**, *7*, 111439–111465. [CrossRef]
52. Lei, X.J.; Fang, M.; Wu, F.X.; Chen, L.N. Improved flower pollination algorithm for identifying essential proteins. *BMC Syst. Biol.* **2018**, *12*, 46. [CrossRef]
53. Shen, L.; Fan, C.Y.; Huang, X.T. Multi-Level Image Thresholding Using Modified Flower Pollination Algorithm. *IEEE Access* **2018**, *6*, 30508–30519. [CrossRef]
54. Gao, M.L.; Shen, J.; Jiang, J. Visual tracking using improved flower pollination algorithm. *Optik* **2018**, *156*, 522–529. [CrossRef]
55. Rodrigues, D.; Silva, G.F.A.; Papa, J.P.; Marana, A.N.; Yang, X.S. EEG-based person identification through Binary Flower Pollination Algorithm. *Expert Syst. Appl.* **2016**, *62*, 81–90. [CrossRef]
56. Guan, C.; Zhang, Z.Q.; Li, Y.P. A flower pollination algorithm for the double-floor corridor allocation problem(dagger). *Int. J. Prod. Res.* **2019**, *57*, 6506–6527. [CrossRef]
57. Kherabadi, H.A.; Mood, S.E.; Javidi, M.M. Mutation: A new operator in gravitational search algorithm using fuzzy controller. *Cybern. Inf. Technol.* **2017**, *17*, 72–86. [CrossRef]
58. Szentesi, S.; Illés, B.; Cservenák, Á.; Skapinyecz, R.; Tamás, P. Multi-Level Optimization Process for Rationalizing the Distribution Logistics Process of Companies Selling Dietary Supplements. *Processes* **2021**, *9*, 1480. [CrossRef]
59. Bányai, Á.; Illés, B.; Schenk, F. Supply Chain Design of Manufacturing Processes with Blending Technologies. *Solid State Phenom.* **2017**, *261*, 509–515. [CrossRef]
60. Hardai, I.; Illés, B.; Bányai, Á. View of the opportunities of Industry 4.0. *Adv. Logist. Syst. Theory Pract.* **2021**, *14*, 5–14. [CrossRef]
61. Kundrák, J.; Mamalis, A.G.; Molnár, V. The efficiency of hard machining processes. *Nanotechnol. Percept.* **2019**, *15*, 131–142. [CrossRef]
62. SAP Data in Plant Simulation. Homepage of Siemens. Available online: https://community.sw.siemens.com/s/article/sap-data-in-plant-simulation (accessed on 17 September 2021).

Article

Blockchain-Empowered Digital Twins Collaboration: Smart Transportation Use Case

Radhya Sahal [1,2,*], Saeed H. Alsamhi [3,4], Kenneth N. Brown [1], Donna O'Shea [5], Conor McCarthy [6] and Mohsen Guizani [7]

[1] SMART 4.0 Fellow, School of Computer Science and Information Technology, University College Cork, T12 E8YV Cork, Ireland; k.brown@cs.ucc.ie

[2] Faculty of Computer Science and Engineering, Hodeidah University, Al Hodeidah 3114, Yemen

[3] SMART 4.0 Fellow, Software Research Institute, Athlone Institute of Technology, N37 W089 Athlone, Ireland; salsamhi@ait.ie

[4] Faculty of Engineering, IBB University, Ibb 70270, Yemen

[5] Department of Computer Science, Cork Institute of Technology, T12 P928 Cork, Ireland; donna.oshea@cit.ie

[6] School of Engineering, University of Limerick, V94 T9PX Limerick, Ireland; conor.mccarthy@ul.ie

[7] College of Engineering, Qatar University, Doha 2713, Qatar; mguizani@ieee.org

* Correspondence: rsahal@ucc.ie

Citation: Sahal, R.; Alsamhi, S.H.; Brown, K.N.; O'Shea, D.; McCarthy, C.; Guizani, G. Blockchain-Empowered Digital Twins Collaboration: Smart Transportation Use Case. *Machines* **2021**, *9*, 193. https://doi.org/10.3390/machines9090193

Academic Editors: Zhuming Bi, Li Da Xu, Puren Ouyang and Pingyu Jiang

Received: 21 July 2021
Accepted: 7 September 2021
Published: 9 September 2021

Publisher's Note: MDPI stays neutral with regard to jurisdictional claims in published maps and institutional affiliations.

Abstract: Digital twins (DTs) is a promising technology in the revolution of the industry and essential for Industry 4.0. DTs play a vital role in improving distributed manufacturing, providing up-to-date operational data representation of physical assets, supporting decision-making, and avoiding the potential risks in distributed manufacturing systems. Furthermore, DTs need to collaborate within distributed manufacturing systems to predict the risks and reach consensus-based decision-making. However, DTs collaboration suffers from single failure due to attack and connection in a centralized manner, data interoperability, authentication, and scalability. To overcome the above challenges, we have discussed the major high-level requirements for the DTs collaboration. Then, we have proposed a conceptual framework to fulfill the DTs collaboration requirements by using the combination of blockchain, predictive analysis techniques, and DTs technologies. The proposed framework aims to empower more intelligence DTs based on blockchain technology. In particular, we propose a concrete ledger-based collaborative DTs framework that focuses on real-time operational data analytics and distributed consensus algorithms. Furthermore, we describe how the conceptual framework can be applied using smart transportation system use cases, i.e., smart logistics and railway predictive maintenance. Finally, we highlighted the future direction to guide interested researchers in this interesting area.

Keywords: blockchain; digital twins; Industry 4.0; smart manufacturing; data analysis; transportation; logistics; railway

1. Introduction

Industry 4.0 revolution is considered to be a new paradigm of digital, autonomous, and decentralized control with the Industrial Internet of Things (IIoT), Machine Learning (ML), big data, and edge computing [1,2]. For smart manufacturing, distributed manufacturing is a form of decentralized manufacturing practiced by enterprises using digitalization [3,4]. It uses an effective collaboration form in terms of information sharing, analytics, and collaborative decision-making in real-time. A Digital Twin technology (DT) is one of the core elements of manufacturing digitalization, representing a real-world system such as production systems in a virtual space [5]. Multiple DTs are used to represent a distributed production system in hierarchical levels [6]: (i) DTs in a flat network represent individual things at the machine level. They exchange information with each other on things and learn about their operation and behavior to build a common understanding of the machine condition, (ii) DTs for things in a tree or a chain represent the sub-system level and the system level where each DT is passing on information to the next level.

Multiple DTs are deployed to represent the up-to-date industrial data of the physical assets in operation, including the asset status and the relevant historical data. The deployed DTs can intelligently collaborate by utilizing the intelligence of DT-driven operational data to predict the potential risks within the distributed manufacturing systems. In particular, the DTs are collaborating by applying predictive data analytics to analyze DT-based historical operating data, learn about their things using shared knowledge and real-time data, and then predict the potential risks in real-time. A better understanding of the predicted potential risks can facilitate consensus decision-making among participants on the management floor within the distributed manufacturing systems. However, the DT paradigm is still at an early stage, and many challenges still exist to adopt DTs collaboration in the distributed manufacturing environment, including:

- Interoperability: The models and strategies of the sharing policies (i.e., internal and external data) need to define the DTs data schema and the collaboration requirements.
- Authentication: In some scenarios in distributed manufacturing systems, the deployed DTs are owned by independent entities that want to collaborate. Therefore, securing a digital distributed manufacturing system needs efficient technology to acquire secure real-time data exchange and analysis across multiple participants.
- Distributed machine learning: A large-scale input data size from multiple participants needs to be analyzed to obtain accurate predictions about the potential risks within the distributed manufacturing system.
- Distributed decision-making: Centralizing suffers from single failure data, while decentralization suffers from lacking global data, so the decision-making consensus is required.
- Scalability and robustness: a system needs to accommodate a large number of DTs which represent multiple participants, e.g., objects, devices, machines, nodes, people, workstations, etc., within manufacturing systems. The distributed manufacturing system also needs to deal with multiple deployed DTs and simultaneously maintain the robustness at a required level, especially with hacked nodes and malfunctioning.

Most works have proposed adopting blockchain with DTs to guarantee transparency, decentralized data storage, data sharing, peer-to-peer communication, secure and trusted traceability, and scalability [7]. Using blockchain, multiple DTs can collaborate in a hierarchical and granular manner, using shared knowledge to manage and trace the product assembly data [8]. A smart contract is used to execute some actions automatically to increase data sharing efficiency, and higher security [9], and provide trusted data provenance, audit, traceability, and tracking transactions initiated by participants involved in the creation of DTs [10]. However, the existing research lacks solutions for collaborative DTs based on operational data analytics because its focus is mainly related to blockchain adoption for DTs. There are still many challenges requiring further investigation to identify, diagnose, and remove the potential risks in distributed manufacturing systems using the intelligence of the DT-driven operational data.

1.1. Contribution

The combination of blockchain and DT technologies represents the key technologies that allow continuous data acquisition in Industry 4.0. Furthermore, the combination of both technologies has significant advantages to address the challenges mentioned above such as traceability, security, the guarantee of ownership rights, decentralization, etc. [7,8,10–12]. However, the combination of blockchain and DT is rather still under exploration. Many research works have proposed simple blockchain model adoption of DT with a focus on centralized production systems. In contrast, these works have fallen short of providing DT-based solutions for distributed manufacturing. Moreover, the blockchain is not designed for DTs collaboration scenarios of risk prediction. On the other hand, the distributed consensus decision-making has been adopted in blockchain technology [13,14].

Various consensus implementations were proposed to make replicas reach an agreement on transactions updating using a distributed ledger. However, the development of many ledger-based DTs with the distributed consensus decision-making for risk prediction is still an unsolved problem. Therefore, more smart and collaborative solutions for DTs based on Distributed Ledger Technology (DLT) and distributed consensus decision-making for risk prediction are required to add progressive value to distributed manufacturing.

Consequently, the main purpose of this work is to develop a conceptual framework for data-driven ledger-based collaborative DT. In particular, the proposed framework targets smart distributed manufacturing to predict the potential risks using the intelligence of sharing operational data. Figure 1 depicts the high level of the blockchain-based collaborative DTs using predictive data analysis (i.e., analysis DT-based operational data).

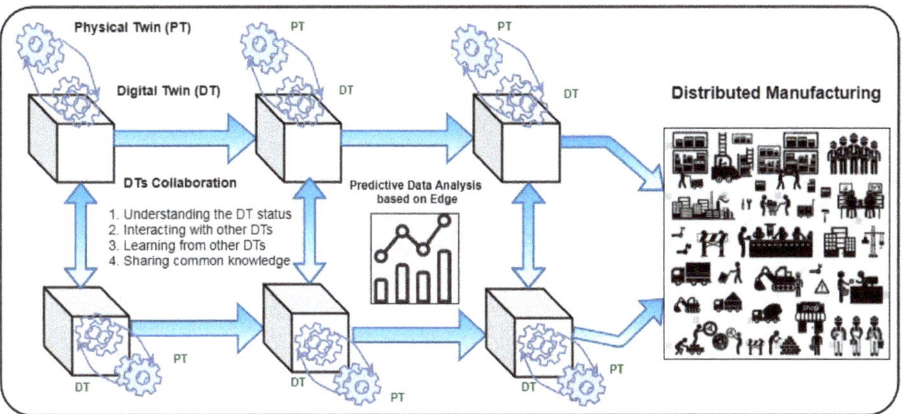

Figure 1. The high-level of the blockchain-based collaborative digital twins (DTs) using predictive data analysis (i.e., analysis DT-based operational data).

Our main contributions in this conceptual framework paper are summarized as follows:

- We explore how blockchain employing in DTs collaboration with highlighting the benefits of the combination.
- We propose the conceptual framework of the data driving-based DTs collaboration with the help of blockchain technology. The proposed framework consists of two components:
    1. The data-driven ledger-based predictive model is used to predict the potential risks using DT-driven operational data. The DLT performs intelligent and secure interoperability, including real-time operational data exchange, querying the real-time operational database, and dynamic interactions among the deployed DTs. At the same time, the distributed predictive model plays a vital role in developing and evaluating DT deployment locally using the DT-driven operational data.
    2. A distributed consensus algorithm to improve the decentralized DTs collaboration. The distributed decision-making algorithm develops based on the essence of the consensus mechanism and the dynamic prediction, which uses real-time DT-driven operational data. The developed distributed consensus algorithm can make most nodes agree on the potential risks and notify the decision-makers within the distributed manufacturing systems.
- We describe how the conceptual framework can be applied in smart transportation systems, i.e., smart logistics and railway predictive maintenance.

1.2. Paper Organization

The remainder of this paper is organized as shown in Figure 2. The comparison with other existing solutions of blockchain empowering digital twins collaboration is introduced in Section 2. The proposed conceptual framework for collaborative DTs is introduced in Section 3. The use case of smart transportation is described in Section 4. The discussion, validation, and future direction are presented in Section 5. Finally, conclusions are presented in Section 6.

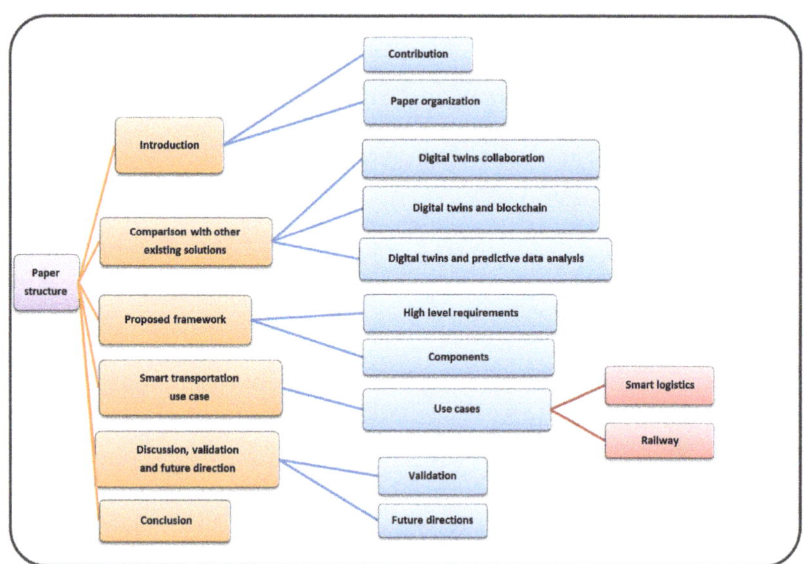

Figure 2. Paper structure.

2. Comparison with Other Existing Solutions of Blockchain Empowering Digital Twins Collaboration

In this section, we compare the contributions of our work with state-of-the-art solutions, particularly solutions based on the use of blockchain and digital twins.

2.1. Digital Twins Collaboration

DTs are merging the virtual worlds and real worlds. It is used to describe the detailed presentation of machines, devices, robots in the warehouse, production, and process. The DTs' advantages in Industry 4.0 include improving data security and quality, reducing cost, and faster decision-making. The authors of [15] described DTs as one virtual replica of a machine, robots, and devices containing data, function, and communication interfaces. The main parts of DTs are the physical entity, virtual entity, and information that connect virtual and physical entities [16,17] as shown in Figure 3 [18].

Collaboration means sharing and exchange information among entities and share tasks to act accordingly. The authors of [19,20] discussed the importance of Artificial Intelligence (AI) and Machine Learning (ML) for robots. The collaboration is based on improving the quality of services, connectivity, and reliability. Furthermore, the collaboration of drones and the Internet of things to enhance smartness of smart cities applications [21] and public safety [22], and for better Quality of Service (QoS) [23]. The collaboration among multi-user and identifying the activities is described in [24]. Collaboration of DTs and humans is described with details in [25]. However, the authors highlighted the challenges of collaboration in industry platform [26–28].

Collaboration is essential for a group of users to perform complex activities effectively and efficiently, while a single can not do [24]. In [29], the authors introduced blockchain technology for heterogeneous multi-robot collaboration to combat COVID-19 in decentralized peer-to-peer networks without human intervention. Furthermore, the authors of [30] applied blockchain for decentralized multi-drone collaboration to combat COVID -19 in delivering goods and monitoring people in the quarantine area. Furthermore, the authors [19] introduced a machine learning technique for multi-robot collaboration based on keeping connectivity, maintaining the quality of services, and improving mobility during task performances. In [22] addressed drones and IoT devices collaboration to improving greener and smarter cities, while the drones and IoT collaboration resulting green IoT [31,32]. However, no one of the above studies addresses DTs collaboration and applications. Therefore, we discuss DTs collaboration for improving transportation and logistics applications.

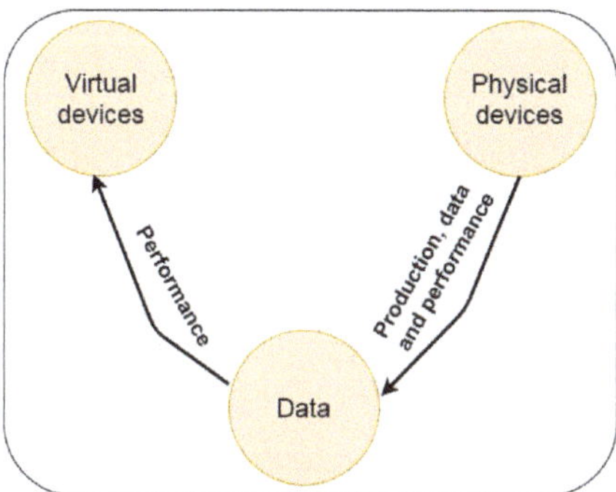

Figure 3. The main parts of DTs [18].

Smart industries depend on gathered data from smart IoT devices or their DTs of the production lines. Collected data can be erratic sensors, RFID, actuators, or their DTs injecting and producing incorrect data for analyzing and taking affecting decision-making. Based on this, the author [18] has proposed a unified interaction mechanism for collaborative DTs to provide an auto-detection using the intelligence of operational data in the cyber-physical production system. The proposed mechanism can detect whether the DT has erratic behavior by interacting with other collaborative DTs within the edge level. The authors [33] introduced architecture for smart factories, relying on service-based DTs. The architecture was presented how automatically DTs combine the corresponding physical processes and sharing analogies with web services. The authors of [34] introduced DT and big data in the smart industry, focusing on applications, manufacturing, production, maintenance prediction, etc. Furthermore, the authors of [35] discussed the enabling technologies for DT in smart industries. He et al. [36] presented DT-driven sustainable smart industries. Moreover, the authors of [37] introduced DTs with Industry 4.0 and data analytics.

2.2. Digital Twins and Blockchain

DT technology has been aligned with blockchain technology in different industrial sectors [7]. ManuChain is an iterative bi-level model proposed based on the incorporation of blockchain into DT on decentralized manufacturing [11]. ManuChain model has defined the lower-level for fine-grained self-organization intelligence and upper-level to iterate the coarse-grained holistic optimization intelligence by utilizing decentralized features of blockchain. Makerchain is another blockchain-driven model based on DTs, which was proposed to handle the cyber-credit of social manufacturing among various makers [38]. The Makerchain model has used DTs to synchronize the updating of data tags and ensure the personalized demands to make manufacturing service transactions among makers more trustworthy. In [39], the authors have proposed a manufacturing blockchain of things (MBCoT) architecture for secure, traceable, and decentralized manufacturing configuration. The authors have defined the data-and knowledge-driven DT manufacturing cell as a reference model for decentralized manufacturing. They have also introduced the consensus-oriented transaction logic of MBCoT for the fault-tolerant protocol to support the autonomous manufacturing process. In [40], the authors have developed ExplorerChain, including online ML, metadata of transaction, and the Proof-of-Information-Timed algorithm to serve as a reference for researchers who would like to implement and deploy blockchain technology in the healthcare/genomic domain. In [12], the authors have used DLT to develop a protocol to guarantee the transfer of values between DTs in economic systems. Furthermore, a framework for secure DT data sharing based on DLT is proposed to track events and provenance information along with an asset's lifecycle and increase transparency for all participants [41].

All of the above literature discussed collaboration in smart industries. Still, none of the studies focused on discussing the collaboration of DTs using blockchain technologies to improve management in the distributed smart manufacturing.

2.3. Digital Twins and Predictive Data Analysis

A predictive DTs methodology has been proposed to enable data-driven physics-based DTs using a library of component-based reduced-order models and interpretable machine learning [42]. The predictive DTs methodology has been performed using a case study of fixed-wing UAVs. The UAV uses structural sensors to detect damage or degradation on one of its wings to inform the UAV's decision-making about performing an aggressive maneuver or a more conservative one to avoid structural failure. In [43], a smart city DT architecture is proposed where knowledge representation, reasoning, and ML formalisms are used to provide complementary and supportive roles in the collection and processing of data, identification of events, and automated decision-making.

Maintenance is one of the predictive data analysis-based applications in Industry 4.0, which is referred to as predictive maintenance 4.0 or PM 4.0. It is one of the recent researched industrial applications as its impact on the manufacturing sectors. The authors in [44] have reviewed the use of DTs technologies to apply maintenance strategies to provide a deeper insight into the synergies between both DTs and maintenance in the industrial sectors. Furthermore, the authors in [1] proposed a set of requirements to enable predictive maintenance with big data for Industry 4.0 applications. The authors have studied the railway industry concerning big data streaming processing platforms, distributed message queue management systems, big data storage platforms, and streaming SQL engines. Moreover, the authors in [45] have proposed a ledger-based DT reference model for predictive maintenance defined in three layers: edge, fog, and cloud. Regarding data analysis of DTs, Song et al., introduced a predictive maintenance model that comprised DTs plurality [46]. DTs corresponded to the plurality of remotely located physical machines. Each DT combined product nameplate data that correspond to a unique physical device with more simulation models. The database contained the run time and logged data gathered from sensors with the associated physical machine. Moreover, modular-based corrective maintenance was used in DTs, which was proposed for automating decision-making in complex systems [47]. The proposed model was corrective maintenance and was relied on DTs development. Aivaliotis et al., proposed the physics approach to predict the maintenance by using DTs [48]. The proposed approach was discussed to reduce modeling efforts and provide the modeling framework of different resources for enabling DTs. The machine modeling was included the machine dynamic behavior using the grey box, black box, and white box. The virtual sensor modeling referred to gather data during the simulation. Furthermore, modeling parameters were included parameters updated with frequency to guarantee machine DT.

Table 1 describes a comparison of existing work and the present work concerning the applications, including blockchain, DTs, collaboration, data analysis, Industry 4.0, and transportation. Because only a few publications exist in blockchain-based DT, DT collaboration has not yet been a focus in the literature to date. Moreover, data interoperability-by-design concepts have not been considered yet.

Table 1. Comparison of existing work and the present work (Industry 4.0 (I4.0), Transportation(Transp)).

Ref	Highlighted	Blockchain	DT	Collaboration	Data Analysis	I4.0	Transp
[11] (2020)	ManuChain model is proposed based on the incorporation of blockchain into DT on a decentralized manufacturing	✓	✓	X	X	✓	X
[38] (2019)	Makerchain model based on DTs is proposed to handle the cyber-credit of social manufacturing among various makers	✓	✓	X	X	✓	X
[40] (2020)	ExplorerChain is a reference for researchers to implement and deploy blockchain technology in the healthcare/genomic domain using online machine learning	✓	X	X	✓	✓	X
[39] (2020)	MBCoT architecture for the configuration of a secure, traceable, and decentralized manufacturing	✓	X	X	X	✓	X
[12] (2019)	A protocol based on DLT to guarantee the transfer of values between DTs in economic systems	✓	✓	X	X	✓	X
[29] (2020)	The homogeneous and heterogeneous of multi-robot collaboration to perform their complex task in decentralized fashion with blockchain technology	✓	X	✓	X	X	X
[45] (2019)	Aledger-based DT reference model for predictive maintenance	✓	✓	X	✓	✓	X
[41] (2019)	A framework for secure DT data sharing based on DLT to track events and provenance information	✓	✓	X	X	✓	X
[42] (2020)	A predictive DTs methodology based on ML using UAV case study	X	✓	X	✓	✓	X
[43] (2020)	A smart city DT architecture using reasoning and ML to automated decision-making	X	✓	X	✓	✓	X

Table 1. *Cont.*

Ref	Highlighted	Blockchain	DT	Collaboration	Data Analysis	I4.0	Transp
[19] (2019)	Machine learning technique for multi-robot collaboration based on keeping connectivity, maintaining the quality of services, and improving mobility during tasks performances	X	X	✓	✓	X	X
[22] (2019)	Drones and IoT devices collaborate to improving greener and smarter cities	X	X	✓	X	X	✓
[49] (2018)	DT monitoring that for monitoring and development of wind farms.	X	✓	X	✓	X	X
[50] (2017)	An approach for identifying the network physical vulnerabilities in industry 4.0 systems.	X	X		X	✓	X
[51] (2014)	AI-based supervisory control and data acquisition method for prediction and fault diagnosis of wind turbines	X	X	X	✓	X	X
[33] (2019)	A conceptual architecture and model for smart manufacturing relying on service-based DTs	X	✓	X	X	✓	X
[48] (2019)	Advanced physics-based modeling approach for predictive maintenance using DTs	X	✓	X	✓	✓	X
[46] (2016)	A model-based machine predictive maintenance based on DTs and a simulation platform	X	✓	X	✓	✓	X

Table 1. *Cont.*

Ref	Highlighted	Blockchain	DT	Collaboration	Data Analysis	I4.0	Transp
[47] (2018)	A modular-based corrective maintenance methodology using DTs to automate decision making in complex systems	X	✓	X	✓	✓	X
[34] (2018)	Discussion of the DT and big data in smart manufacturing in terms of applications, production, manufacturing, maintenance prediction	X	✓	X	✓	✓	X
[37] (2019)	A tool and technologies for DT in smart manufacturing	X	✓	X	X	✓	X
[36] (2020)	A DT-driven sustainable technique smart manufacturing	X	✓	X	X	✓	X
[35] (2020)	Focusing on DTs with Industry 4.0 and data analytics	X	✓	X	✓	✓	X
[52] (2018)	Investigation wind farm and power consumption for smart manufacturing using IoT and DTs to perform wind turbines maintenance.	X	✓	X	✓	✓	X
[23] (2020)	Collaboration of drone and IoT to enhance smartness of smart cities applications	X	X	✓	X	X	X
[18] (2021)	A conceptual framework for DTs collaboration to provide an auto-detection of erratic operational data by utilizing the intelligence of operational data in the manufacturing systems.	X	✓	✓	✓	✓	X
Our work	A conceptual framework for blockchain based DTs collaboration to improve DTs collaboration in transportation systems and focuses on real-time operational data analytics.	✓	✓	✓	✓	✓	✓

3. Proposed Conceptual Framework of Data-Driven Blockchain-Based Collaborative Digital Twins

The conceptual framework of data-driven blockchain-based collaborative DTs is proposed to empower more intelligent and collaborative solutions for DTs based on DLT and distributed consensus decision-making. The proposed framework is considered one level higher than the adoption of blockchain with DT in production systems that could integrate blockchain and operational data analysis. Moreover, the proposed framework could be developed and implemented on top of the DT platform, which exploits blockchain capabilities to guarantee transparency, decentralized data storage, data sharing, peer-to-peer communication, secure and trusted traceability, and scalability.

3.1. High-Level Requirements for Digital Twins Collaboration

In this subsection, we have identified a set of requirements regarding DTs collaboration in distributed manufacturing. Table 2 summarizes nine criteria to fulfill the requirements for digital twins collaboration. For any manufacturing process, the data is collected from the data sources, which are DTs of physical things such as devices, machines, people . . . , etc., across the entire network within the manufacturing units. This data is frequently updated in real-time, beneficial to the organization's decision-making process (R1 & R2) Interestingly, the collected data from the collaborative DTs are data-driven learning systems In particular, DTs provide the data analysis engine by continuously updating data fed into the learning models to enable advanced predictions of the potential risks (R3).

Basically, simulation models help to understand what may happen when changes occur on the physical assets. However, DTs help understands what is currently happening on the physical asset and what could happen in the future (R4). For the collaborative environment, the virtual visibility of the potential risks in the future within the physical assets can help to refine the product design, real-time troubleshooting, and implement new ideas. Furthermore, DTs networks are used to enhance the connectivity of the network participants (e.g., devices, machines, people, departments, organizations). The interaction between the participants through the DTs network is used for reliable data exchange to allow internal and external data sharing (R5). However, the connected DTs network needs to be authenticated to maintain trust among network peers (R6). So, authentication provides a trust level that can keep secure collaboration and interactions among the DTs network. DT is a virtual representation of a real thing and transparently visualizes the physical things and their behavior within the collaborative environment. In particular, the transparent visibility of things through the DT model allows accurate traceability across the DT network (R7).

For any centralized network, all nodes are connected under a single authority. However, the decentralized network has not a single authority server that controls the nodes, where all nodes have individual entities. Substantially, centralizing suffers from single failure data, while decentralization suffers from lacking global data, so decision-making is required for DTs collaboration (R8 & R9). Substantially, the consensus-based distributed decision-making process provides insightful and delivers efficient and reliable collaborative solutions.

Table 2. Requirement of digital twins collaboration.

Req. No	Requirement	Reason
R1	Data collection	supporting data-driven decision making
R2	Data update frequency	providing realtime update of the physical twin
R3	Data analysis	enabling advanced predictions of the potential risks
R4	Simulation capabilities	enabling virtual visibility of the products
R5	Data exchange	allowing internal and external data sharing
R6	Authentication	maintaining trust among network peers
R7	Transparency	allowing traceability across the entire network
R8	Distributed decision making capabilities	providing insightful consensus-based decision making process
R9	Decentralization	delivering efficient and reliable solutions

3.2. Components of the Proposed Framework

The developing blockchain can overcome the safety and security that have prevented DT initiatives. Because of blockchain characteristics like immutability and decentralization, DTs initiatives can evolve more effectively and quickly in their environments [53]. A decentralized blockchain network can help trust DTs with a data track and digital identity. Furthermore, decentralization, secure and safe data transportation is more related to authenticate DTs industry environments cases. The authors of [10] introduced blockchain for DTs processing which guarantees safe, secure, and reliable transactions without data accessibility, traceability, and immutability. Smart contracts are used to track and manage the transactions in the developed DTs. Blockchain technology is used to exchange and store DTs data to exchange information between DTs in a decentralized fashion.

Consequently, to implement DTs collaboration in distributed manufacturing, the aforementioned high-level requirements could be fulfilled using a combination of DTs, blockchain, and artificial intelligence. Based on these requirements, two main components are required to equip the conceptual framework of blockchain-based DTs collaboration. Figure 4 depicts the components including (1) data-driven ledger-based collaborative DTs for predictive analytics and, (2) consensus-based decision making. These two parts will be elaborated flowingly. Furthermore, A more detailed architecture with an in-depth discussion of every component is beyond the scope of this paper.

3.2.1. Data-Driven Ledger-Based Collaborative DTs for Predictive Analytics

This component is used to develop a methodology for creating and updating data-driven ledger-based collaborative DTs. It demonstrates the predictive analytics approach by developing offline and online predictive models using the data ledger-based historical DTs data and live streaming DTs data. Two main sub-components are required to equip the data-driven ledger-based collaborative DTs for predictive analytics as follows:

Ledger-Based DTs Model

Multiple DTs could be connected through the blockchain network using DLT to secure distributed operational data management and analytics across multiple participants. Figure 5 describes the ledger-based DT model. At the technical level, the ledger-based DT model needs to define the five components [10,41]: (1) registered DT owner, (2) DT status, (3) timestamp, (4) transaction, and (5) ledger database. The information that maintains a physical object's specifications is stored within the ledger, such as the DT owner. The communication mechanism that transfers bi-directional data between a DT and its physical counterpart will generate data that is considered DT status within timestamp, which is used to create a transaction. DLT is used to store the transactions, DTs data, and actions.

Moreover, DLT guarantees the transfer of values between DTs for collaborative DTs-based applications. In doing so, the ledger-based running database synchronizes the updated DTs' status within the manufacturing systems in real-time, which leads to an increase in real-time prediction accuracy, such as the potential risks to improve the quality of the decision making. The requirements of the DTs in terms of data schema and collaborations, including information and analytics sharing, will be identified based on the collaborative DTs-based applications [54]. Accordingly, the participants represented by DTs (i.e., if they are the same type or different types) should be defined. The type of collaborations in the participants' communications activities should also be determined based on the collaboration scenario from the beginning.

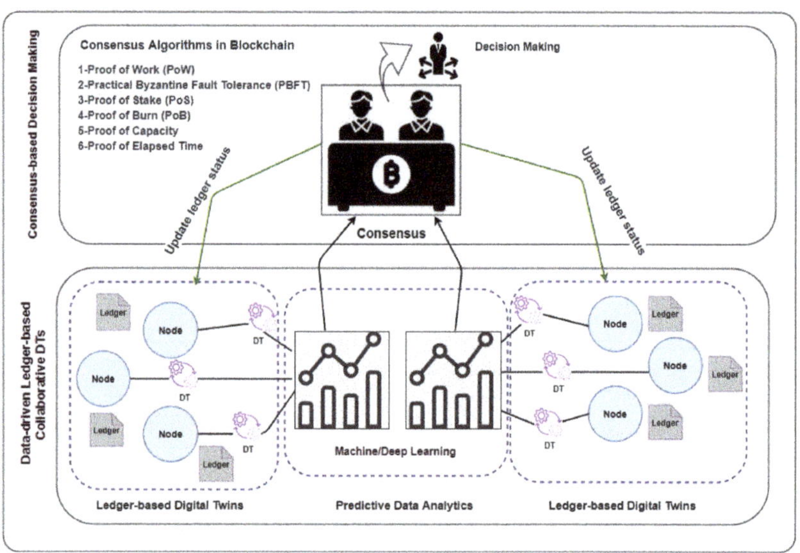

Figure 4. The architecture of the conceptual framework of data-driven blockchain-based for DTs collaboration.

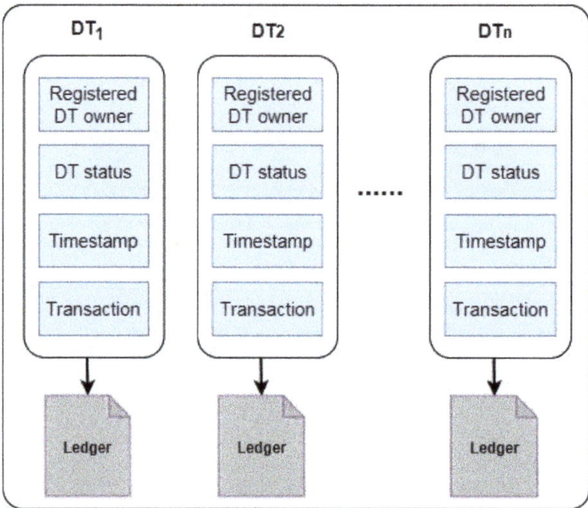

Figure 5. Ledger-based DTs.

Data-Driven DTs Based Predictive Analytics

For data-driven DTs collaborations, the DTs collaboration concept within the proposed framework will help to understand the DT status, interact with other DTs at the edge level, learn from other DTs, and share common semantic knowledge within industrial systems [18]. Figure 6 shows the workflow description of building a predictive model for data-driven ledger-based collaborative DTs. The workflow consists mainly of two phases: building an offline predictive model and deploying an online predictive model [55,56]. For the offline predictive model, we will use the distributed predictive model (i.e., classifier) by applying the distributed machine learning frameworks like Apache Spark MLlib. Apache Spark MLlib is a scalable library that implements many machine learning algorithms (i.e., regular machine learning and deep learning) [57]. An offline predictive model will be developed and trained using ledger-based historical DT operational data. For the online predictive model, the developed predictive model (i.e., classifier) could be evaluated in the smart contracts. In particular, the developed model will be used to predict the potential risks online using DT-driven real-time operational data. It will be run until it reaches a consensus. The smart contract will make decisions based on these output predictions and then store the decisions in the blockchain ledger. The smart contract executed the developed model and applied it using the DT-driven live streaming data stores in the ledgers. One example of the predicted potential risk with production systems is detecting the early faults indicated by degraded performance or damaged physical counterpart (e.g., node, device). Consequently, the proposed framework can help the decision-makers dynamically replan a set of safety precautions and take the proper action to decrease downtime within the production systems.

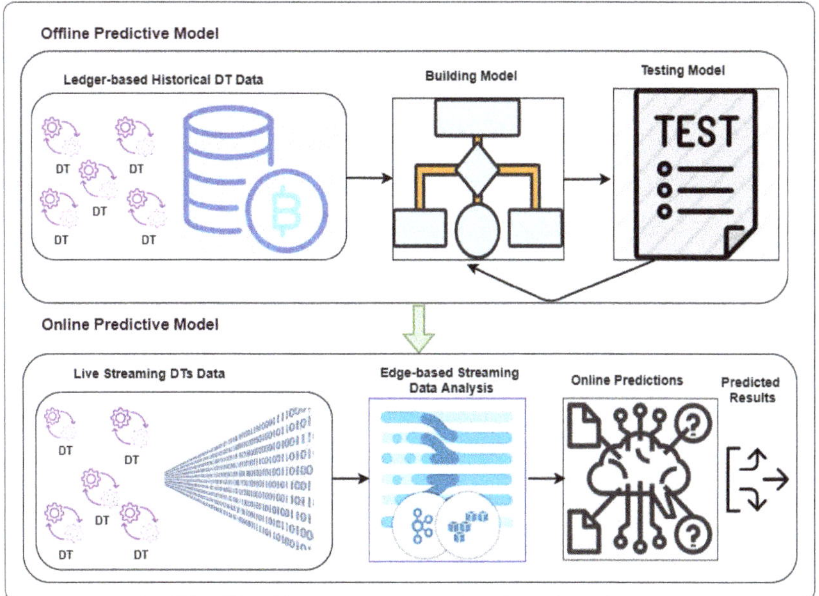

Figure 6. The workflow of building a predictive model for data-driven ledger-based collaborative DTs.

3.2.2. Consensus-Based Decision Making

This component is used to develop a distributed consensus algorithm to improve the DTs collaboration. The proposed conceptual framework aims to develop a collaborative DTs system to provide distributed decision-making to avoid potentially threatening the production system. Therefore, a distributed decision-making algorithm will be developed based on the essence of the consensus mechanism and the dynamic prediction, which uses real-time DT-driven operational data [13,14]. The developed distributed consensus algorithm will provide evidence that most nodes agree on the potential risks to notify the decision-makers within the distributed manufacturing systems. Some examples of the use of the consensus algorithms include Proof of Work (PoW), Practical Byzantine Fault Tolerance (PBFT), Proof of Stake (PoS), Proof of Burn (PoB), Proof of Capacity, and Proof of Elapsed Time.

To do so, the Directed Acyclic Graph (DAG) structured DLT solution will be considered as part of its consensus mechanism that would make it possible to have a fully decentralized manufacturing environment [45,58]. Furthermore, DLT can ensure that all the participants (i.e., deployed DTs) share identical knowledge to allow the possibility of prediction that needs to reach a global agreement regarding the object of interest (e.g., fault diagnosis) [41] This agreement will be reached using the consensus mechanism, which utilizes the local interaction of deployed DTs within a manufacturing node, and then globally using the ledger-based running database [59].

4. Smart Transportation Use Case

This section introduces an overview of the smart transportation industry, followed by the three selective use cases, including smart logistics, railway predictive maintenance, and their mapping to our proposed framework.

4.1. Overview of Smart Transportation Industry

The concept of Smart Transportation Systems (STS) includes a wide variety of advanced technologies, such as communication, sensing, and power, which have been utilized to process massive volumes of information to overcome the issues in the urban region [60] We present an overview of the literature by considering the DTs' role and data analytics in transportation systems.

The authors in [61] have addressed the data fusion generated from the connected vehicles in STS. The authors have introduced a multi-variate data fusion (MVDF) technique The proposed MVDF aims to handling asynchronous and discrete data from the environment and streamlining them into continuous and delay-less inputs for the applications. They have used regression learning to identify the errors, and they have used network simulator experiments for the metrics error, data utilization ratio, and computation time. In the context of the Internet of Vehicles and security, the authors in [62] have proposed two models to enhancing the security and performance of nodes on the Internet of Vehicles. The first model has been proposed to detect the data fascination attack using the hashing technique, while the second model has been proposed to reduce the travel time in case of traffic congestion. Furthermore, the authors in [63] have introduced the hybridized cryptographic-integrated steganography (HCIS) algorithm. The HCIS algorithm is used with auxiliary data inputs for secured data sharing in IoT assisted cloud environment for the urban transportation system.

In the context of logistics, the authors in [64] have proposed an IoT-assisted intelligent logistics transportation management framework to design an optimized logistics plan, improve customer service and reduce transportation costs. The authors focused on identifying the optimal routes for the directed autonomous vehicle, considering different vehicles.' necessities, renewable generations, logistic requests, and the essential transportation systems. In [65], the authors have explored the potential of using DT technology in synchromodal transport. They have introduced a proof-of-concept for long-distance DTs solution. The DT-based solution aims to connect real-time data generated from the physical system to a virtual GIS environment and then utilize this data in real-time synchromodal deliveries.

For the predictive maintenance in the transportation context, the authors in [44] have reviewed the use of DTs technologies to apply maintenance strategies to provide a deeper insight into the synergies between both DTs and maintenance in the industrial sectors. Furthermore, the authors in [1] proposed a set of requirements to enable predictive maintenance with big data for Industry 4.0 applications. The authors have studied the railway industry concerning big data streaming processing platforms, distributed message queue management systems, big data storage platforms, and streaming SQL engines. For the railway, the authors in [66] have also explored the potential use of blockchain technology in the railway industry by providing a simple analysis of the possible adoption rate. The authors have introduced a prototype for speech recognition and a mobility-based data collection solution to enhance the technology adoption rate.

In the context of adoption DTs in crane operations, the authors in [67,68] have proposed a maintenance model called Integrated Maintenance Decision Making Model (IDMM) for cranes operating in control terminals. The IDMM model targets to improve crane operations using DTs technologies by introducing crane maintenance based on Monte Carlo Markov Chain and Particle Swarm Optimization. Furthermore, Autiosalo et al.,[69] have introduced a multi-component DT for an industrial overhead crane. The authors have proposed a prototype called Ilmatar based on DTs technologies. The Ilmatar prototype aims to provide a maintenance service for cranes' daily tasks.

On the other hand, the authors in [70] have introduced the Robotic Process Automation (RPS) solutions to deliver service for the patients. They have proposed a simulation-based framework using RPA solution, development, 3-D building information collection, supply chain simulation, and optimization. The authors have also studied the Greenfield hospital in Singapore and then used DTs technology to visualize the operational logistics supply chain. The authors in [71] have explored the potential of using DTs technologies to manage the COVID-19 pandemic by supporting flexible decision making. The authors have discussed various challenges, including modeling and data-driven analysis for pandemic management and modeling and predictive analysis. Finally, the authors have introduced a framework using DTs and AI tools to improve the control of the COVID-19 pandemic.

Table 3 describes a comparison of existing work and the present work concerning the STS use cases, including blockchain, DTs, collaboration, DT, data analysis, IoT, and the three selected transportation use cases; logistics and railway.

Table 3. Comparison of existing work and the present work for smart transportation use cases.

Ref	Highlighted	Blockchain	DT	Collaboration	Data Analysis	IoT	Selected Use Cases Logistics	Railway
[61] (2021)	MVDF technique for handling asynchronous generated from the connected vehicles in STS	X	X	X	✓	✓	X	X
[62] (2021)	Enhancing the security and performance of nodes on the Internet of Vehicles.	X	X	X	X	✓	X	X
[63] (2020)	Hybridized cryptographic integrated steganography algorithm for secured data sharing in IoT in cloud environment	X	X	X	X	✓	X	X
[64] (2021)	IoT-assisted intelligent logistics transportation management framework to optimize logistics	X	X	X	X	✓	✓	X
[65] (2020)	Exploring potential of using DT in synchromodal transport	X	✓	X	X	✓	X	X
[44] (2020)	Using DTs to apply maintenance strategies in the industrial sectors.	X	✓	X	X	✓	X	X
[1] (2020)	Identify a set of requirements to enable predictive maintenance for Industry 4.0 including railway.	X	X	X	X	✓	X	✓
[66] (2018)	Exploring and analysis of adoption of blockchain in the railway	✓	X	X	X	X	X	✓
[67] (2019) [68] (2020)	Proposing a DT-based maintenance model for cranes operating in control terminal.	X	✓	X	X	✓	✓	X

Table 3. *Cont.*

Ref	Highlighted	Blockchain	DT	Collaboration	Data Analysis	IoT	Selected Use Cases	
							Logistics	Railway
[69] (2021)	Proposing multi-component based on DTS called Ilmatar for overhead cranes.	X	✓	X	X	✓	✓	X
[70] (2020)	Proposing a robotic process automation solution to deliver service for the patients	X	✓	X	✓	X	✓	X
[71] (2020)	Study the potential of using DTs to manage the COVID-19	X	✓	X	✓	X	X	X
[29] (2020)	A blockchain framework for heterogeneous multi-robot collaboration to combat COVID-19	✓	X	✓	X	✓	X	X
[22] (2019)	The collaboration between drone and IoT devices for improving Industry 4.0 applications such as smart city, smart healthcare	✓	X	✓	X	✓	X	X

4.2. Selective Use Cases for Smart Transportation Industry

Today, many transportation systems connect their information systems using new technologies, including IoT, big data, DTs, and AI. DTs are used to visualize the transportation infrastructure to support collaborations for accelerating the transportation process. In particular, DTs represent the physical assets in the transportation system to understand the assets' status and model their performances. They are continuously updated in real-time from multiple transportation systems, including sensors, vehicles, CCTV, people, road networks, etc. These DTs can be collaborated by sharing their operational data to provide insightful information about the assets throughout the lifecycle within the transportation system.

Consequently, DT collaboration is considered the backbone of the transportation system. It provides up-to-date information to implement the approaches based on predictive analytics for making decisions. In particular, decisions will be taken based on the predicted potential risks within the transportation system to avoid delays and optimize transportation asset performance.

This section introduces two use cases of smart transportation that use blockchain-based collaborative DTs: (1) smart logistics and (2) railway predictive maintenance. The proposed framework, data-driven blockchain-based collaborative DTs, could be applied in actual use cases for smart transportation. A set of concepts can be discussed using the proposed framework for designing smart transportation use cases, which are the need for DTs collaborations, DLT, predictive data analysis, and distributed consensus decision making. To guide this research work, we stated these five questions derived from the objectives of the proposed framework.

1. Why do we have to use collaborative DTs in this use case?
2. What are the data schema and requirements which DTs will represent?
3. How could a DLT be used for data sharing to support collaborative DTs?
4. How can the DTs-based operational data intelligence help gain insight to enhance the prediction about the potential risks?
5. How could a distributed consensus algorithm be used to ensure a consensus of the decision-making based on the predicted potential risks?

4.2.1. Smart Logistics

In this subsection, an overview of smart logistics is presented together with a detailed mapping of our proposed framework.

Overview

In the supply chain context, the transportation process refers to the movement of products from one location to another to deliver them. Smart logistics makes supply chain transportation more effective and efficient at each step. This means that logistics are becoming increasingly challenged, and the transport of large items is becoming a huge issue. The efficient logistics system can purchase, transportation, and store the raw materials until they are delivered, making more profits for the business and ensuring reasonable customer satisfaction about timely delivery. The connected devices in the transportation system are used to visualize the logistics process and track the movement of the products. For example, the sensors within the containers are used to track each stage of shipment, weather conditions, temperature, and humidity to give companies real-time visibility of the product movement through the logistics life cycle.

One of the state-of-the-art research works of logistics collaboration, Jabeur et al., has addressed the problem of collaboration within logistics [72]. The authors have proposed a multi-agent-based solution for collaboration between logistics objects. Three types of logistics applications could be considered: shipment alert, dynamic routing, and predictive maintenance. Dynamic routing in logistics operation is the key to a successful logistics company. Therefore, optimizing the dynamic routing is essential in logistics for the most efficient routes allocated to delivery fleets. Due to the COVID-19 pandemic, logistics companies face challenges providing high logistics services for customers who prefer a safer and faster delivery method rather than venturing out for the same.

Consequently, the dynamic route plays a vital role in ensuring online deliveries to the quarantine areas. This motivates us to discuss the dynamic routing scenario within the logistics use case concerning our proposed framework. Based on this, Figure 7 depicts the high level of mapping of our proposed framework to logistics systems. Further details are elaborated following:

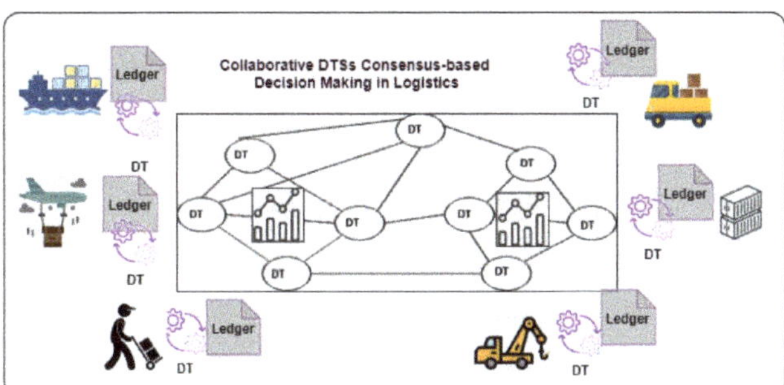

Figure 7. Blockchain-based DTs collaboration for logistics system.

DTs Collaboration in Logistics

Recently, Logistics 4.0 technologies have emerged as one of the dimensions of Industry 4.0, including smart robotics, self-driving vehicles, and automated systems for managing the movement of products among warehouses and factories. Logistics 4.0 solutions aim to create interoperable and connected logistics chains to become more innovative. The logistics process generates big data generated by tracking the movement of goods. According to the Industry 4.0 principles, Logistics 4.0 can be described as collaborative cyber-physical systems. Therefore, collaborative digital twins are used to represent the logistics data. The DTs-based logistics data is used for potential logistics optimization by monitoring physical assets and other equipment to eliminate downtime within the logistics system. The physical asset could be a fleet, truck, ship, container, robots, warehouse, and people with a set of sensors that can collect real-time data and operational status about the logistics supply chain. The digital logistics supply chain twin is used for end-to-end product tracking and identifying issues by visualizing goods' digital movements over time. Furthermore, the decentralized digital logistics supply chain twin comprises the DTs representing the geographical logistics warehouses, logistics centers, and participants.

In particular, the data generated from the elements within the logistics supply chain could be represented in DTs. The collaborative DTs are adopted to visualize the logistics supply chain, which could track products and provide end-to-end service from unloading at the quayside to shipping goods to their destinations. The proposed data-driven blockchain-based collaborative DTs framework will provide smart logistics service for a faster flow of goods, real-time analysis of comprehensive supply chain data, better synchronization of dynamic routing logistics processes, unbroken shipment tracking to improve distribution planning and delivery reliability.

Data Schema and Requirements for DTs

The collaborative DTs model consists of three components; (1) digital model, (2) data analytics, and (3) knowledgebase. The logistics data within the digital supply chain is generated from sensors attached to containers, fleets, warehouses, and robots to capture real-time data about logistic items and report on-time data about environmental changes. For warehouse management, the sensors monitor the weather, e.g., temperature and humidity in the warehouse for storing safety items. For containers, sensors are used to monitor the environmental conditions during deliveries. For fleet management, sensors attached to ships, trucks, buses, airplanes are used to collect the operating parameters of the fleet. Finally, for logistics workers, sensor data are used to track staff's physical safety.

DLT and Logistics Operational Data Sharing

The complexity of the logistics system comes from the distribution of goods, from raw materials to finished products. The logistic supply chain is divided into hundreds of geographically distributed stages globally, among multiple warehouses and logistic centers. These distributed logistic processes have massive data, making monitoring and analyzing targeted actions in response difficult. Another relevant aspect of complexity is represented by the security, integrity, sharing, and interoperability of logistic data sources. Consequently, DLT can implement collaborative DTs that allow data sharing among multiple DTs in a decentralized supply logistic chain. Therefore, the ledger-based data sharing for DTs collaboration will be considered for DT data, sharing data, and collaboration-by-communication within logistic participants.

Data-Driven DTs Based Predictive Analytics

With dynamic routing, logistics systems have a flexible and powerful scheduling capability to deliver the products to the customers on time to keep their reputation and quality of services aligned with the customers' stratification. The logistics companies can use real-time information such as weather or construction delays to change carrier routes on the fly. They can plan for future shipping routes based on collected historical logistics operational data. They can also increase their profits by taking place the dynamic routing to continue delivering a flawless shipping experience.

To assess the potential risks within the decentralized logistics manufacturing (e.g., whether it is harmful to the products such as medicines and frozen food), a prediction is needed to estimate how long a refrigerated truck will require to arrive at one or more processing plants. Therefore, dynamic routing is essential to direct the fleets based on past experiences and real-time tracking of the on-road performance of the fleet. When any problem occurs due to weather or roadblocks, the dynamic rerouting feature helps decision-makers suggest alternate and efficient routes for delivery. Consequently, many variables are operating within trucks that are needed to be monitored (i.e., temperature and humidity inside the containers, driver's road time, and the route conditions). The data generated from the elements regarding the refrigerated truck could be represented into DTs at hierarchical levels: (1) With local DT at each container, (2) intermediate more powerful DTs on the refrigerated truck at the network edge, and (3) much more powerful when DTs represent the logistics units in the cloud. The refrigeration unit's collaborative DTs system can predict the product state and queue length using the DT-driven operational data. Based on these predictions, the decision can be made in a consensus-based manner to direct the truck to the best plant to avoid potential risks.

Consensus-Based Decision Making

A consensus is a decision-making process in which members of a group of logistics centers agree to develop and support decisions to speed up the logistics supply chain considering mutual logistics traceability. Using collaborative DTs provides a better understanding of potential risks for logistics supply chains and facilitates consensus-building among participants involving the decision-makers. In doing so, many nodes involved in logistics information that is represented in DTs are divided into multiple consensus sets. The consensus mechanism which could be used is the PBFT algorithm. According to the PBFT algorithm, some amount of fault (damaged objects, hijacking, and theft, climate change, fleet failure, CCTVs failure, staff's physical safety, transportation conditions) can be tolerated without affecting the integrity of the network. The PBFT algorithm is used in Hyperledger in the transaction approval process to avoid malicious decisions among participants with logistics supply chains [73].

4.2.2. Railway Predictive Maintenance

In this subsection, an overview of predictive railway maintenance is presented together with a detailed mapping of our proposed framework.

Overview

Railway 4.0 is one of Industry 4.0 dimensions using new digital technologies including big data, IoT, DT, AI, and cloud computing [1,74,75]. The railway companies compete to provide high and attractive service to the passengers by utilizing automation and emerging technologies. One of the significant challenges that the rail industry faces is avoiding delays to meet passengers' satisfaction and maximize their profits. To do that, the rail companies start to deploy predictive maintenance applications to early diagnoses the fault and perform maintenance actions.

Nowadays, railway companies use DTs to improve railway performance by utilizing railway DT-based operational data analysis intelligence. In particular, they use the DTs collaboration to gain improved information visibility and better understand the past, present, and future predictions. With the DT-based prediction of the potential risks, the decision-makers in rail companies can support the transformation of rail track maintenance and deliver safe, reliable, and resilient service.

As the result of digitalization in the railway sector, blockchain has been adopted to provide security, scalability, traceability, transparency, and decentralization [66]. However, the adoption rate of the blockchain can be seen as slow due to the lack of technology stability, maturity, and developers' skills. Despite that, the combination of blockchain and DTs technologies in the railway sector has a beneficial role in providing high quality and safe service by (1) representing the complete railway supply chain, (2) automating of internal accounting process and passenger traveling, (3) Conducting contracts between machines and objects, and (4) monitoring the railway assets including station, train, track, switches, point machine, and sensors, and (5) managing signaling, passenger information systems, physical flows, ticketing, and goods delivery. One of the biggest challenges for the rail companies is making their stations service with a minimum maintenance cost by avoiding unnecessary expenses for the maintenance company. As the late railway maintenance can result in failure and additional costly repair, using simulation by utilizing DTs capabilities, predictive data analysis, and blockchain technologies can capture the early fault and notify the decision-makers to take the proper actions. Based on this, Figure 8 depicts the high level of mapping of our the proposed framework for fault diagnosis in railway systems. Further details are elaborated following. Furthermore, Table 4 summarizes a comparison of some current research work in predictive maintenance.

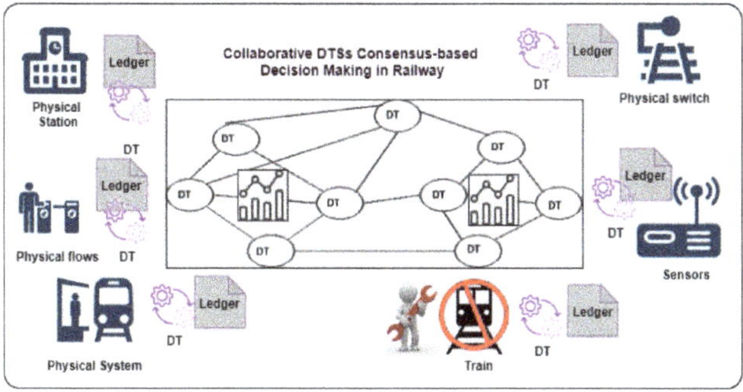

Figure 8. Blockchain-based DTs collaboration for railway.

Table 4. Comparison of some existing research work in predicative maintenance.

Ref	Highlighted	Technologies	Advantages	Limitations
[1] (2020)	Identify a set of requirements to enable predictive maintenance for Industry 4.0 including railway	Big data streaming technologies including -distributed queuing management, -big data stream processing, -big data storage, -streaming SQL engines	Provide a breadth-first mapping of predictive maintenance use-case requirements to the capabilities of big data streaming technologies focusing on open-source tools	Using the capabilities of blockchain and DTs Collaboration
[74] (2016)	Discussing the possibility of applying predictive maintenance in the railway transportation industry	-RFID technologies -Business intelligence	Explore the most important positive effects of applying predictive maintenance that affect the organization, the economy, and the people in the whole system	Using the capabilities of blockchain and DTs Collaboration
[75] (2018)	Proposing a cloud-based system for cost-effective and reliable real-time data collection, processing, and analysis from the shop floor with a multi-criteria decision-making algorithm and a condition-based maintenance strategy	-Cloud computing -Wireless sensor network -Data acquisition technologies	-Increasing awareness on both machine and shop-floor level condition -Effective and accurate maintenance of machine tools -Accurate decisions though condition-based maintenance and adaptive scheduling. -Increasing interoperability, automation and communication	Using the capabilities of blockchain and DTs Collaboration

Table 4. *Cont.*

Ref	Highlighted	Technologies	Advantages	Limitations
[76] (2018)	Reviews the overall framework to develop a DT coupled with the industrial IoT technology to advance aerospace platforms autonomy	-Industrial IoT -DTs	Discuss the role of data fusion in predictive maintenance using DT	Using the capabilities of blockchain and DTs Collaboration
[77] (2020)	Investigating of creating an automation cell for the fan-blade reconditioning component of maintenance, repair, and overhaul services to ensure that fleets of aircraft are in airworthy conditions	-Vision sensor -DTs -Robotic technologies	Track and remove the coating material of a fan blade in a closed-loop approach	Using the capabilities of blockchain and DTs Collaboration

DTs Collaboration in Railway

The rail sector is a complex network of assets and systems that come together to enable people and goods to travel safely, in a timely way at various speeds and distances. The rail assets are including station, process, rail system, and individual assets such as train, switch, sensors, etc. These rail assets could be represented in interoperable and collaborative DTs to show high visibility of the rail supply chain. The DTs representing the rail supply chain of the transportation industry can collaborate to diagnose the railway's fault, whether caused by hardware failures (e.g., rail, train, switch) or weather conditions that affect the rail, for example temperature and humidity. The DTs collaboration can understand each DT status, interact with other DTs, learn from other DTs, and share common semantic knowledge across geographical railways.

Data Schema and Requirements for DTs

In the rail industry, the physical assets (e.g., trains, track, switches, point machines) generate a vast amount of machine data from sensors such as temperature, light, vibration, and GPS. The rail data can be used to identify potential railway failures. To do so, the physical rail assets are being monitored by collecting their sensory data. As DTs represent the railway industry, the DT model is defined based on the data required for data-driven analytics for fault diagnosis, holding data fields that could be fitted in the predictive data models. Consequently, the collaborative DT model is defined based on the basic DT model, and it is consists of three components, including digital model, data analytics, and knowledgebase [78]. These components are integrated to investigate the DTs collaboration, locating the fault within the rail industry.

The Digital model of the operational data contains semi-structured content, e.g., JSON and XML. According to the domain of the rail industry, the data is generated from rail assets such as station, process, train, and weather sensors. For example, the rail data describes the physical system of the rail station, and the weather sensors are used to monitor climate change. Their data represent the weather variables, including temperature and humidity. Furthermore, the model is used to semantically model the data, reflecting the DT features and their relations using object-oriented concepts. Some semantic work could describe the relationships between the models by using model-to-model, e.g., OOP, RDF, and OWL in the case of the complex DT systems with heterogeneous DTs types.

DLT and Railway Operational Data Sharing

Distributed ledgers provide a novel technology for multi-party data sharing that emphasizes data sharing and integrity. To implement a collaborative DTs system that allows data sharing among multiple DTs in a decentralized railway supply chain, we consider DLT. The proposed framework for DLT-based architecture for DT data sharing to support DTs collaborations [41] could be extended to adopt DTL for collaborative DTs. Regarding data sharing, both the communication between lifecycle parties and the bidirectional communication between the DT and its real-world railway asset counterpart need to be considered. Therefore, the proposed ledger-based data sharing for DTs collaboration should consider: (1) DT data, (2) sharing data, and (3) collaboration-by-communication. Regarding using the DLT for the rail industry, each ledger should share data for the rail station with all parties.

Data-Driven DTs Based Predictive Analytics

Predictive data analytical models are used to support decision-making by utilizing the intelligence of the DTs-based operational data. For instance, data analytics can be used within DTs' interaction and communication to describe, diagnose, predict, and prescribe the behavior of the physical rail system for fault diagnosis. The outcome of the data analytics will be used as inputs for the consensus algorithm to make the best decision for abnormal data or warn the decision-makers in case of potential failure over the rail industry supply chain. The knowledgebase also contains the set of knowledge learned through relevant machine learning techniques from historical maintenance.

The component of data-driven ledger-based collaborative DTs for predictive analytics in our proposed framework can perform the potential failure risk using rail-based operational data including heterogeneous data sources (i.e., sensor data and historical data) [79]. To do so, the offline predictive model will be trained using ledger-based historical DT operational data. For the online predictive model, the developed predictive model (i.e., classifier) could be evaluated to predict the future faults with the railway. The outcome of the data analytics will be used as inputs for the consensus algorithm to make the best decision for abnormal data or warn the decision-makers in case of potential failure over the rail industry supply chain.

Consensus-Based Decision Making

The consensus algorithms will be used to provide the agreement of the diagnosed fault provided by collaborative DTs in the railway. This agreement will be reached locally utilizing the interaction of deployed DTs within a railway and then globally using the running database within the distributed ledger, which synchronizes the updated DTs' status within the supply chain [54]. The average consensus about the fault within the rail (e.g., fleet, train, station), which relies on a belief consensus technique, will advise the decision-makers about the fault within the rail industry supply chain.

5. Discussion, Validation, and Future Direction

In this section, we discuss the validity of our proposed framework concerning the requirements and future directions.

5.1. Validation of the Proposed Framework

This section validates the proposed framework, which aims to apply a data-driven blockchain-based for DTs collaboration. For this purpose, we also consider who the aforementioned high-level requirements were fulfilled by using the industrial technologies and informative concepts. Table 5 shows the overview of mapping the industrial technologies to the identified requirements for the proposed framework. For the data collection, IoT technologies are used to allow various data sources, such as physical things like devices, machines, people, etc. These collected data are stored into DTs, which are considered as the image of physical things. These data are also frequently updated to inform the current status of the physical things. To consider this rapid update of the physical things, the concept of timely updating can be offered by using DTs technology. DTs technology can provide an AI-based system for the data analysis requirement by continuously updating data to give timely predictions that help the decision-making process. Furthermore, the DTs technology has been adopted for its dynamic simulation capability to understand what is currently happening on the physical asset and what could happen in the future.

Besides that, the update frequency of the data needs to be exchanged among the DTs network in a secure, trust, authenticated, and transparent process. Furthermore, collaboration means sharing and exchange information among entities and share tasks to act accordingly. Blockchain technology is beneficial for DTs collaboration to (1) maintain the trust among peer to peer network [10], which DTs represent, (2) allow traceability across the entire DTs network [7], (3) provide insightful consensus-based decision-making process [80], and (4) deliver efficient solutions by utilizing the decentralization feature of blockchain technology [81]. Additionally, with a decentralized infrastructure of physical nodes represented in DTs, the blockchain particularly, DLT technologies can help relieve the risk of the point of failure. The blockchain and DLT technologies can overcome the safety and security that have prevented DT initiatives. The decentralized blockchain network can help trust DTs with a data track and digital identity. For a reliable decision-making process, the Consensus algorithms are used to improve the DTs collaboration in terms of the agreement of the majority of nodes about the potential risks to notify the decision-makers within the distributed manufacturing systems.

Table 5. Validation of the proposed framework.

Req. No	Requirement	Enabled by
R1	Data collection	IoT technology
R2	Data update frequency	DT technology
R3	Data analysis	AI techniques
R4	Simulation capabilities	DT technology
R5	Data exchange	Blockchain DLT technology
R6	Authentication	Blockchain technology
R7	Visibility and transparency	Blockchain technology
R8	Distributed decision making capabilities	Consensus algorithms
R9	Decentralization	Blockchain DLT technology

Validation of the Smart Logistics Use Case

A usual way to assess the validity of the conceptual frameworks is to define a set of criteria and then compare the capability of the proposed framework with the specified criteria. The criteria used above (see Table 5) may serve as a starting point. As the proposed conceptual framework implementation is a work in progress, the smart logistics use case scenario has been obtained to be validated [72]. Concerning that the proposed framework is implemented-dependent, Figure 9 describes the workflow of blockchain-based DTs collaboration in the logistics system. Being implementation-dependent, the participants in the blockchain network of the logistics system (i.e., represented in DTs) are detailed as follows:

- The factory is responsible for transporting the loads to the suppliers. Each factory DT checks the smart contact to meet the load requirements. Then factory DT has all the data about the loading status, certificates, suppliers' locations, and the number of batches written into the ledgers. Once the factory sends the load to the supplier, the blockchain network is updated.
- The supplier is responsible for transporting the load to the warehouse. Each supplier DT has frequently updated data about the loaded products, warehouses locations, and shipments date. The collected data about the products and shipments are written into ledgers, and then the blockchain network is updated.
- The logistics operator is responsible for updating all necessary records, including a packing list, order number, batch number, production data, etc. Each logistics operator DT has frequently updated data about the corresponding shipment data recorded by the operator. The recorded data are written into ledgers, and then the blockchain network is updated.
- The long haul carrier is used to carry the heavy shipment and transport them to the warehouses. Each carrier DT checks the smart contact to meet the rules of shipment transportation. As a result, the carrier DT frequently has the updated bill of loading, shipment details, the destination warehouses location, and diver details. The shipment data are written into ledgers, and then the blockchain network is updated about the shipment movement track.
- The warehouse is used to store the shipments. Each warehouse DT has the data about the stored products, including location, temperature, humidity, product items, etc. These updated warehouse DT data are used to check the product storing conditions

concerning the smart contract rules to avoid product damages. The warehouse DT data, including the product quantity, are also used to check the smart contract for new orders. The stored product data are written into ledgers, and then the blockchain network is updated about the stored products.

- The delivery process is responsible for delivering the orders to the customers. Each delivery DT has updated data about the warehouse location, customer address, routing instruction, packing details, driver details, bills, etc. The delivering product data are written into ledgers, and then the blockchain network is updated about the products being delivered.
- A customer is a person who orders the product and who receives it. The customer DT has the data about the customer delivery address, customer ID, etc. The smart contract will check the delivery data based on the customer data, and then the delivering information is recorded into the ledgers. Once the delivery process is successfully completed, all the blockchain network participants are updated about completing the delivery process.
- The decision-making unit is responsible for deciding in case of potential risk for the products during the loading, storing, and delivering processes.

Based on the criteria above and the participant in the blockchain network, the validity of the proposed conceptual framework are discussed as follows:

Tracking. Blockchain network allows efficient tracking of the changes along the logistics process. Using the combination of emerging technologies like the blockchain, AI, and DTs can improve the productivity of the logistics process with effective tracking. The proposed framework can incorporate blockchain with DTs to get accurate data about every step in the shipping process. Once the products are loaded and shipped, the logistics participants' DTs collaborate to exchange the logistics data. The logistics data is stored along with the information on the product movement [65]. Therefore, the blockchain network can provide the participants with the logistics by-product data like showing the person handling the product at that time. For example, the logistics system can track the product damage using the ledger-based logistics records in case of product damage. The smart contracts are settled, the logistics data will be stored in the public ledgers. All logistics records are stored to track the changes (i.e., what is the change, why it is done, and who made the changes). Furthermore, sharing information about logistics tracking with the customers across the blockchain network can increase the transparency of the logistics system.

Delivery. To assess the potential risks within the decentralized logistics manufacturing delivery process (e.g., whether it is harmful to the products such as medicines and frozen food), a prediction is needed to estimate how long a refrigerated truck will require to arrive at one or more processing plants. The proposed framework can incorporate blockchain with AI and DTs to predict the potential risks in advance and then take the appropriate action like routing redirection [64]. On the other hand, the proposed framework can improve the secure delivery process by reducing fraud and theft issues. To do this, the smart contracts check the detailed rules, such as requiring government-approved photo IDs to access the goods for pickup or delivery.

Performance monitoring. Based on the components of the proposed framework (see Section 3.2), the predictive data analysis component is used to monitor the performance of the logistics process by analyzing the product data which are collected from logistics participants DTs including factory DT, supplier DT, long haul carrier DT, logistics operator DT, warehouse DT, and delivery DT. These DTs are collaborating and interacting to feed the learning models to predict the potential risk of the product, such as harmful product, damage, theft, and so on. The decision is making based on the consensus to avoid the risk such as logistics process delay.

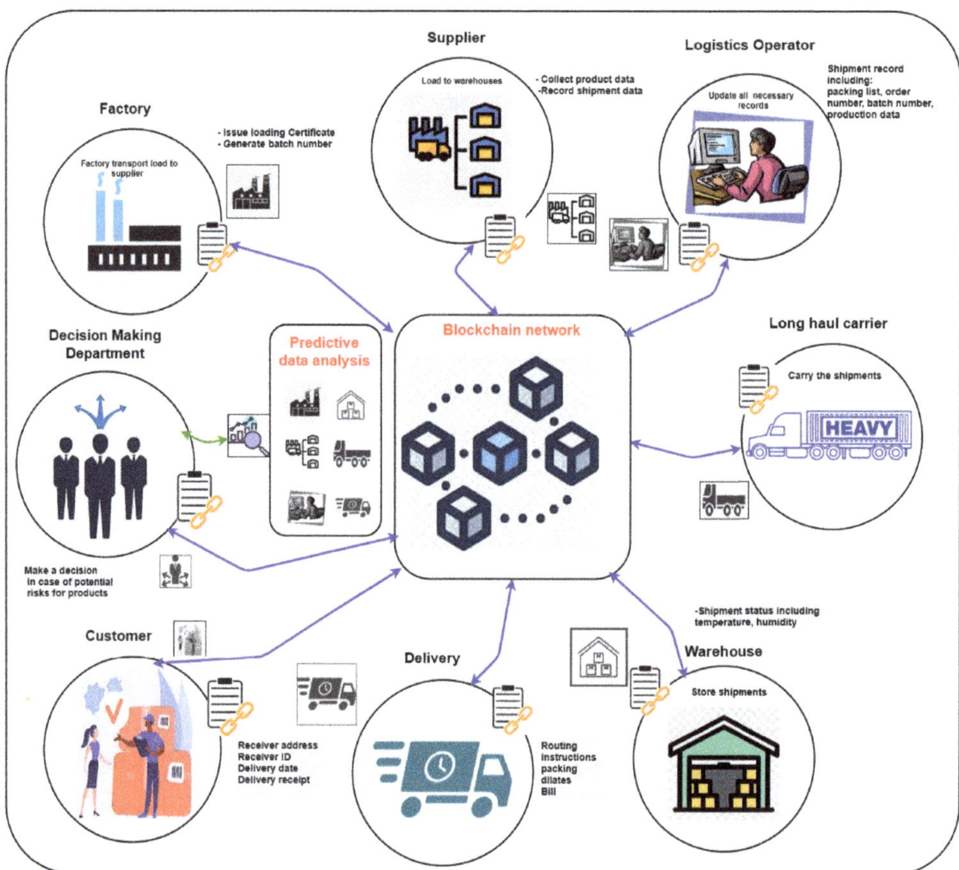

Figure 9. The workflow of blockchain-based DTs collaboration for logistics system.

5.2. Future Directions

5.2.1. Security and Privacy

The security and privacy associated with DTs are challenging within smart transportation because of the massive amount of data and the risk of sensitive data created from smart transportation systems. Therefore, IoT devices should analyze DTs data locally using federated learning and then share only the model to the blockchain instead of sending the raw data. Thus, the issue of security can be solved by using blockchain technology, while privacy can be solved by using federated learning. The combination of both techniques can significantly enhance the security and privacy of DTs in transportation systems.

5.2.2. Connectivity

With the growth of smart devices in smart transportation systems, connectivity is still a challenge for these smart devices to perform the goal in real-time. The massive number of smart devices in smart transportation needs for advanced communication technologies like Beyond Fifth Generation (B5G) or Sixth Generation (6G). If any smart devices get disconnected, blockchain may help devices borrow data from neighboring devices to keep the transportation system working efficiently. Running machine learning at the edge may ensure full connectivity, high accuracy and prevent missing data.

5.2.3. Global Logistic Networks

DTs in logistic networks play a vital role in improving logistics such as highways, railways, streets, oceans and smart cars can create their data from surrounding resources and make their map accordingly. They can then share their data with others to reduce traffic. The shared data can be traffic speed, parking, road closures, etc. and blockchain technology collaborates with them quickly in the decentralized network. While federated learning can improve processes locally and sharing only the model, this will help solve blockchain storage issues and resources. Therefore, digital twins of smart cars can collaborate in the real-time location of a specific car or people with the help of blockchain and federated learning for creating compelling, optimize, and efficient logistics.

5.2.4. Timing, Speed, and Response

Beyond logistics and transportation, DTs timing and speed are challenging. Timing and speed can lead to several changes in logistics and transportation. First, time improves decision-making and response of taking decisions and taking actions for the customer service demand, which needs high accuracy and fast responses. Businesses based on transportation and logistics do not want data, but they want visibility to be quick and timely.

5.2.5. Packaging and Containers in Transportation

In logistics, most of the transportation is in the packaging form. Therefore, the development, management, and monitoring of containers and packaging face many logistics sector challenges. Currently, due to COVID-19, increasing demand for containers and packaging can be noted due to growth in E-commerce. Furthermore, varieties of packaging also need to be taken into consideration. These results significantly reduce operation efficiency and waste due to poor utilization. A container model can be created using DTs in conjunction with computer vision, and such problems can be automatically detected. Historical information storage in the blockchain of the containers moments for starting DT can influence decision-making about container status. The decision-making should be repaired, used, and maintained as a fault in the container. DTs with blockchain technology can develop a lighter, more robust, and eco-friendly environment for packaging goods and managing containers effectively and efficiently.

5.2.6. Decision-Making Process

The proposed framework work could be extended by integrating two levels of the decision-making process to derive alternative system configurations. The decision-making process levels are 1) decision to avoid the potential risks and 2) dynamic planning to reconfigure the system in cases of unexpected events. For example, in the case of the railway transportation industry, the system can predict failure then diagnose and trigger maintenance by using IoT. In another instance, in the transportation logistics industry, the system can predict the potential risks of the product state within a truck. Based on these predictions, the decision can be made in a consensus-based manner to direct the truck to the best plant to avoid potential risks.

6. Conclusions

This conceptual framework shows how blockchain technology-empowered DTs collaboration in smart distributed manufacturing. DTs collaboration supports the interaction mechanism to understand the DT status, sharing a goal, exchange information, interacting with each other, mutual learning, and mutual adaption. Based on the literature, we present the challenges that DTs collaboration is suffering and then how blockchain technology solves these challenges. The proposed framework can improve DTs collaboration and analysis data in real-time. Furthermore, we discuss how the conceptual framework can be applied in smart transportation, i.e., smart logistics and railway predictive maintenance. Finally, we highlighted the future direction to guide interested researchers in this interesting area.

Author Contributions: R.S.: Conceptualization, Data curation, Formal analysis, Methodology, Writing—original draft, Writing—review & editing. S.H.A.: Conceptualization, Methodology, Formal analysis, Writing—review & editing. Brown, K.N.B.: Conceptualization, Supervision, Investigation, Project administration, D.O.: Supervision C.M.: Conceptualization, reviewing and Project administration M.G.: Conceptualization and reviewing. All authors have read and agreed to the published version of the manuscript.

Funding: This research has emanated from research supported by a research grant from Science Foundation Ireland (SFI) under Grant Number SFI/16/RC/3918 (CONFIRM), and Marie Skłodowska-Curie grant agreement No. 847577 co-funded by the European Regional Development Fund.

Institutional Review Board Statement: Not applicable.

Informed Consent Statement: Not applicable.

Data Availability Statement: No Data available online. For further query email to corresponding author (rsahal@ucc.ie, radhya.sahal@gmail.com).

Conflicts of Interest: The authors declare no conflict of interest.

References

1. Sahal, R.; Breslin, J.G.; Ali, M.I. Big data and stream processing platforms for Industry 4.0 requirements mapping for a predictive maintenance use case. *J. Manuf. Syst.* **2020**, *54*, 138–151. [CrossRef]
2. Tao, F.; Qi, Q.; Liu, A.; Kusiak, A. Data-driven smart manufacturing. *J. Manuf. Syst.* **2018**, *48*, 157–169. [CrossRef]
3. Srai, J.S.; Kumar, M.; Graham, G.; Phillips, W.; Tooze, J.; Ford, S.; Beecher, P.; Raj, B.; Gregory, M.; Tiwari, M.K.; et al. Distributed manufacturing: Scope, challenges and opportunities. *Int. J. Prod. Res.* **2016**, *54*, 6917–6935. [CrossRef]
4. Zhong, R.Y.; Xu, X.; Klotz, E.; Newman, S.T. Intelligent Manufacturing in the Context of Industry 4.0: A Review. *Engineering* **2017**, *3*, 616–630. [CrossRef]
5. Lu, Y.; Liu, C.; Wang, K.I.K.; Huang, H.; Xu, X. Digital Twin-driven smart manufacturing: Connotation, reference model, applications and research issues. *Robot. Comput.-Integr. Manuf.* **2020**, *61*, 101837. [CrossRef]
6. Benkamoun, N.; ElMaraghy, W.; Huyet, A.L.; Kouiss, K. Architecture framework for manufacturing system design. *Procedia CIRP* **2014**, *17*, 88–93. [CrossRef]
7. Yaqoob, I.; Salah, K.; Uddin, M.; Jayaraman, R.; Omar, M.; Imran, M. Blockchain for digital twins: Recent advances and future research challenges. *IEEE Netw.* **2020**, *34*, 290–298. [CrossRef]
8. Altun, C.; Tavli, B.; Yanikomeroglu, H. Liberalization of digital twins of IoT-enabled home appliances via blockchains and absolute ownership rights. *IEEE Commun. Mag.* **2019**, *57*, 65–71. [CrossRef]
9. Huang, S.; Wang, G.; Yan, Y.; Fang, X. Blockchain-based data management for digital twin of product. *J. Manuf. Syst.* **2020**, *54*, 361–371. [CrossRef]
10. Hasan, H.R.; Salah, K.; Jayaraman, R.; Omar, M.; Yaqoob, I.; Pesic, S.; Taylor, T.; Boscovic, D. A blockchain-based approach for the creation of digital twins. *IEEE Access* **2020**, *8*, 34113–34126. [CrossRef]
11. Leng, J.; Yan, D.; Liu, Q.; Xu, K.; Zhao, J.L.; Shi, R.; Wei, L.; Zhang, D.; Chen, X. ManuChain: Combining permissioned blockchain with a holistic optimization model as bi-level intelligence for smart manufacturing. *IEEE Trans. Syst. Man. Cybern. Syst.* **2019**, *50*, 182–192. [CrossRef]
12. Obushnyi, S.; Kravchenko, R.; Babichenko, Y. Blockchain as a Transaction Protocol for Guaranteed Transfer of Values in Cluster Economic Systems with Digital Twins. In Proceedings of the 2019 IEEE International Scientific-Practical Conference Problems of Infocommunications, Science and Technology (PIC S&T), Kyiv, Ukraine, 8–11 October 2019; pp. 241–245.
13. Li, S.; Oikonomou, G.; Tryfonas, T.; Chen, T.M.; Da Xu, L. A distributed consensus algorithm for decision making in service-oriented internet of things. *IEEE Trans. Ind. Inform.* **2014**, *10*, 1461–1468.
14. Gramoli, V. From blockchain consensus back to Byzantine consensus. *Future Gener. Comput. Syst.* **2020**, *107*, 760–769. [CrossRef]
15. Schluse, M.; Priggemeyer, M.; Atorf, L.; Rossmann, J. Experimentable digital twins—Streamlining simulation-based systems engineering for industry 4.0. *IEEE Trans. Ind. Inform.* **2018**, *14*, 1722–1731. [CrossRef]
16. Avventuroso, G.; Silvestri, M.; Pedrazzoli, P. A networked production system to implement virtual enterprise and product lifecycle information loops. *IFAC-PapersOnLine* **2017**, *50*, 7964–7969. [CrossRef]
17. Nikolakis, N.; Alexopoulos, K.; Xanthakis, E.; Chryssolouris, G. The digital twin implementation for linking the virtual representation of human-based production tasks to their physical counterpart in the factory-floor. *Int. J. Comput. Integr. Manuf.* **2019**, *32*, 1–12. [CrossRef]
18. Sahal, R.; Alsamhi, S.H.; Breslin, J.G.; Brown, K.N.; Ali, M.I. Digital Twins Collaboration for Automatic Erratic Operational Data Detection in Industry 4.0. *Appl. Sci.* **2021**, *11*, 3186. [CrossRef]
19. Alsamhi, S.H.; Ma, O.; Ansari, M.S. Convergence of Machine Learning and Robotics Communication in Collaborative Assembly: Mobility, Connectivity and Future Perspectives. *J. Intell. Robot. Syst.* **2020**, *98*, 541–566. [CrossRef]

20. Alsamhi, S.H.; Ma, O.; Ansari, M.S. Survey on artificial intelligence based techniques for emerging robotic communication. *Telecommun. Syst.* **2019**, *72*, 483–503. [CrossRef]
21. Alsamhi, S.H.; Ma, O.; Ansari, M.S.; Almalki, F.A. Survey on collaborative smart drones and internet of things for improving smartness of smart cities. *IEEE Access* **2019**, *7*, 128125–128152. [CrossRef]
22. Alsamhi, S.H.; Ma, O.; Ansari, M.S.; Gupta, S.K. Collaboration of drone and internet of public safety things in smart cities: An overview of qos and network performance optimization. *Drones* **2019**, *3*, 13. [CrossRef]
23. Gupta, A.; Sundhan, S.; Gupta, S.K.; Alsamhi, S.H.; Rashid, M. Collaboration of UAV and HetNet for better QoS: A comparative study. *Int. J. Veh. Inf. Commun. Syst.* **2020**, *5*, 309–333. [CrossRef]
24. Li, Q.; Gravina, R.; Li, Y.; Alsamhi, S.H.; Sun, F.; Fortino, G. Multi-user activity recognition: Challenges and opportunities. *Inf. Fusion* **2020**, *63*, 121–135. [CrossRef]
25. Wang, T.; Li, J.; Kong, Z.; Liu, X.; Snoussi, H.; Lv, H. Digital twin improved via visual question answering for vision-language interactive mode in human–machine collaboration. *J. Manuf. Syst.* **2020**, *58*, 261–269. [CrossRef]
26. Andreassen, T.W.; Lervik-Olsen, L.; Snyder, H.; Van Riel, A.C.R.; Sweeney, J.C.; Van Vaerenbergh, Y. Business model innovation and value-creation: The triadic way. *J. Serv. Manag.* **2018**, *29*, 883–906. [CrossRef]
27. De Reuver, M.; Sørensen, C.; Basole, R.C. The digital platform: A research agenda. *J. Inf. Technol.* **2018**, *33*, 124–135. [CrossRef]
28. Eloranta, V.; Orkoneva, L.; Hakanen, E.; Turunen, T. Using platforms to pursue strategic opportunities in service-driven manufacturing. *Serv. Sci.* **2016**, *8*, 344–357. [CrossRef]
29. Alsamhi, S.H.; Lee, B. Blockchain for Multi-Robot Collaboration to Combat COVID-19 and Future Pandemics. *arXiv* **2020**, arXiv:2010.02137.
30. Alsamhi, S.H.; Lee, B.; Guizani, M.; Kumar, N.; Qiao, Y.; Liu, X. Blockchain for decentralized multi-drone to combat COVID-19 and future pandemics: Framework and proposed solutions. *Trans. Emerg. Telecommun. Technol.* **2021**, e4255. [CrossRef]
31. Alsamhi, S.H.; Afghah, F.; Sahal, R.; Hawbani, A.; Al-qaness, A.A.; Lee, B.; Guizani, M. Green Internet of Things using UAVs in B5G Networks: A Review of Applications and Strategies. *Ad Hoc Netw.* **2021**, *117*, 102505. [CrossRef]
32. Almalki, F.; Alsamhi, S.; Sahal, R.; Hassan, J.; Hawbani, A.; Rajput, N.; Saif, A.; Morgan, J.; Breslin, J.; et al. Green IoT for Eco-Friendly and Sustainable Smart Cities: Future Directions and Opportunities. *Mob. Netw. Appl.* **2021**, 1–25.
33. Catarci, T.; Firmani, D.; Leotta, F.; Mandreoli, F.; Mecella, M.; Sapio, F. A Conceptual Architecture and Model for Smart Manufacturing Relying on Service-Based Digital Twins. In Proceedings of the2019 IEEE International Conference on Web Services (ICWS), Milan, Italy, 8–13 July 2019; pp. 229–236.
34. Qi, Q.; Tao, F. Digital twin and big data towards smart manufacturing and industry 4.0: 360 degree comparison. *IEEE Access* **2018**, *6*, 3585–3593. [CrossRef]
35. Qi, Q.; Tao, F.; Hu, T.; Anwer, N.; Liu, A.; Wei, Y.; Wang, L.; Nee, A.Y.C. Enabling technologies and tools for digital twin. *J. Manuf. Syst.* **2019**, *58*, 3–21. [CrossRef]
36. He, B.; Bai, K.J. Digital twin-based sustainable intelligent manufacturing: A review. *Adv. Manuf.* **2020**, *9*, 1–21. [CrossRef]
37. Fuller, A.; Fan, Z.; Day, C.; Barlow, C. Digital Twin: Enabling Technologies, Challenges and Open Research. *IEEE Access* **2020**, *8*, 108952–108971. [CrossRef]
38. Leng, J.; Jiang, P.; Xu, K.; Liu, Q.; Zhao, J.L.; Bian, Y.; Shi, R. Makerchain: A blockchain with chemical signature for self-organizing process in social manufacturing. *J. Clean. Prod.* **2019**, *234*, 767–778. [CrossRef]
39. Zhang, C.; Zhou, G.; Li, H.; Cao, Y. Manufacturing Blockchain of Things for the Configuration of a Data-and Knowledge-Driven Digital Twin Manufacturing Cell. *IEEE Internet Things J.* **2020**, *7*, 11884–11894. [CrossRef]
40. Kuo, T.T. The anatomy of a distributed predictive modeling framework: Online learning, blockchain network, and consensus algorithm. *JAMIA Open* **2020**, *3*, 201–208. [CrossRef] [PubMed]
41. Dietz, M.; Putz, B.; Pernul, G. A distributed ledger approach to digital twin secure data sharing. In *IFIP Annual Conference on Data and Applications Security and Privacy*; Springer: Berlin/Heidelberg, Germany, 2019; pp. 281–300.
42. Kapteyn, M.G.; Knezevic, D.J.; Willcox, K. Toward predictive digital twins via component-based reduced-order models and interpretable machine learning. In Proceedings of the AIAA Scitech 2020 Forum, Orlando, FL, USA, 6–10 January 2020; p. 0418.
43. Austin, M.; Delgoshaei, P.; Coelho, M.; Heidarinejad, M. Architecting smart city digital twins: Combined semantic model and machine learning approach. *J. Manag. Eng.* **2020**, *36*, 04020026. [CrossRef]
44. Errandonea, I.; Beltrán, S.; Arrizabalaga, S. Digital Twin for maintenance: A literature review. *Comput. Ind.* **2020**, *123*, 103316. [CrossRef]
45. Altun, C.; Tavli, B. Social Internet of Digital Twins via Distributed Ledger Technologies: Application of Predictive Maintenance. In Proceedings of the 2019 27th Telecommunications Forum (TELFOR), Belgrade, Serbia, 26–27 November 2019; pp. 1–4.
46. Song, Z.; Canedo, A.M. Digital Twins for Energy Efficient Asset Maintenance. U.S. Patent Application 15/052,992, 25 February 2016.
47. Vathoopan, M.; Johny, M.; Zoitl, A.; Knoll, A. Modular fault ascription and corrective maintenance using a digital twin. *IFAC-PapersOnLine* **2018**, *51*, 1041–1046. [CrossRef]
48. Aivaliotis, P.; Georgoulias, K.; Arkouli, Z.; Makris, S. Methodology for enabling digital twin using advanced physics-based modelling in predictive maintenance. *Procedia CIRP* **2019**, *81*, 417–422. [CrossRef]

49. Pargmann, H.; Euhausen, D.; Faber, R. Intelligent big data processing for wind farm monitoring and analysis based on cloud-technologies and digital twins: A quantitative approach. In Proceedings of the 2018 IEEE 3rd International Conference on Cloud Computing and Big Data Analysis (ICCCBDA), Chengdu, China, 20–22 April 2018; pp. 233–237.

50. DeSmit, Z.; Elhabashy, A.E.; Wells, L.J.; Camelio, J.A. An approach to cyber-physical vulnerability assessment for intelligent manufacturing systems. *J. Manuf. Syst.* **2017**, *43*, 339–351. [CrossRef]

51. Wang, K.S.; Sharma, V.S.; Zhang, Z.Y. SCADA data based condition monitoring of wind turbines. *Adv. Manuf.* **2014**, *2*, 61–69. [CrossRef]

52. Sivalingam, K.; Sepulveda, M.; Spring, M.; Davies, P. A review and methodology development for remaining useful life prediction of offshore fixed and floating wind turbine power converter with digital twin technology perspective. In Proceedings of the 2018 2nd International Conference on Green Energy and Applications (ICGEA), Singapore, 24–26 March 2018; pp. 197–204.

53. Valdivia, L.J.; Del-Valle-Soto, C.; Rodriguez, J.; Alcaraz, M. Decentralization: The failed promise of cryptocurrencies. *IT Prof.* **2019**, *21*, 33–40. [CrossRef]

54. George, J.; Elwin, M.L.; Freeman, R.A.; Lynch, K.M. Distributed fault detection and accommodation in dynamic average consensus. In Proceedings of the 2018 Annual American Control Conference (ACC), Milwaukee, WI, USA, 27–29 June 2018; pp. 5019–5024.

55. Alharbi, A.; Alosaimi, W.; Sahal, R.; Saleh, H. Real-Time System Prediction for Heart Rate Using Deep Learning and Stream Processing Platforms. *Complexity* **2021**, *2021*, 5535734. [CrossRef]

56. Saleh, H.; Younis, E.M.; Sahal, R.; Ali, A.A. Predicting systolic blood pressure in real-time using streaming data and deep learning. *Mob. Netw. Appl.* **2021**, *26*, 326–335. [CrossRef]

57. Verbraeken, J.; Wolting, M.; Katzy, J.; Kloppenburg, J.; Verbelen, T.; Rellermeyer, J.S. A survey on distributed machine learning. *ACM Comput. Surv. (CSUR)* **2020**, *53*, 1–33. [CrossRef]

58. Hutton, N.; Maloberti, J.; Nickel, S.; Rønnow, T.F.; Ward, J.J.; Weeks, M. *Design of a Scalable Distributed Ledger*; Technical Report; 2018. Available online: https://staging3.fetch.ai/wp-content/uploads/2019/10/Fetch.AI-Ledger-Yellow-Paper.pdf (accessed on 6 September 2021).

59. Chaudhry, N.; Yousaf, M.M. Consensus algorithms in blockchain: Comparative analysis, challenges and opportunities. In Proceedings of the 2018 12th International Conference on Open Source Systems and Technologies (ICOSST), Lahore, Pakistan, 19–21 December; pp. 54–63.

60. Wang, F.Y.; Li, Y.; Zhang, W.; Bennett, G.; Chen, N. Digital Twin and Parallel Intelligence Based on Location and Transportation: A Vision for New Synergy Between the IEEE CRFID and ITSS in Cyberphysical Social Systems [Society News]. *IEEE Intell. Transp. Syst. Mag.* **2021**, *13*, 249–252. [CrossRef]

61. Manogaran, G.; Balasubramanian, V.; Rawal, B.S.; Saravanan, V.; Montenegro-Marin, C.E.; Ramachandran, V.; Kumar, P.M. Multi-Variate Data Fusion Technique for Reducing Sensor Errors in Intelligent Transportation Systems. *IEEE Sens. J.* **2020**, *21*, 15564–15573. [CrossRef]

62. Vijayarangam, S.; Chandra Babu, G.; Ananda Murugan, S.; Kalpana, N.; Malarvizhi Kumar, P. Enhancing the security and performance of nodes in Internet of Vehicles. *Concurr. Comput. Pract. Exp.* **2021**, *33*. [CrossRef]

63. Bi, D.; Kadry, S.; Kumar, P.M. Internet of things assisted public security management platform for urban transportation using hybridised cryptographic-integrated steganography. *IET Intell. Transp. Syst.* **2020**, *14*, 1497–1506. [CrossRef]

64. Abosuliman, S.S.; Almagrabi, A.O. Routing and scheduling of intelligent autonomous vehicles in industrial logistics systems. *Soft Comput.* **2021**, *25*, 11975–11988. [CrossRef]

65. Ambra, T.; Macharis, C. Agent-Based Digital Twins (ABM-Dt) In Synchromodal Transport and Logistics: The Fusion of Virtual and Pysical Spaces. In Proceedings of the 2020 Winter Simulation Conference (WSC), Orlando, FL, USA, 14–18 December 2020; pp. 159–169.

66. Naser, F. The Potential Use Of Blockchain Technology In Railway Applications: An Introduction Of A Mobility And Speech Recognition Prototype. In Proceedings of the 2018 IEEE International Conference on Big Data (Big Data), Seattle, WA, USA, 10–13 December 2018; pp. 4516–4524.

67. Szpytko, J.; Duarte, Y.S. Digital Twins Model for Cranes Operating in Container Terminal. *IFAC-PapersOnLine* **2019**, *52*, 25–30. [CrossRef]

68. Szpytko, J.; Duarte, Y.S. A digital twins concept model for integrated maintenance: A case study for crane operation. *J. Intell. Manuf.* **2020**, *32*, 1863–188. [CrossRef]

69. Autiosalo, J.; Ala-Laurinaho, R.; Mattila, J.; Valtonen, M.; Peltoranta, V.; Tammi, K. Towards Integrated Digital Twins for Industrial Products: Case Study on an Overhead Crane. *Appl. Sci.* **2021**, *11*, 683. [CrossRef]

70. Liu, W.; Zhang, W.; Dutta, B.; Wu, Z.; Goh, M. Digital twinning for productivity improvement opportunities with robotic process automation: Case of greenfield hospital. *Int. J. Mech. Eng. Robot. Res.* **2020**, *9*, 258–263. [CrossRef]

71. Mezzour, G.; Boudanga, Z.; Benhadou, S. Smart Pandemic Management Through a Smart, Resilient and Flexible Decision-Making System. *Int. Arch. Photogramm. Remote Sens. Spat. Inf. Sci.* **2020**, *44*, 285–294. [CrossRef]

72. Jabeur, N.; Al-Belushi, T.; Mbarki, M.; Gharrad, H. Toward leveraging smart logistics collaboration with a multi-agent system based solution. *Procedia Comput. Sci.* **2017**, *109*, 672–679. [CrossRef]

73. Li, X.; Lv, F.; Xiang, F.; Sun, Z.; Sun, Z. Research on key technologies of logistics information traceability model based on consortium chain. *IEEE Access* **2020**, *8*, 69754–69762. [CrossRef]

74. Kans, M.; Galar, D.; Thaduri, A. Maintenance 4.0 in Railway Transportation Industry. In Proceedings of the 10th World Congress on Engineering Asset Management (WCEAM 2015), Tampere, Finland, 28–30 September 2015; Springer: Berlin/Heidelberg, Germany, 2016; pp. 317–331.

75. Mourtzis, D.; Vlachou, E. A cloud-based cyber-physical system for adaptive shop-floor scheduling and condition-based maintenance. *J. Manuf. Syst.* **2018**, *47*, 179–198. [CrossRef]

76. Liu, Z.; Meyendorf, N.; Mrad, N. The role of data fusion in predictive maintenance using digital twin. In *AIP Conference Proceedings*; AIP Publishing LLC: Melville, NY, USA, 2018; Volume 1949, p. 020023.

77. Oyekan, J.; Farnsworth, M.; Hutabarat, W.; Miller, D.; Tiwari, A. Applying a 6 DoF Robotic Arm and Digital Twin to Automate Fan-Blade Reconditioning for Aerospace Maintenance, Repair, and Overhaul. *Sensors* **2020**, *20*, 4637. [CrossRef] [PubMed]

78. Wang, J.; Ye, L.; Gao, R.X.; Li, C.; Zhang, L. Digital Twin for rotating machinery fault diagnosis in smart manufacturing. *Int. J. Prod. Res.* **2019**, *57*, 3920–3934. [CrossRef]

79. Li, H.; Parikh, D.; He, Q.; Qian, B.; Li, Z.; Fang, D.; Hampapur, A. Improving rail network velocity: A machine learning approach to predictive maintenance. *Transp. Res. Part C: Emerg. Technol.* **2014**, *45*, 17–26. [CrossRef]

80. Oyinloye, D.P.; Teh, J.S.; Jamil, N.; Alawida, M. Blockchain Consensus: An Overview of Alternative Protocols. *Symmetry* **2021**, *13*, 1363. [CrossRef]

81. Dong, Z.; Luo, F.; Liang, G. Blockchain: A secure, decentralized, trusted cyber infrastructure solution for future energy systems. *J. Mod. Power Syst. Clean Energy* **2018**, *6*, 958–967. [CrossRef]

Article

Application of IoT-Aided Simulation to Manufacturing Systems in Cyber-Physical System

Yifei Tan [1], Wenhe Yang [2,*], Kohtaroh Yoshida [2] and Soemon Takakuwa [2]

[1]   Faculty of Commerce, Chuo Gakuin University, 451 Kujike, Abiko, Chiba 270-1196, Japan; yftan@cc.cgu.ac.jp
[2]   Department of Industrial and Systems Engineering, Chuo University, 1-13-27 Kasuga, Bunkyo-ku, Tokyo 112-8551, Japan; a14.ynns@g.chuo-u.ac.jp (K.Y.); takakuwa@indsys.chuo-u.ac.jp (S.T.)
*   Correspondence: yangwh@indsys.chuo-u.ac.jp; Tel.: +81-3-3817-1939

Received: 26 October 2018; Accepted: 17 December 2018; Published: 3 January 2019

Abstract: With the rapid development of mobile and wireless networking technologies, data has become more ubiquitous and the IoT (Internet of Things) is attracting much attention due to high expectations for enabling innovative service, efficiency, and productivity improvement. In next-generation manufacturing, the digital twin (DT) has been proposed as a new concept and simulation tool for collecting and synchronizing real-world information in real time in cyber space to cope with the challenges of smart factories. Although the DT is considered a challenging technology, it is still at the conceptual stage and only a few studies have specifically discussed methods for its construction and implementation. In this study, we first explain the concept of DT and important issues involved in developing it within an IoT-aided manufacturing environment. Then, we propose a DT construction framework and scheme for inputting data derived from the IoT into a simulation model. Finally, we describe how we verify the effectiveness of the proposed framework and scheme, by constructing a DT-oriented simulation model for an IoT-aided manufacturing system.

Keywords: IoT; cyber-physical systems; digital twin; simulation approach; smart factory

---

1. Introduction

With the rapid development of mobile and wireless networking technologies, the Internet of Things (IoT) has turned the vision of a more connected world into reality by emerging volume of data and numerous services offered through heterogeneous networks. The IoT, also called the Internet of Everything or the Industrial Internet, is a new technology paradigm envisioned as a global network of machines and devices capable of interacting with each other [1], realizing innovative service, efficiency and productivity improvement.

Unlike other excessive marketing terms in the information technology industry, the IoT is now widely accepted as a technology already having a major impact on industrial manufacturing systems. The IoT has the potential of drastically changing the competitive domain in various industries. Gartner [2] predicted that by the year 2017, 8.4 billion connected things—up by 31 percent from 2016—will be in use worldwide and the figure will reach 20.4 billion by the year 2020. According to IDC [3], the global annual spending in 2017 on the IoT was estimated to exceed US$800 billion, which was an increase of 17% as compared to the previous year. By 2021, the figure is expected to reach US$1.4 trillion. As a representative movement of the application of the IoT to the manufacturing industry, the German government proposed the Industrie 4.0 strategy in 2011. Immediately thereafter, countries around the world, including the Industrial Internet Consortium established in the United States, the "Made in China 2025" plan announced by the Chinese government, and the Robot Revolution Initiative jointly initiated by the Japanese industry, government, and academia, have launched their own IoT manufacturing strategies.

When referring to the IoT and Industrie 4.0, some concepts and terms, including big data, cyber-physical systems (CPS), and digital twin (DT), have recently appeared and evolved. Regarded as the convergence of the physical and digital worlds, a CPS implies a system that includes gathering data in the real world (physical space) through the IoT, automatically analyzing the data using large-scale data processing technologies in cyberspace and feeding the results back to physical space to solve problems in the real world. Each CPS includes smart machines, storage systems, and production facilities that can exchange information with autonomy and intelligence, make decisions and trigger actions, and control each other independently. To implement a CPS in industry, the virtual simulation of products and processes before and during operations is a key aspect of achieving critical goals for product design and production flexibility [4]. In this context, there is a need for a new concept or methodology to address these new issues in the CPS environment.

The concept of *twin* (which later evolved into the concept of DT) was initially proposed and adopted by NASA for the monitoring and optimizing of the safety and reliability of spacecraft [5,6]. Recently, the scope of DT application has extended beyond the scope of aerospace equipment to the field of manufacturing, including product life-cycle management (PLM). The DT has been proposed as a tool for collecting data from the IoT and synchronizing real-world information in real time on the cyber side. This tool facilitates the cyber-physical integration of manufacturing, which is a critical bottleneck to achieving the next-generation manufacturing concept of smart manufacturing [7]. The research firm, Gartner, highlighted the DT as number four in the top 10 strategic technology trends for 2018 and predicted that these trends would mark the course of the next decade [8].

Although the DT is considered a challenging technology, it is still at the conceptual stage [9] and only a few studies have specifically discussed methods for their construction and implementation in the manufacturing domain. Yang et al. [10] focused on simulation experiments with real-time data and constructed a prototypical DT-driven simulation using data derived from a distributed train model equipped with sensors. Zhang et al. [5] presented a DT-based approach to production design and optimization. In [5], DT-based simulation was proposed and utilized for the individualized design of a hollow glass production line. These previous studies provided some conceptual frameworks and case studies that guide the approach to DT implementation. However, issues such as the reception of real-time data from the IoT, as well as the conversion and inputting of the data into a simulation model have not yet been completely solved. Even though some general roadmaps and frameworks have been proposed, it is still not clear what kind of data and information must be integrated. When constructing a DT, a specific framework to guide the process of extracting the necessary data from the physical system and a scheme for entering these data into the cyber-side simulation model are required.

In response to these issues, in this study, we first explain the concept of a DT and highlight the important issues of developing a DT remaining in an IoT-aided manufacturing environment. Then, we propose a framework for the construction of a DT and a scheme for inputting data derived from the IoT into a simulation model. Finally, we discuss how we verified the effectiveness of our proposed framework and scheme by constructing a DT-oriented prototype simulation model for a CPS-based manufacturing system.

2. Simulation Approach and Digital Twin

Simulation is a method to replicate a real-world system (or conceptual scenarios) on a computer, conducting experiments to understand the behavior of the system, and/or evaluating various strategies for the operation of the system. Simulation-based optimization has been widely implemented and validated in various industries.

In the IoT environment, an enormous amount of data can be collected in real time from the network of sensors and devices. However, only less than one percent of the data is being utilized today [11]. Additional value can be obtained by using the remaining 99 percent of valuable IoT data for predictive maintenance or optimizing operational management. On the other hand, traditional simulation approaches have limitations in processing large amounts of real-time data. Novel simulation techniques are required to enhance scalability and permit the real-time execution of a largescale IoT environment.

From the perspective of the Industrie 4.0, the DT can be considered as a simulation approach utilizing data collected from sensors attached to industrial equipment used for smart manufacturing [10]. Using the real-time data generated from the equipment, a computer-based real-time simulation can model events, such as part locations and machine states, in the physical world. This simulation model acts as a twin of the real world in cyberspace and behaves in the same way as the physical space.

Different from the traditional simulation architecture, the simulation in DT is used as a validation tool for optimizing solutions rather than only visually displaying the simulation of random events [5]. Although the concept of the DT was born in the field of aerospace engineering for the re-engineering of structural life prediction and management, the concept has also been adopted in manufacturing contexts. The DT paves the way for cyber-physical integration and serves as a bridge between the physical world and the cyber world, providing manufacturing enterprises with a new way of implementing smart production and precision management [7]. A DT consists of a virtual representation of a production system that is able to use sensory data, connected smart devices, mathematical models, and real-time data elaborations. The DT can be run on different simulation disciplines that are characterized by the synchronization of virtual and real systems [9]. With its capabilities of conceptualization, comparison, and collaboration, the DT frees us from the physical realm, where humans operate relatively inefficiently [12]. The concept of the DT and its relation to the CPS considered in this study are shown in Figure 1.

In industrial practice, the definition and understanding of DT may be different by industry. For example, General Electric focuses on forecasting the health and performance of their products over their lifetimes [6], whereas some product developers regard DT as a set of augmented reality (AR) tools and integrate DT with other computer-aided design software to improve product development efficiency. By integrating these definitions of DT in various industries, we think that a DT should include the functions of real-time synchronization, prediction, and testing. In order to achieve these three functions, the DT should be executed separately in three modes, that is, synchronization mode, evaluation mode, and experimental mode.

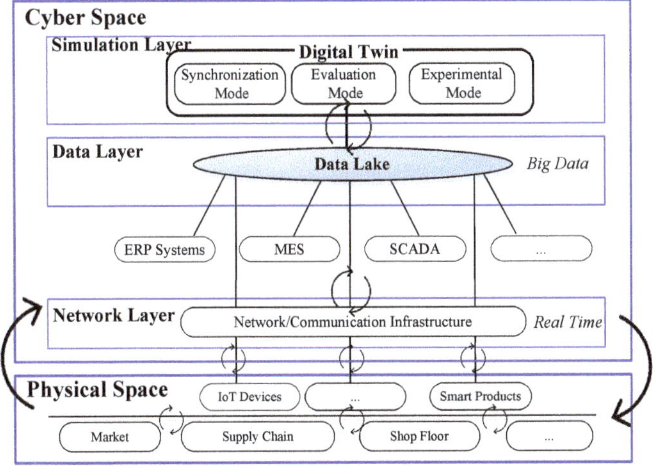

Figure 1. Digital twin with a cyber-physical system.

A DT is realized through simulation software, hardware, and data transmission and processing techniques in addition to other components. From the perspectives of simulation granularity and scope of analysis, a DT can be classified into three levels from the bottom to the top: phenomenon-based, device- and product-based, and social and process-based. Figure 2 shows a hierarchical structural diagram of a DT.

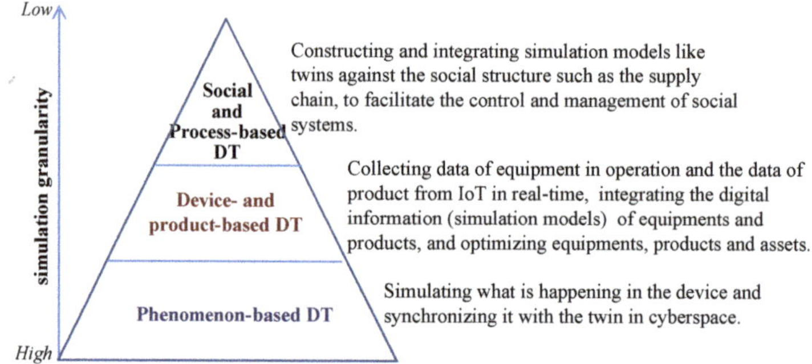

Figure 2. Hierarchical structure of digital twins.

In the conventional simulation model, stochastic or deterministic parameters are assigned to a specific simulation model all at once and the results are analyzed on the basis of these data. The model itself is usually fixed and seldom changed. However, manufacturing and logistics in the real world are affected by their internal and external environments, so their characteristics and functions may change frequently and quickly, possibly leading to an increasing gap between the simulation model and reality. Contrary to this weakness in conventional simulation models, the DT always uses the latest information to synchronize the mirroring of real-world behaviors so that the virtual models can be closer to reality and perform more complex simulations.

With the utilization of DTs, it is possible to visually check and inspect the states of manufacturing plants even from remote areas. Furthermore, by virtualizing the processes from product design to production, operation, and maintenance, the DT integrates the whole factory into the simulation models in cyberspace, facilitating the control and management of the factory in physical space to achieve efficient smart manufacturing.

3. Digital Twin and Monitoring Systems

The DT may not be a new concept if we only consider it as a tool for monitoring things and behaviors in the real world through the digital world, because digital mock-up technologies based on monitoring systems and 3D models are being widely used. However, with technological innovations, such as IoT, artificial intelligence (AI), and big data, the necessity for DTs is increasing more than ever and their applications continue to expand.

Even though a monitoring system may observe movements in the real world, it cannot be used to predict the occurrence of a failure or irregularity. Moreover, it cannot be used for hypothetical experiments or predictive analytics but only for gathering data from the real world and compiling and displaying them in the digital world. It does not provide a mechanism for feeding the decision support information back to the real world to allow for optimization.

Addressing the above problems, the DT is a simulation method that can be used for evaluation and optimization. As shown in Figure 1, the DT continuously acquires data from a data lake and incorporates the data into a real-time simulation model. In real time, the data lake gathers and stores data from machines and sensors connected to industrial equipment in the real world. After the DT processes and analyzes the data, the results are fed back to physical space to optimize the real-world system. The DT is a concept that applies a CPS to the field of manufacturing. A DT compares and analyzes sensory data obtained from the IoT with computer-aided design data derived from a product design, and then incorporates the data into a module inside a synchronous simulation model for monitoring. Compared to traditional monitoring systems, the DT not only represents the real world but also makes predictions, provides solutions, and supports decision-making before a failure or anomaly occurs. Thus, the DT simulates possible future events. As a result, it can prevent unplanned shutdowns and their incurred costs. The DT has been listed by major industrial equipment manufacturers, such as General Electric and Siemens, as an important technology for supporting the next generation of Industrie 4.0.

4. Framework for Constructing Digital Twin-oriented Simulations

As mentioned above, the DT is still at the conceptual stage and only a few studies specifically mention model construction and implementation methods. In this paper, we propose a framework for constructing a DT in an IoT-aided manufacturing environment. The proposed framework is shown in Figure 3.

In this study, we propose that a DT should include the functions of real-time synchronization, prediction, and testing. To perform these three functions, the DT should be executed separately in three modes, which are synchronization mode, evaluation mode, and experimental mode. The evaluation mode and the experimental mode are actually provided in conventional simulation models which are often used for what-if analysis and optimization. Unlike the evaluation mode and the experimental mode, the synchronization mode is a mode unique to the DT. The DT is constantly updating of the simulation model with data being drawn from the IoT.

As shown in Figure 3, the first step in building a DT model is to design the concept. Next, similarly to the typical process of constructing a conventional simulation model, after performing the input analysis, a simulation model using stochastic data is constructed and validated. This simulation model can be used for the evaluation mode and experimental mode for prediction and optimization. After constructing a model using stochastic parameters, a simulation model is built for the synchronization mode. To use the real-time data of the IoT to synchronize the physical world, it is necessary to convert the stochastic parameters in the model into deterministic parameters to reconfigure the model. We can use the synchronization mode to monitor behaviors in physical space and save the historical data gathered from the IoT. Those historical data can be converted into stochastic distributions assigned to the simulation models of the evaluation and experimental modes, of which the results are fed back to the model of the synchronization mode.

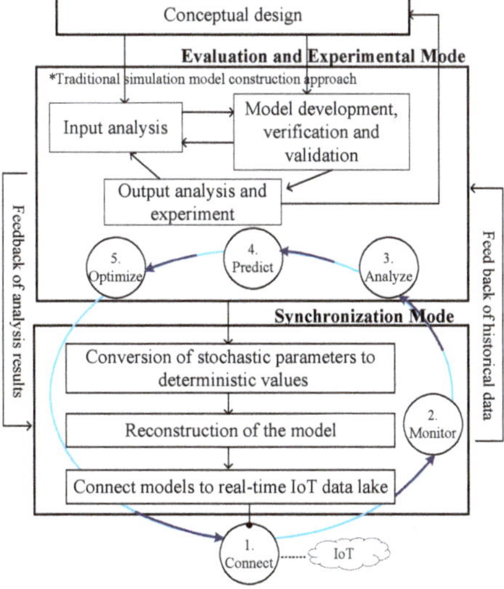

Figure 3. Framework for construction of a DT-oriented simulation model.

5. Proposal for Data Input Scheme for Simulation Model Construction of a Digital Twin

Generally, in constructing a simulation model, data inputs with randomness are employed to model the uncertainty events. When entering the parameters into a simulation model, instead of directly inputting the observed data, the probability distributions estimated from the observation value are utilized, and the model is executed. Distributions such as the triangular distribution and the normal distribution together with the associated parameters are specified as the input of the simulation model, especially for processing time and interval time, and so on.

Naturally, entering parameters with randomness into the model also leads to the randomness of execution results (outputs). Therefore, to obtain statistically significant results, it is necessary to perform statistical analyses on the repetitive execution results of the model. However, when constructing the synchronization mode (the real-time simulation) in DT, to express the behavior in the physical world as it is, the simulation needs to be carried out at the actual-time progressing speed. Therefore, it is necessary to avoid entering parameters with randomness, which causes random outputs, into the model as much as possible.

For this reason, we propose a data input scheme, as shown in Figure 4, for simulation models in a DT. When constructing a synchronization mode in DT, we enter the data into a simulation model by using a numerical sheet such as an Excel sheet, rather than probability distribution parameters. The digital sheets in the database collect and record the status data of the machines (e.g., instant processing time) in real-world production systems in real time via IoT. Then, the simulation model reads the status data from the digital sheet in real time and executes the simulation. Figure 4 indicates the differences between the conventional and DT simulation models in the data entry scheme.

As shown in Figure 4, these real-time data collected from the machines and equipment written to the digital sheet is aggregated and stored in a database. The evaluation and experimental modes of the DT will then read the historical data stored in the database and enter the fitted probability distribution parameters into the simulation model.

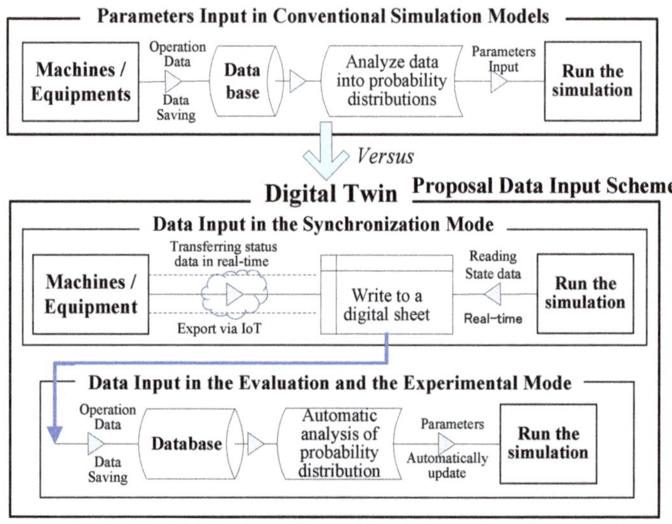

Figure 4. Proposal of data input scheme for simulation models in a DT.

6. Application of Digital Twin Construction

To validate the proposed DT construction framework and data input scheme, a DT-oriented simulation model for an IoT-aided manufacturing factory model was constructed.

6.1. General Description of a Factory Simulation Model

The distributed model used in this study is an adoption of the Fischertechnik® Factory Simulation 9V model [13]. Table 1 shows a process schedule and processing parameters for this factory model. An overhead view and overlapped processing map of the factory is shown in Figure 5. The letters *A* to *H* in the circle represent the task codes shown in Table 1.

As shown in Figure 5, this factory model consists of three main parts: the Controller Automated High-Bay Warehouse with a Vacuum Gripper Robot (VGR), Controller Multiprocessing Station with Oven, and Controller Sorting Line with Detection. This system simulates a series of basic processes, including material storage, multiprocessing, and product sorting.

Table 1. Process schedule and processing parameters in the factory model.

Proc. Seq.	Task Code	Processes	Processing Time (Unit: Second)	Resources
1	A	Picking and transporting workpieces from high-bay rack to identification station	*TRIA(13.4, 18.1, 26.3)*	High-bay rack feeder
2	B	Scanning and carrying workpieces by conveyor belt to vacuum gripper robot	*TRIA(3.60, 3.66, 3.71)*	Conveyor system with identification
3	C	Picking and carrying workpieces to the multiprocessing station with oven	*TRIA(20.7, 20.8, 21.0)*	3D Vacuum Gripper Robot A
4	D	Processing by the oven	*TRIA(17.6, 17.8, 17,9)*	Conveyor belt, Oven
5	E	Picking and transporting workpieces to saw station	*TRIA(9.1, 9.3, 9.4)*	Vacuum Gripper Robot B
6	F	Processing by the saw	*TRIA(9.73, 9.77, 9.89)*	Saw, ejector
7	G	Color sorting line with detection	*TRIA(7.3, 7.96, 9.37)*	Color detection, conveyor belt, storage locations
8	H	Conveyor belt carries workpieces to storage location	*TRIA(4,6,8)*	Conveyor belt, storage locations

Note: *TRIA(Min, Mod, Max)* indicates the triangle distribution.

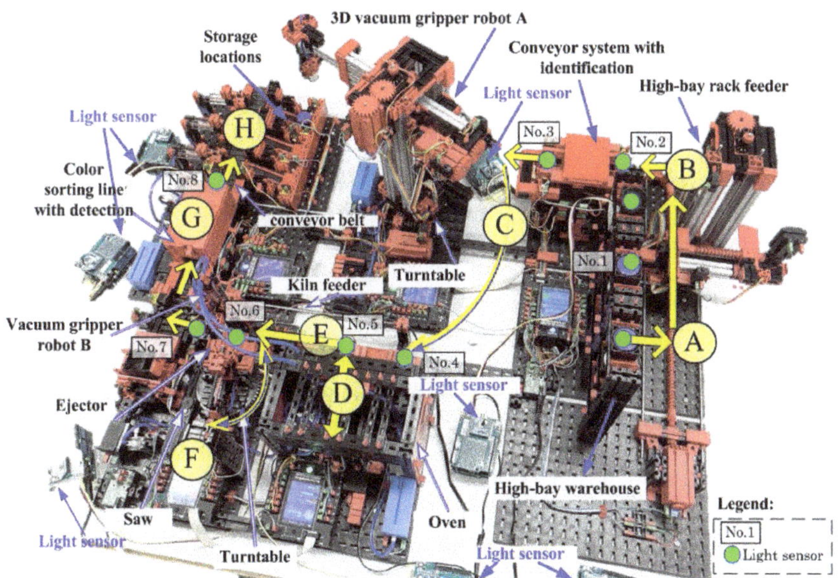

Figure 5. Overhead view and the sequence of processes of the factory model.

Workpieces are stored at the high-bay warehouse, which is designed as a pallet rack storage system for storing and retrieving goods. These are handled by a rack feeder that moves in a lane between three rows of racks. This area is part of the receiving station, where the identification of goods takes place. This is known as a conveyor system with identification, as the workpieces are transferred by a conveyor belt and identified by a barcode reader. Then, the workpieces are handled and transported by VGR A from the storage area to the processing area, where the workpiece automatically runs through several stations.

Processing begins with the oven, which represents the firing process. The light barrier is interrupted when a workpiece is placed on the kiln feeder, triggering the opening of the oven's door and drawing in the kiln feeder. At the same time, VGR B is requested to transport the workpiece to the turntable after the firing process is complete. The turntable positions the workpieces under the saw for processing and conveys them to the ejector, which slides the workpieces onto the conveyor belt to carry to the sorting area.

The sorting line is used for the automated separation of different colored workpieces by color detection, which is handled by a color sensor that detects color on the basis of different surface reflections. With the help of a pulse switch, the workpieces are transferred to the correct position and pushed to the appropriate storage locations by the ejector. Meanwhile, several light barriers monitor the fill level of the storage locations. The system's control logic, such as the warehousing principle, VGR movements, machine processing time, and sorting into the storage locations, are programmed by ROBO Pro and controlled with TXT controllers.

6.2. The IoT-Aided Manufacturing System in the Factory Model

Although Fischertechnik's Factory Simulation 9V model is equipped with some sensors, such as color sensors, it is not connected to the Internet and does not have the structure or functionality of the IoT. To convert this factory model into an IoT-aided manufacturing system, we added eight sets of light sensors wirelessly connected to the Internet. Figure 6 shows an example of a light sensor unit with a wireless module. As an example of the IoT system, the high-bay warehouse area with built-in light sensors is shown in Figure 7. The positions where other light sensors are installed are marked in Figure 5. The sets of the light sensor units are numbered. With the light sensor, we can obtain the working status and operating time of each machine.

To construct a DT model, we need to receive data from the sensors connected to the IoT and save the data to a data lake (see Figure 1). We receive data wirelessly from the light sensors in real time and utilize the data to build a DT-driven simulation model. To wirelessly receive and record data from the factory model, we use an Arduino microcomputer with sensors and Microsoft Excel VBA programs.

The Arduino has an original application development environment based on C/C++. For the sensors, we used Grove Light Sensor v1.2 and Grove Ultrasonic Ranger (Seeed Studio). The former sensor sent a changed value depending on the light intensity to Arduino as a signal, while the latter sensed the distance of an object away from it. A total of 8 sets (16 units) of sensors were installed and assigned numbers from 1 to 8 according to their locations, as shown in Figure 5.

To wirelessly send the data, an XBee ZB (S2C) (manufactured by Digi) was attached to the Arduino (Figure 6). Along with the defined algorithm in Arduino, the Arduino received signals from the sensors and sent the sensor number to the computer if the condition set by the algorithm was satisfied.

An Excel VBA program was also developed. When Excel received a signal from the Arduino, the time of the reception was recorded in a worksheet in real time. If the reaction sensor was not Sensor No. 1, the program calculated the interval between the current and the previous time, then wrote the interval times onto a specific worksheet. Then, a VBA module, EasyComm, received the signal via serial communication. Figure 8 shows the flow of the processing logic of the Excel VBA program.

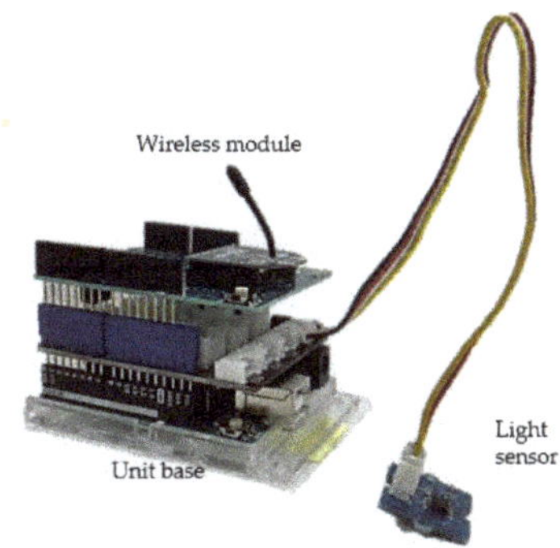

Figure 6. A light sensor unit body with a wireless module attached.

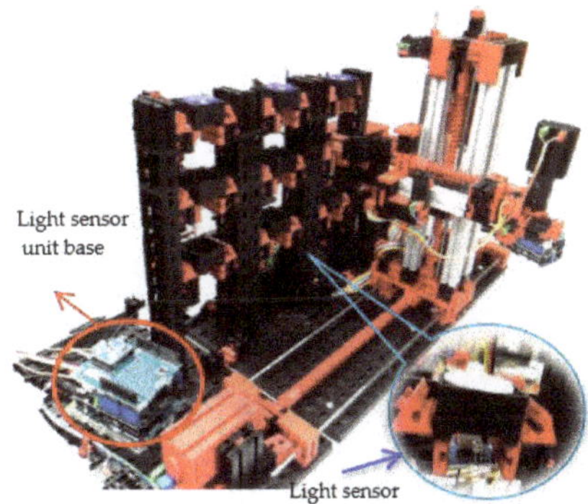

Figure 7. High-bay warehouse area with built-in light sensors.

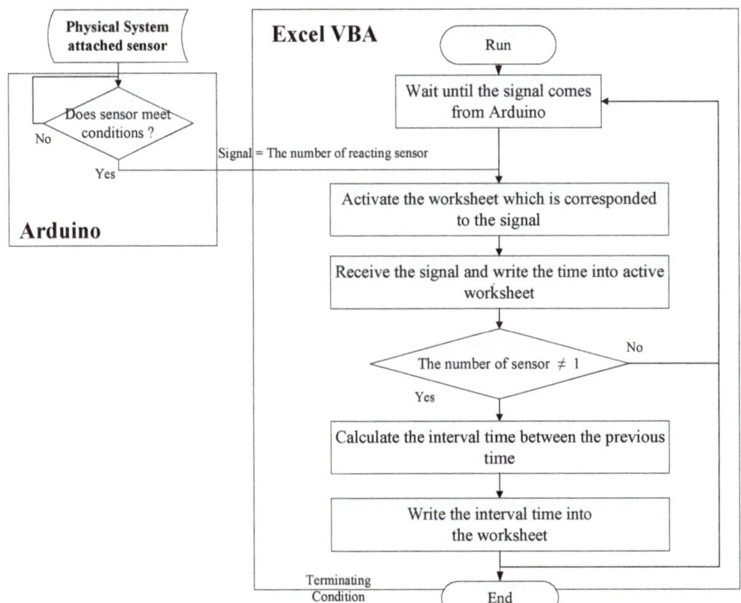

Figure 8. Flow of processing logic of the Excel VBA program.

6.3. Construction of Prototypical Model for DT-Oriented Simulation

To validate the proposed DT construction framework and data input scheme, we developed a DT-oriented simulation model for the factory model. The simulation model was programmed in Arena [14] and overlaid on a scaled layout. The main stochastic parameters were the processing time and resources used by each process, as shown in Table 1.

According to the framework proposed in Figure 3, we firstly constructed a conventional simulation, using the stochastic parameters for evaluation and optimization. Subsequently, we converted the stochastic parameters into deterministic parameters and rebuilt the model. When the model was executed, Excel acquired the data from the IoT in real time according to the processing logic shown in Figure 8, and then the simulation model continuously updated the parameter values by reading the data from Excel. This model was used to monitor and synchronize the behavior in physical space. The historical data were fed back to the simulation models for evaluation and experimental mode to calculate the stochastic distribution of the stochastic parameters (Figure 4). A DT schematic diagram of this factory is shown in Figure 9. Now, there are three modes for the DT-oriented simulation model, namely the evaluation model, the experimental mode and the synchronization mode. We can use these simulation models according to our needs.

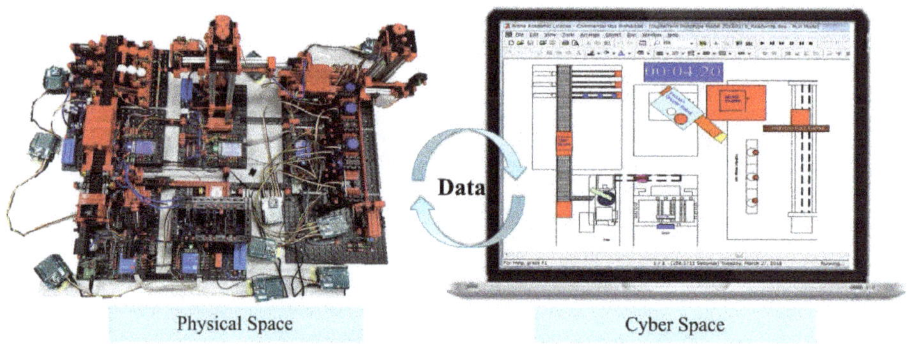

Figure 9. A DT schematic diagram of the factory.

6.4. Experiment and Results

After building the DT model, it was validated by an interactive process between the factory model and the simulation model modelers. This process compared the model's output with the actual sensory data. After confirming the reliability of the model, the simulation models were run and the results were analyzed. Figure 10 shows a synchronized display of experimental results on the input and output of workpiece interval time between physical and cyber sides. As shown in Figure 10, the time interval received from sensors as input from the real world is exactly the same as the interval time output from the simulation model on cyberspace. It was confirmed that the synchronization mode of the DT simulation could accurately mirror the actual behavior of the factory model.

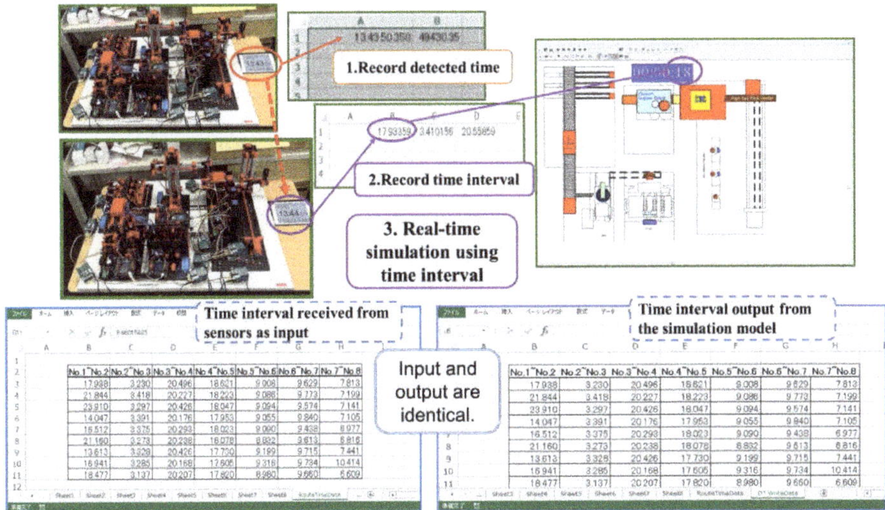

Figure 10. Synchronized display of input and output of interval time between physical and cyber sides.

The main research objectives of this study are to propose a framework and a data input scheme for the construction of a DT, as well as develop a prototype model for the DT based on the proposed framework. Therefore, this article does not focus on the what-if analysis of the experiment. However, from the operational experiment, we obtained the following findings:

- The DT's synchronization mode can be used to remotely observe the factory's operation and the statuses of the machines via a network. However, the DT differs from the conventional monitoring system in that it can quantitatively grasp an entity's time-related performance indicators, resource utilization, etc.
- Like the conventional simulation model, the DT's evaluation mode can be used to predict the system's behaviors, such as machine failures and entity cycle times.
- The DT's optimization mode can be used to optimize a system just like the conventional simulation model. This mode can be used to seek and find a solution, such as shortening the lead time of orders, by scenario comparison analysis.

7. Conclusions

The IoT is recognized as a technology that has already had economic impacts and created high expectations for drastically changing the competitive domain in various industries. In the manufacturing field, sensor-equipped machines can collect data from the production system in real time. On the cyber side, the DT has been proposed as a tool for collecting and synchronizing real-world information in real time. By updating data in real time and comparing cyberspace with physical space in parallel, the DT can be continuously improved and synchronized with the real world. Although the DT is considered a challenging technology, it is still at the conceptual stage.

In this study, in order to develop the DT, we proposed a framework for constructing a DT. We proposed that the DT should contain three execution modes, that is, the evaluation mode, the experimental mode, and the synchronization mode. The evaluation mode and experimental mode can be developed as conventional simulation models using stochastic parameters for prediction and optimization. Therefore, the DT's evaluation mode can be used to predict machine failures and test failure avoidance methods. The synchronization mode is used to observe the factory's operation and the machine status remotely via a network, but differs from a conventional monitoring system.

Traditionally, randomness is adopted to execute simulation experiment. However, when constructing the synchronization mode in DT, it is necessary to avoid entering parameters with randomness. For this reason, we proposed a data input scheme for inputting data derived from the IoT into a simulation model of a DT.

In order to verify the effectiveness of the framework as a prototype of a DT-oriented simulation model, a DT-oriented model for an IoT-aided bench-scale factory system was constructed. In this factory model system, to simulate the IoT environment, we modified a simple IoT system by attaching some sensor units. With the proposed framework for DT construction and proposed data input scheme, we were able to receive real-time data from an IoT-aided manufacturing system and construct a DT model that successfully reflected the real situation of the physical system. A simulated factory model was developed for research and education.

In this study, a digital twin was constructed for a bench-scale model of the factory. In actual factories, increasing the number of devices and processes may increase the complexity of the model. In that case, further discussion is still needed on how to efficiently input large amounts of data into the simulation model. In addition, when considering the implementation of a smart factory, there remains a problem of connection experiments between ERP (Enterprise Resources Planning), MES (Manufacturing Execution System) and simulation. Furthermore, what-if analysis using digital twin should be done, following this study. Forecasting, experimentation and optimization are other important features of DT to predict the future status or performance. These issues should be developed and examined in future researches.

Author Contributions: Conceptualization, Y.T., W.Y. and S.T.; methodology, Y.T. and W.Y.; simulation models development, Y.T., K.Y. and W.Y.; IoT-aided factory model construction, K.Y.; model validation, Y.T., W.Y., K.Y. and S.T.; writing—original draft preparation, Y.T., W.Y. and K.Y.; writing—review and editing, S.T. and W.Y.; supervision, S.T.; funding acquisition, S.T. and W.Y.

Funding: This research was partly supported by JSPS KAKENHI, grant number JP17K12984.

Conflicts of Interest: The authors declare no conflict of interest.

References

1. Lee, I.; Lee, K. The Internet of Things (IoT): Applications, investments, and challenges for enterprises. *Bus. Horiz.* **2015**, *58*, 431–440. [CrossRef]
2. Gartner, Inc. Gartner Says 8.4 Billion Connected "Things" will be in Use in 2017, up 31 Percent from 2016. Available online: https://www.gartner.com/newsroom/id/3598917 (accessed on 25 March 2018).
3. IDC. Worldwide Semiannual Internet of Things Spending Guide. Available online: https://www.idc.com/getdoc.jsp?containerId=IDC_P29475 (accessed on 25 March 2018).
4. Posada, J.; Toro, C.; Barandiaran, I.; Oyarzun, D.; Stricker, D.; de Amicis, R. Visual Computing as a Key Enabling Technology for Industrie 4.0 and Industrial Internet. *IEEE Comput. Graph. Appl.* **2015**, *35*, 26–40. [CrossRef] [PubMed]
5. Zhang, H.; Liu, Q.; Chen, X.; Zhang, D.; Leng, J. A digital twin-based approach for designing and multi-objective optimization of hollow glass prosution line. *IEEE Access* **2017**, *5*, 26901–26911. [CrossRef]
6. Schleich, B.; Anwer, N.; Mathieu, L.; Wartzack, S. Shaping the digital twin for design and production engineering. *CIRP Ann. Manuf. Technol.* **2017**, *66*, 141–144. [CrossRef]
7. Qi, Q.; Tao, F. Digital Twin and Big Data towards Smart Manufacturing and Industry 4.0: 360 Degree Comparison. *IEEE Access* **2018**, *6*, 3585–3593. [CrossRef]
8. Gartner, Inc. Gartner Identifies the Top 10 Strategic Technology Trends for 2018. Available online: https://www.gartner.com/newsroom/id/3812063 (accessed on 25 March 2018).
9. Negri, E.; Fumagalli, L.; Macchi, M. A Review of the Roles of Digital Twin in CPS-based Production Systems. *Procedia Manuf.* **2017**, *11*, 939–948. [CrossRef]
10. Yang, W.; Tan, Y.; Yoshida, K.; Takakuwa, S. Digital Twin-Driven Simulation for A Cyber-Physical System in Industry 4.0. In *DAAAM International Scientific Book 2017*; DAAAM International Publishing: Vienna, Austria, 2017; pp. 227–234. ISBN 978-3-902734-12-9.

11. Kecskemeti, K.; Casale, G.; Jha, D.N.; Lyon, J.; Ranjan, R. Modelling and Simulation Challenges in Internet of Things. *IEEE Cloud Comput.* **2017**, *4*, 62–69. [CrossRef]
12. Grieves, M. Digital Twin: Manufacturing Excellence through Virtual Factory Replication. Available online: https://research.fit.edu/media/site-specific/researchfitedu/camid/documents/1411.0_Digital_Twin_White_Paper_Dr_Grieves.pdf (accessed on 25 March 2018).
13. Fischertechnik. Factory Simulation 9V–Education. Available online: https://www.fischertechnik.de/en/products/teaching/training-models/536629-edu-factory-simulation-9v-education (accessed on 25 March 2018).
14. Kelton, D.; Sadowski, R.; Swets, N.B. *Simulation with Arena*, 5th ed.; McGraw-Hill, Inc.: New York, NY, USA, 2010; ISBN 978-0071267717.

 *machines*

Article

Smart Hybrid Manufacturing Control Using Cloud Computing and the Internet-of-Things [†]

Jonnro Erasmus *, Paul Grefen, Irene Vanderfeesten and Konstantinos Traganos

School of Industrial Engineering, Eindhoven University of Technology, 5612 AZ Eindhoven, The Netherlands; p.w.p.j.grefen@tue.nl (P.G.); i.t.p.vanderfeesten@tue.nl (I.V.); k.traganos@tue.nl (K.T.)
* Correspondence: j.erasmus@tue.nl; Tel.: +31-40-247-2685
† This paper is an extended version of our conference paper published in proceedings of the 5th IEEE International Conference on Cloud Networking, Pisa, Italy, 3–5 October 2016: Supporting Hybrid Manufacturing: Bringing Process and Human/Robot Control to the Cloud (Short Paper).

Received: 31 October 2018; Accepted: 28 November 2018; Published: 3 December 2018

Abstract: Industry 4.0 is expected to deliver significant gains in productivity by assimilating several technological advancements including cloud computing, the Internet-of-Things, and smart devices. However, it is unclear how these technologies should be leveraged together to deliver the promised benefits. We present the architecture design of an information system that integrates these technologies to support hybrid manufacturing processes, i.e., processes in which human and robotic workers collaborate. We show how well-structured architecture design is the basis for a modular, complex cyber-physical system that provides horizontal, cross-functional manufacturing process management and vertical control of heterogenous work cells. The modular nature allows the extensible cloud support enhancing its accessibility to small and medium enterprises. The information system is designed as part of the HORSE Project: a five-year research and innovation project aimed at making recent technological advancements more accessible to small and medium manufacturing enterprises. The project consortium includes 10 factories to represent the typical problems encountered on the factory floor and provide real-world environments to test and evaluate the developed information system. The resulting information system architecture model is proposed as a reference architecture for a manufacturing operations management system for Industry 4.0. As a reference architecture, it serves two purposes: (1) it frames the scientific inquiry and advancement of information systems for Industry 4.0 and (2) it can be used as a template to develop commercial-grade manufacturing applications for Industry 4.0.

Keywords: smart manufacturing; Industry 4.0; human-robot collaboration

---

1. Introduction

Industry 4.0 is a trend in automation and digitization that promises significant gains in production output, product customizability, and manufacturing flexibility [1]. This new industrial age stems from the coincident rise of cloud computing, the Internet-of-Things (IoT), and smart devices [2]. It is expected that mass customized products will be produced by smart robotics in dynamic processes managed in the cloud [3]. It is even conceptually understood how these technologies should work together to achieve smart manufacturing and deliver on the promises of Industry 4.0 [4,5]. Computation that is not time-critical is relegated to the cloud. The IoT facilitates commands and responses to and from devices and teams of humans and smart robotics perform sophisticated operations. Figure 1 gives an overview of the technologies and their roles in smart manufacturing.

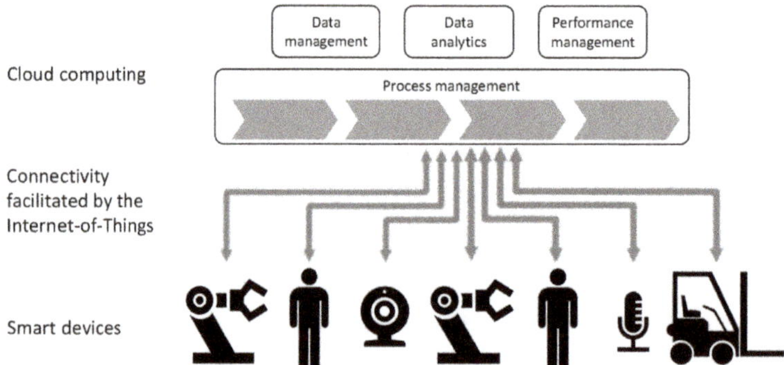

Figure 1. Roles of selected technologies in smart manufacturing.

The technologies that underpin Industry 4.0 are increasingly appealing and accessible to manufacturing enterprises. Cloud computing and internet connectivity is prevalent in industrialized countries and robots are becoming increasingly intelligent and affordable [6,7]. Future scenarios are proposed where humans and robots harmoniously collaborate to perform irregular and complicated tasks [8]. Even small and medium enterprises (SMEs) now consider user-friendly automation solutions. For example, a robot that can be programmed by a demonstration is significantly easier and inexpensive to introduce in a relatively low-tech environment [9].

Although more accessible, affordable, and available, these technologies are developed independently and remain largely detached. The separate technologies can be acquired and even implemented, but it is unclear how to unlock the promised benefit of an integrated solution. Monostori [10] argues that these new technologies threaten the traditional automation hierarchy, which further distorts our understanding of the manufacturing system and its various elements. Thus, the adoption of smart manufacturing technology is hindered by utilization and integration instead of acquisition. In fact, the absence of out-of-the-box solutions that combine the different technologies is considered a primary impediment on the path towards smart manufacturing [6].

The symptoms of the problem manifest most fervently on a factory floor with humans and robots. The activities of humans and robots are controlled differently. Humans receive written, oral, or visual instructions while machines are compelled to action via their control systems. These control regimes function independently [11], which makes it difficult to transfer tasks between humans and robots even if their capabilities would allow this [12,13]. Furthermore, robot control is often poorly integrated with cross-functional processes management [14]. Robot control follows a vertical orientation focused on the operations within a work cell. Process management follow a horizontal orientation focused on the operations across work cells and in the context of enterprise information processing. Thus, current robot control does not support simple reassignment of robots to different work cells. The most apparent symptom is the increased safety hazards introduced by automation. Robots must be equipped with extensive safety precautions to allow close collaboration with humans (the accident in a car factory [15] is well known in the domain). To compensate, human and robot working spaces are usually physically separated. These symptoms and concerns hamper mainstream adoption of human-robot collaboration technology [13,16].

The contribution of this paper is an architecture model of an information system that utilizes modern manufacturing technologies to deliver seamless integration between human and automated activities. The model is proposed as a reference architecture for a manufacturing operations management system for Industry 4.0. As reference architecture, it serves two purposes: (1) it frames the scientific inquiry and advancement of information systems for Industry 4.0 and (2) it can be used as a template or at least a starting point to develop commercial-grade manufacturing applications for Industry 4.0.

The information system model is developed as part of the HORSE Project, which is a European research and innovation project in the Horizon 2020 program [17]. The project brings together 22 organizations including research institutions, technology vendors, and manufacturing enterprises. The primary objective of the project is to make advanced manufacturing technology more accessible to SMEs. These technologies are packaged in a modular, integrated "HORSE System" and include human intention detection, robot force control, teaching-by-demonstration, augmented reality, dynamic allocation of tasks to humans and robots, and manufacturing process management. The intended SME context of the HORSE System often implies limited availability of on-site information technology resources. Therefore, the use of cloud services can be an important enabling factor in the application of HORSE concepts and technology in an SME.

This paper starts with a detailed explanation of the research approach derived from the Reference Architecture for Industry 4.0 (RAMI4.0) and the Kruchten 4 + 1 software engineering framework. In Section 3, the logical system architecture is presented at three levels of aggregation to give insight into the function and structure of the HORSE System. Thereafter, Section 4 presents the physical architecture of the HORSE System in two stages: first, determining which parts of the system can be located 'in the cloud' and, second, presenting the HORSE System as an IoT application. The consideration of cloud-support expands on earlier work of Grefen et al. [18]. In Section 5, three real-world scenarios are presented as proof of concept of the HORSE System. Section 6 considers the possibilities of cloud-based management of inter-organizational manufacturing processes and supply chains. Lastly, conclusions and findings are discussed in Section 7.

2. Research Approach

The information system architecture presented in this paper is the result of rigorous design and science research. The purpose of design science research is to generate prescriptive knowledge that can be used to solve practical problems [19]. As such, problems from the industry are studied to ensure practical relevance and an artefact is created to help solve similar problems. The artefact in this paper is in the form of a conceptual design of an information system [20] to serve as reference architecture to develop and build an information system for the management of smart manufacturing operations.

To structure the research, the design science research framework of Hevner et al. [21] is adopted and discussed in Section 2.1. More importantly, the design approach is thoroughly reported to ensure repeatability. A similar result should be achieved given the same problem and context. The design approach is documented in the form of principles and process. The principles that guide the design are based on information systems architecture theory and discussed in Section 2.2. The research process is based on the widely adopted Kruchten 4 + 1 framework for software engineering [22], which is explained in Section 2.3.

2.1. Research Framework

The HORSE System must be both practical and relevant for the typical challenges faced in SMEs. To achieve this goal, the design science research framework of Hevner et al. [21] is adopted. The framework emphasizes practical relevance by advocating for consideration of business need during the development and evaluation of the results in a realistic environment. The research framework, based on the Hevner et al. framework [21], is shown in Figure 2.

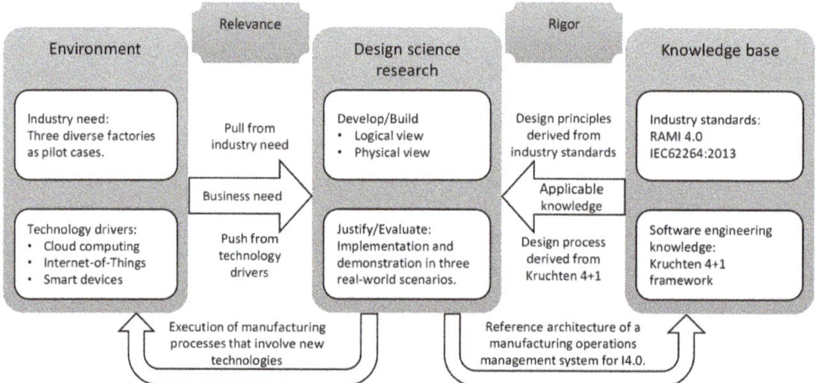

Figure 2. Research framework used in the HORSE Project.

The design science research column of Figure 2 consists of development and evaluation. It produces validated reference architecture. The reference architecture is presented in a logical and physical view in Sections 3 and 4, respectively, according to the Kruchten 4 + 1 framework. The remaining development and process views of the Kruchten framework are omitted in the interest of brevity. The evaluation is articulated in scenarios that demonstrate application of the information system. For relevance, three pilot cases act as the organizations that have the business need while three core concepts of Industry 4.0. provide the technology push driving the artifact development. The artifact is applied in the applicable environments of the three pilot cases, as articulated in the scenarios. For rigor, design principles are derived from industry standards and the design process is based on the Kruchten 4 + 1 framework. Lastly, the contribution toward the knowledge base is a validated information system model.

2.2. Design Principles

RAMI4.0 was established and defined in DIN SPEC 91345:2016-04 [23] to give some structure to the rapidly developing and changing technologies in manufacturing. According to the standard, "the fundamental purpose of Industrie 4.0 is to facilitate cooperation and collaboration between technical objects, which means they have to be virtually represented and connected." The reference architecture brings together the business, life cycle, and hierarchical views of an asset by relating the concepts on three orthogonal dimensions [24]:

- The layers dimension is more formally referred to as the architecture axis. This axis "represents the information that is relevant to the role of an asset." It covers the business-to-technology spectrum by relating different aspects of an asset to layers of the enterprise architecture.
- The life cycle and value stream dimension "represents the lifetime of an asset and the value-added process." This axis distinguishes between the type and instance of the factory and its elements. For example, the digital design of a product and its instantiation as a manufactured product.
- The hierarchy levels dimension is used to "assign functional models to specific levels" of an enterprise. This axis uses aggregation to establish enterprise levels that range from the connected world (i.e., networks of manufacturing organizations in their eco-systems) via stations (manufacturing work cells) to devices and products.

The life cycle and value stream dimension of RAMI4.0 distinguishes between the type and instance of a product and its value-added processes [23]. Type can be equated to the design of the product and processes while instance is the execution of processes to produce a product. This separation emphasizes the importance of consistency across the product life cycle.

The hierarchy levels dimension of RAMI4.0 references the international standard IEC62264:2013 [25]. More specifically, the physical hierarchy of IEC62264:2013 is referenced. The physical hierarchy establishes a naming convention for the sections in the factory. Enterprise is the highest level of the hierarchy and work cell is the lowest for a discreet manufacturing facility. Figure 3 shows an illustrative physical hierarchy of a hypothetical manufacturing enterprise.

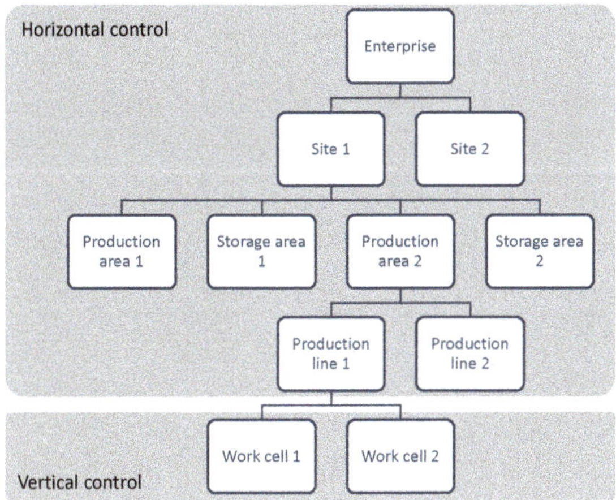

Figure 3. Illustrative physical hierarchy of a manufacturing enterprise showing the different control regimes applied at different levels of the enterprise.

The separation of concerns is widely used to manage complexity in system design [26–28]. The technique allows the designer to consider some aspect of the system separately from the rest of the system, which decreases local complexity. We apply the technique to create two separations in the HORSE System derived from RAMI4.0 and apply these as design principles below.

1. Separation between design-time and run-time system functions based on the life cycle and value stream dimension.
2. Separation between horizontal and vertical control based on the hierarchy levels dimension.

The first separation between design-time and run-time is applied to both the manufacturing processes and the participants in those processes. Manufacturing processes and agents (the participants in the processes) are identified and described during the design period and then instantiated and activated to perform activities during the run-time.

The second separation between the horizontal and vertical control is derived from the physical hierarchy, which is shown in Figure 3. Horizontal control is concerned with the sequence of activities performed by several participants to transform materials into products, i.e., management of the manufacturing processes across multiple work cells. Vertical control is concerned with the actions performed by a single participant or a team of participants within a single work cell of the factory. The different control regimes are also indicated in Figure 3. Thus, horizontal and vertical control is separated to account for different control regimes. Horizontal control is concerned with the coordination of activities that may be spatially dispersed while vertical control is concerned with the sub-second synchronization of actions.

2.3. Design Process

The HORSE System includes several disparate technologies and stresses the need for systematic development and sound architectural principles. The Kruchten 4 + 1 framework [22] is used to deal with the various views of stakeholders and their sequencing in time. The framework, as shown in Figure 4, employs phased development resulting in four views with their respective primary stakeholders.

1. The logical view is concerned with what the system should do. It specifies the functionality of the system in the form of modules and relationships between modules. The main stakeholders are the end users of the system.
2. The development view is concerned with good software management. It specifies how the software system is organized in a developmental environment. The main stakeholders are the software engineers.
3. The process view is concerned with the performance and scalability of the system. It specifies the concurrency and synchronization of the system modules. The main stakeholders are the integrators of the system.
4. The physical view is concerned with realization and deployment of the system. It specifies the allocation of hardware resources to software modules. The main stakeholders are the engineers who are responsible for installing and maintaining the system.

Separate views with different stakeholders can result in a divergence of ideas and an understanding about the system. To avoid such a divergence, the four views are reconciled by a fifth concept:

5. Scenarios represent user cases of the system that demonstrate system functionality and performance. The scenarios should be specific and practical enough to facilitate discussion about the expected operation of the system in its intended context.

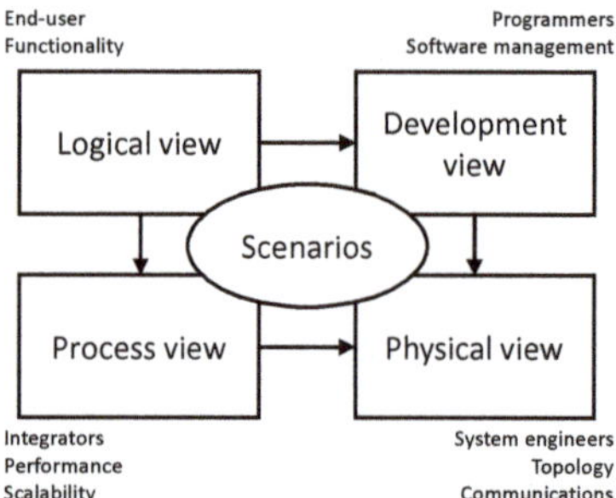

Figure 4. Kruchten 4 + 1 framework [22].

The Kruchten 4 + 1 framework is used to sequence the development of the HORSE System. First, the logical architecture is used to specify the functional structure of the system without reference to specific implementation techniques, technologies, or deployment. The main input into this design is a clear description of the scenarios and the problems faced in those scenarios. The output of this phase is a logical architecture design with five aggregation levels along with one context level [29]. We discuss four of these six levels in Section 3 of this paper. The development and process views are concerned with a good software engineering practice and are, thus, omitted from this paper.

In the physical view, it is determined how and where the software resulting from the previous two views will run. The exact software and hardware may be different for each deployment of the HORSE System, but the type of technology remains the same. The exact software and hardware used in the HORSE Project is specified in this case study [30]. This research paper is more concerned with the separation between cloud-based and on-premise deployments. Thus, the cloud-supported parts of the HORSE System are identified, which leads to a clear division between the modules running 'in the cloud' and those running on premise. This division is discussed and justified in Section 4 of this paper based on performance and security considerations. The scenarios are located within the three industrial pilot cases of the project and discussed in Section 5.

3. Logical View of the HORSE System

The logical system architecture of the HORSE System is a hierarchical, multi-level view. The complete design comprises six levels of aggregation, but only the first four levels are discussed in this paper in the interest of brevity. The full system design report [29] elaborates on the remaining two levels of aggregation, but it limits the discussion to system structure and design decisions. Conversely, this research paper emphasizes scientific rigor and presents the HORSE System in the context of Industry 4.0. The four levels discussed in this paper are labeled Context Architecture and Architecture Levels 1, 2, and 3. The Level 2 architecture is divided into the design time half and the execution time half. Only the execution time subsystem on the local layer is discussed because this subsystem is the most complicated and is responsible for the main interface between software and hardware on the factory floor.

3.1. HORSE Context Architecture

The HORSE System is primarily concerned with highly configurable, flexible manufacturing processes involving human and robotic participants. All three pilot cases in the HORSE Project, which are discussed in Section 5, feature processes with cooperation between humans and robots. Therefore, the HORSE System interfaces with a variety of humans, robots, and sensors that participate in the manufacturing processes. These processes do not exist in isolation. Any manufacturing enterprise also performs other business processes such as purchasing, product development, sales, and customer support. Consequently, the HORSE system must be contextualized in the existing hardware and software systems of the enterprise.

An illustrative enterprise architecture is shown in Figure 5 with the HORSE System positioned as a central hub amongst typical systems found in factories [14]. Business management systems are represented with an enterprise resource planning system (ERP), which is a manufacturing execution system (MES) and a product life-cycle management system (PLMS) at the top of Figure 5. Toward the lower end of Figure 5, the HORSE System is connected to a robot, human, and sensor, which promotes human-robot co-existence. In practice, the situation is typically more complicated, but this simplified view shows the context of the HORSE System.

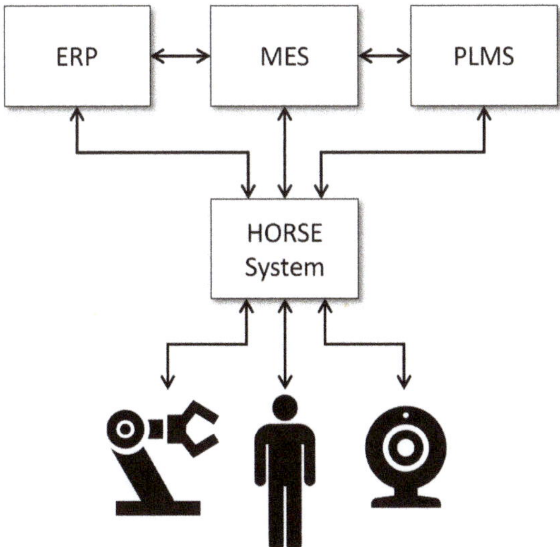

Figure 5. Context of the HORSE System simplified to only show typical systems.

3.2. HORSE System Architecture Level 1

Level 1 of the architecture model unpacks the HORSE System box of Figure 5 to refine its contents. This refinement applies the two design principles described in Section 2.2: separation of design-time and run-time system functions and separation of horizontal and vertical control. The first design principle calls for system functionality dedicated to design and control, respectively. This consideration gives the architecture a columned style with a design-time column and an execution-time column. The second design principle calls for separation between functionality that is aimed at the support of activities within a single manufacturing work cell and functionality that is aimed at synchronizing activities across multiple work cells. This consideration gives the architecture a layered style with global and local control layers. The global layer interfaces with business management systems at the top of Figure 5 while the local layer interfaces with humans and robot controllers, which is shown at the bottom of Figure 5.

Applying the two design principles results in a system architecture with four sub-systems and two data stores, as shown in Figure 6. This is a columned architecture embedded in a layered architecture [31]. The design-time and execution-time columns are connected via databases that contain specifications of manufacturing activities and the participants involved in the activities. During design-time, the global and local layers are indirectly connected via the data stores since these subsystems are used to create and edit the manufacturing activities and actors. However, for execution time, HORSE Exec Global and HORSE Exec Local are directly coupled to pass messages directly between global and local control. Note that the design-time subsystem on the local layer is labeled as configuration instead of design, since this fits better with the configuration of equipment and tools within work cells.

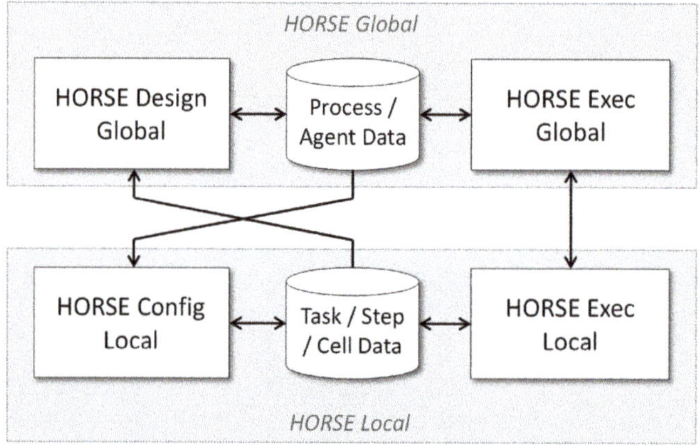

Figure 6. HORSE architecture, aggregation level 1.

In the two subsections below, the HORSE system architecture is further elaborated. The design period and execution time columns of Figure 6 are discussed separately to manage the inherent complexity of the system.

3.3. HORSE Design Time Architecture Level 2

This section elaborates on the design-time column of the HORSE System, which is shown in Figure 7. On the global layer, the HORSE Design Global subsystem provides functionality to design manufacturing processes across multiple work cells populated by multiple, possibly heterogenous, actors. As such, the subsystem contains two modules dedicated to the design of processes and agents (terminology used to denote any independent process participant), respectively. For the Process Design module, existing business process management (BPM) technology is used as the basis and extended to accommodate the physical nature of manufacturing processes as opposed to the administrative processes for which this technology is traditionally used [32]. Most significantly, the location of manufacturing operations must be considered to accommodate the time it takes for material to flow between locations. The Agent Design module is a graphical user interface that enables the user to create new agent profiles or edit existing agent profiles. Such a profile comprises attributes that describe the agent including abilities, skills, authorization, cost, and performance.

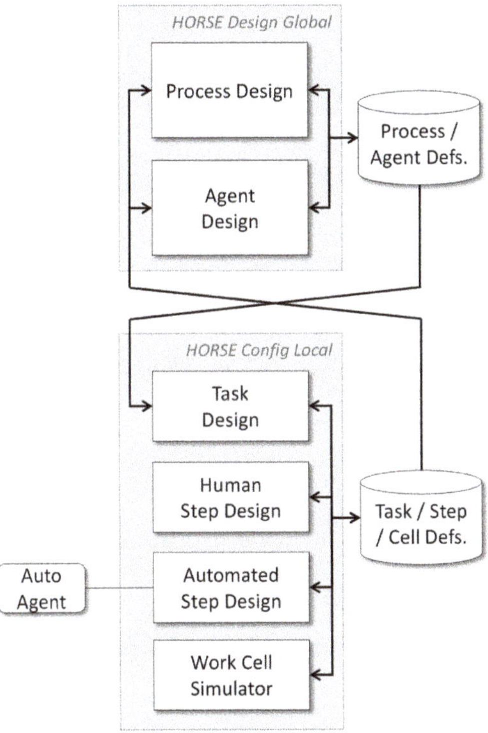

Figure 7. HORSE architecture, design time aspect, aggregation level 2.

HORSE Config Local provides functionality for defining the operations performed within a work cell. The terminology of 'task' and 'step' is used here to distinguish between the actions of multiple and single agents. The task design module is used to specify the synchronization of a team of agents within a work cell like the interplay between a human worker and a robot. The actions performed by the individual members of a team are specified using the two-step design modules. The human step design module is used to create work instructions. These work instructions may range from simple textual descriptions to technology-supported guidance like augmented reality. The automated agent step design module is used to create execution scripts for robot, automated guided vehicles or any other non-human agents. Several different instances of this module may exist in the same enterprise, which corresponds to the different types of automated agents used and supports textual scripting, graphical scripting, and scripting by physical manipulation (physically showing the robot what to do, which is also called programming by demonstration [9]). The latter requires a direct connection to the involved robot, which is shown in Figure 7. Lastly, the work cell simulator module is used to digitally define and evaluate the physical constraints within which a task will be executed. Physical constraints may include inter alia, the space available for agents to move, the location of the work cell, and the mounted position of the automated agents.

3.4. HORSE Execution Time Architecture Level 2

This section elaborates the execution-time column of the HORSE System, as shown in Figure 8. Within the global and local layers, each contain an execution and an awareness module. HORSE Exec Global provides the functionality to enact and monitor manufacturing processes across multiple work cells. Analogous to global design subsystem, the HORSE Exec Global subsystem is based on BPM technology with extensions to make it suitable for manufacturing processes [33].

HORSE Exec Local provides the functionality to control and monitor the activities of a team of agents in a single work cell (a team may be only one agent). This subsystem directly interacts with agents by sending commands and receiving responses. The interaction differs between humans and automated agents. Work instructions are displayed on handheld devices or fixed monitors and responses are received via physical or virtual buttons. For example, an operator may receive the instruction to perform a task via an instant message. Once at the station, the augmented reality guidance is displayed via the local awareness module and hand movements are detected via a sensor. Automated agents receive execution scripts and respond according to predefined parameters. The Local Awareness module is concerned with regular activity monitoring and exception detection.

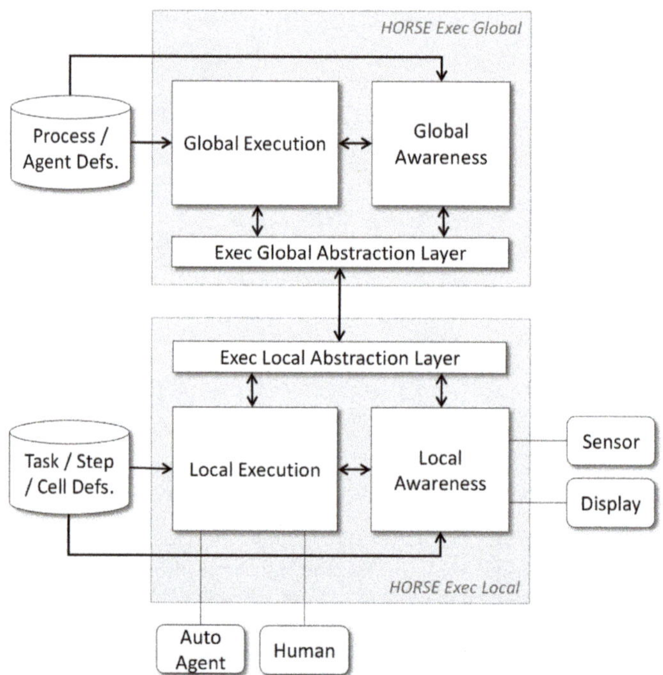

Figure 8. HORSE architecture, execution time aspect, aggregation level 2.

An instance of HORSE Exec Local is created for each team formed to perform a task. Therefore, multiple instances of the subsystem may exist at the same time and may even build on different technology platforms. The abstraction layers facilitate the communication between the HORSE Exec Global and multiple instances of HORSE Exec Local. Integration between the global and local layers, as facilitated by the abstraction layers, is illustrated with a video available on the project website (http://www.horse-project.eu/Media). The HORSE Exec Local subsystem is discussed in more detail in Section 3.5.

3.5. HORSE Exec Local at Aggregation Level 3

Figure 9 shows the refinement of the HORSE Exec Local system module at aggregation level 3. It is, in this part of the HORSE architecture, that the cyber-physical character of the system becomes most apparent because this subsystem is responsible for the real-time control of physical agents.

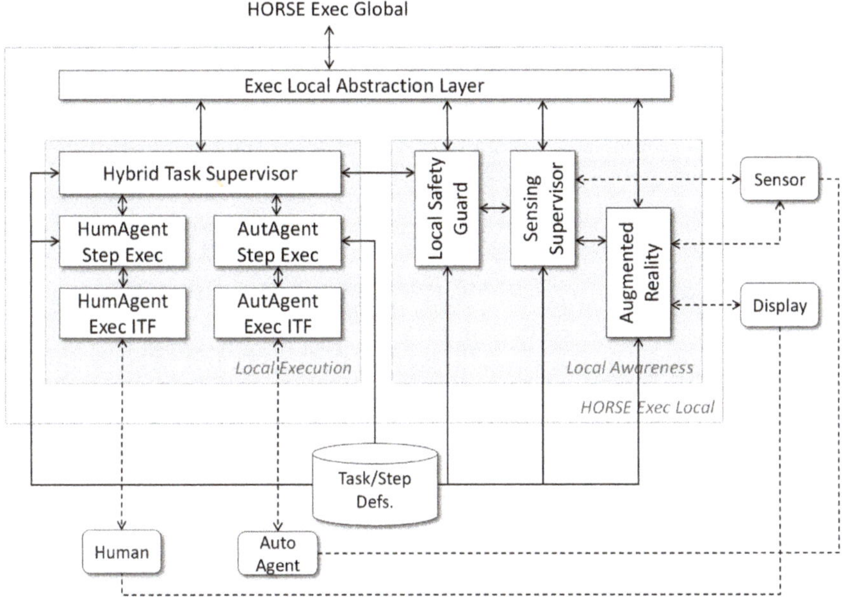

Figure 9. HORSE architecture, execution time aspect, aggregation level 3.

The Local Execution module, as shown on the left side of Figure 9, is responsible for driving the execution of manufacturing tasks. The Hybrid Task Supervisor module delivers the human-robot collaboration capability of the HORSE System. The module controls synchronize the actions of multiple human and/or robotic workers (in HORSE terminology, human agents and automated agents, respectively). These workers receive their instructions via Step Execution modules that manage individual work steps for individual agents and Execution Interfaces that abstract from specific agent control characteristics (such as specific robot control interfaces). Local Execution implements actuation controls from an IoT perspective.

The Local Awareness module, on the right side of Figure 9, is responsible for monitoring the work cell. This module is coupled to sensors and cameras in the work cell that provide real-time information about the status of the work cell such as the position of robotic arms and manipulated products. These sensors and cameras may be attached to robots such as for measuring torque or applying pressure sensors on robot arms. Part of this information can be fed to displays that provide information to human workers. Local Awareness implements sensing controls from an IoT perspective.

4. Physical View of the HORSE System

The logical view of the HORSE System architecture is explained in Section 3. In this section, the physical view of the Kruchten 4 + 1 framework is applied. The goal of this section is to analyze the role of Industry 4.0 technologies in the HORSE System instead of specifying the actual hardware, infrastructure, and physical systems. Given the primary context of small and medium manufacturing enterprises, it is first determined, in Section 4.1, which modules of the system can be situated in the cloud. Since the design time sub-system and the execution time sub-system of the HORSE system have very different technical requirements, we perform this analysis per sub-system. This analysis naturally establishes the divide between the cloud-supported modules of the system and the Things on the factory floor. The divide is used to construct an IoT-view of the HORSE System, which is presented in Section 4.2.

4.1. Cloud Support for the HORSE System

To determine which modules of the HORSE System can be hosted in the cloud, we consider the performance expectations of various modules. For example, modules used during the design period generally do not require guaranteed sub-second response times. Hence, such modules can be hosted as cloud applications and be subjected to the normal performance impact of distance and Internet traffic. To add some nuance to the discussion, we distinguish between two cloud-based models and a non-cloud model for software deployment [34].

- The Software-as-a-Service (SaaS) model provides a software application as a hosted service on the Internet, eliminating the need to install and run the application on local computers.
- The Platform-as-a-Service (PaaS) model provides an application environment in which users can create their own application that will run on the cloud.
- The on-premise model represents traditional computing with no part of the software or computer hardware hosted in the cloud.

Rimal et al. [35] advocates for scalability, performance, multi-tenancy, configurability, and fault-tolerance as the primary considerations for cloud support of applications. Table 1 lists the advantages and disadvantages inherent to each cloud-support model in relation to the five considerations.

Table 1. Advantages and disadvantages of SaaS, PaaS, and on-premise models in relation to the five considerations for cloud-support.

Consideration	SaaS	PaaS	On-Premise
Scalability	Software and computing resources can be scaled quickly.	Computing resources can be scaled quickly.	Scaling requires installation of new software and hardware.
Performance	Cannot be guaranteed because it is subject to network quality, traffic, and distance.	Cannot be guaranteed because it is subject to network quality, traffic, and distance.	Best response-time performance attainable.
Multi-tenancy	Tenancy can easily be extended to any agent with Internet access.	Additional tenants added with additional software installation or expansion.	Only within its local environment.
Configurability	Software and computing resource changes are subject to the agreements with the service provider.	Computing resource changes are subject to agreements with the service provider.	All changes are under the control of the user.
Fault-tolerance	Fault-tolerance is defined as part of the quality-of-service agreements.	Fault-tolerance is defined as part of the quality-of-service agreements.	Fault-tolerance is under the control of the user.

We apply the five considerations listed in Table 1 to determine which modules of the HORSE System can be hosted with the SaaS or PaaS cloud computing models. The results of the analysis are listed in Table 2.

Table 2. Application of the five considerations for cloud-support for the modules of the HORSE System.

System Module or Data Store	Cloud Support	Rationale
Process Design Agent Design Process/Agent Data Global Execution Global Awareness Task/Step/Cell Data	SaaS	The scalability and multi-tenancy afforded by SaaS is an opportunity to extend service to additional production areas, sites, or even enterprises. Singular deployment implies no configurability requirements. No strict performance guarantees required. Fault-tolerance can be addressed with the quality-of-service agreements with the service provider.
Exec Global Abstraction Layer Exec Local Abstraction Layer	SaaS	The scalability, multi-tenancy, and configurability of a SaaS-based abstraction layer makes it possible to extend the global functionality across multiple production areas, sites, or enterprises. No strict performance guarantees required. Fault-tolerance can be addressed with quality-of-service agreements with the service provider.
Task Design Human Step Design	PaaS	The platform can be scaled to additional production areas, sites, or enterprises. Tenancy can be extended with additional software deployments on the same platform. Technology-heterogeneity requires extensive configuration of software on the same platform. No strict performance guarantees required. Fault-tolerance can be addressed with the quality-of-service agreements with the service provider.
Automated Step Design	PaaS or On-premise	PaaS or On-premise model depends on the direct connection to an automated agent. For textual execution scripts, the PaaS model can provide scalability and multi-tenant support for technology-heterogenous deployments with no strict performance or fault-tolerance requirements. Programming-by-demonstration requires on-premise hardware and software to support immediate response to the movements performed by the human.
Work Cell Simulator	PaaS	The platform can be scaled to additional production areas, sites, or enterprises. Tenancy can be extended with additional software deployments on the same platform. Technology-heterogeneity requires extensive configuration of software on the same platform. No strict performance guarantees required. Fault-tolerance can be addressed with quality-of-service agreements with the service provider.
Local Execution Local Awareness	On-premise	Multiple, technology-heterogeneous realizations for each local deployment requires no scalability or multi-tenancy. The configuration is done during the design-time. The control of multiple, interacting agents require strict performance guarantees in the millisecond domain and near-zero fault tolerance.

Figure 10 shows the result of the five considerations as an overlay on the logical view of the HORSE System architecture. The modules at the global layer support process design, agent design, process enactment, and monitoring. Process and agent design are interactive modules but do require any guaranteed performance. A further advantage of the SaaS model is simplified versioning and upgrade of the software since manufacturing is becoming more flexible [36,37]. Process enactment and monitoring has a real-time character but without very strict timing requirements. Consequently, it is possible to deploy all global layer modules and both the data stores in the cloud. An important requirement for the cloud environment is a very high quality of service (QoS) in terms of availability: unavailability of global functionality typically brings a process-oriented manufacturing plant to a halt within minutes if not seconds.

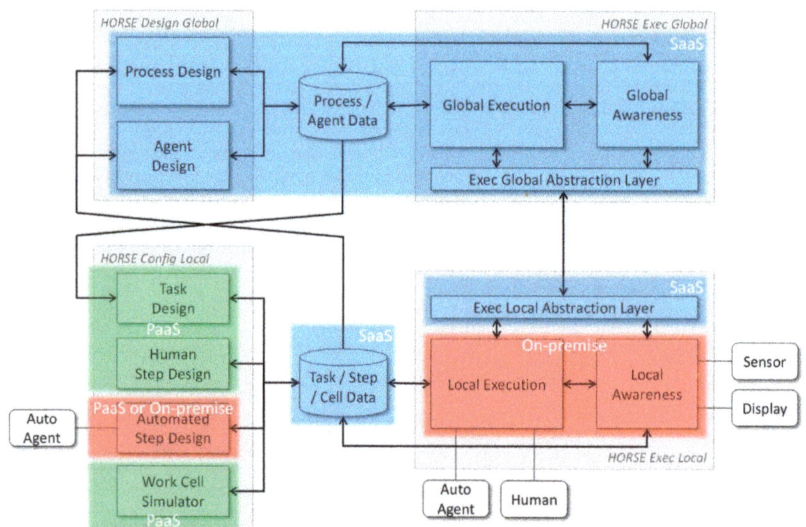

Figure 10. Overview of cloud support for the HORSE System.

The modules at the local layer control the activities of agents—either individually or in teams—are very much real-time cyber-physical systems. This means that part of their functionality has very strict response-time requirements. A good example is safety management, which requires fast synchronization between sensors, local awareness functionality, local execution functionality, and agents (humans and robots). Figure 11 shows the communication path in the system (as a simplified view of Figure 9 to make things clearer) when a human agent enters the operating space of a robotic agent and the robotic agent must immediately stop.

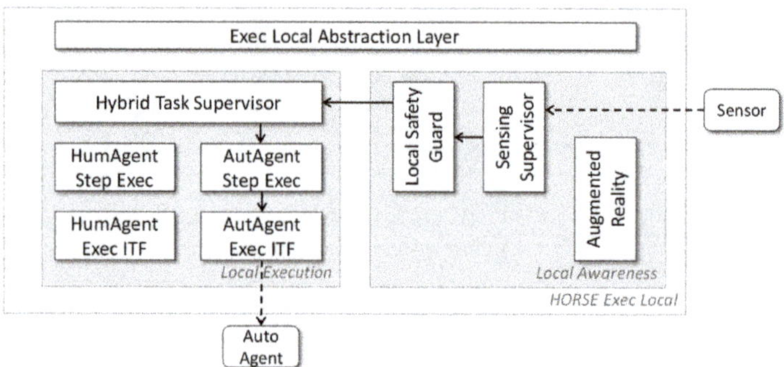

Figure 11. Local communication path in case of an observed safety breach.

4.2. The HORSE System as an IoT Application

With HORSE Exec Global in the cloud and HORSE Exec Local on-site, the HORSE System relies heavily on communication between the cloud and the Things that perform manufacturing activities.

Significant disagreement exists regarding the nature and scope of the IoT [38], but Wortmann and Flüchter [39] offer a complete overview of the concept by bringing together computing, connectivity, and devices in a single technology stack. The technology stack follows the concept of functional abstraction [31], i.e., each layer builds on the functionality of the layer below. For example, the control software of a thing/device makes use of the actuating and sensing components to exert control over the hardware of the thing/device.

The technology stack draws attention to the division between the things or devices (including its embedded control system) on the factory floor and software situated in the IoT cloud. The connectivity layer facilitates communication between the IoT Cloud and one or more things/devices.

Functional abstraction and the division between the cloud and the device is applied to construct a technology stack for the HORSE System. The resulting technology stack, which is shown in Figure 12, serves two purposes: (1) to describe how modern technologies contribute to realize the HORSE System and (2) to justify the designation of the HORSE System as an IoT application.

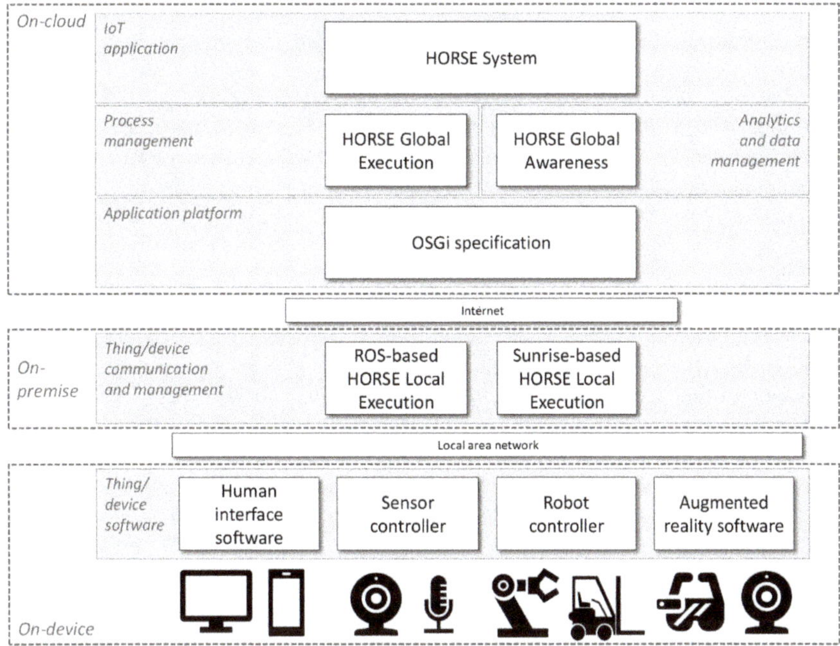

Figure 12. Technology stack showing the HORSE System as an IoT application.

Starting from the bottom of Figure 12, to simplify the explanation of the assimilation of technologies, the various things and devices are shown as simple icons. Human interfaces are represented with displays and handheld devices. The sensors are represented with a camera and microphone while robots are represented with a robotic arm and a vehicle. Lastly, augmented reality is represented with wearable glasses and a sensor to track human movements. The distinction between hardware and components is omitted here because it adds no value to the current discussion. The displayed things or devices are only representative since the actual set may differ in a factory. On the second layer from the bottom, the various devices and things are controlled by their respective software systems. These software systems are left purposefully abstract to signify the open nature of the HORSE System. Many different technologies can be utilized in conjunction with the HORSE System. The software systems of the devices are connected to the HORSE System via the local area network of the factory.

The HORSE Local Execution subsystem serves as thing/device communication and management. This is the most significant deviation from the IoT technology stack of Wortmann and Flüchter [39]. The thing/device communication and management is not included in the Cloud grouping of the technology stack because of the requirement for guaranteed, sub-second communication and synchronization between teams of agents, which is discussed in Section 3.4. The HORSE System currently has two realizations of the Local Execution subsystem based on the Robot Operating System (ROS) and Kuka Sunrise software, respectively. Both these variations can control various technologies involved in smart manufacturing processes and can even be deployed simultaneously in the same factory.

A singular block named Internet is shown to represent the connectivity between Local and Global features. This implies all standard infrastructure and protocols because the current realization of the HORSE System uses standard internet protocols to transport JavaScript Object Notation (JSON) messages. These messages are directed via the message bus of the Open Services Gateway initiative (OSGi) application platform, which allows all subsystems of the HORSE System to subscribe to it and publish messages. The OSGi application platform ties in well with the PaaS model, which is explained in Section 4.1. With an OSGi-based platform and accompanying computing resources hosted in the cloud, the user can install a variety of thing/device communication and management systems based on different technologies to control the activities of different robots and sensors.

HORSE Global Execution, which represents the process management technology, publishes messages and subscribes to listen for responses. Similarly, the HORSE Global Awareness subsystem represents the data analytics technology by listening for various pre-defined events that may indicate unwanted or unexpected factory floor conditions. Thus, the HORSE System is an IoT application with multiple layers of functionality covering the complete technology stack of Wortmann and Flüchter [39].

5. Scenarios as Proof of Concept

The HORSE System is positioned as a manufacturing operations management system for Industry 4.0. Therefore, it is claimed that the HORSE System can be used to manage operations that involve modern manufacturing technologies. To substantiate this claim, manufacturing processes that utilize the smart technologies are demonstrated to serve as proof of concept of the HORSE System. For each process, the involvement and role of various technologies are highlighted in the process models and are subsequently discussed.

The three pilot cases considered in the HORSE Project represent a wide range of problems encountered in the manufacturing industry. These pilot cases are analyzed to articulate clear scenarios where smart manufacturing technology is applied to solve common problems. Each scenario is described to give context, represented as a process model, and demonstrated with video footage.

5.1. Pilot Case 1: Tool Assembly

The first pilot site of the HORSE Project is a medium-sized factory in The Netherlands that produces highly configurable metal products used in furniture assembly. The pilot case features two processes including tool assembly and surface treatment of metal profiles. The first process involves a human operator who attaches tool parts to a base-plate to assemble a configurable tool used for deformation operations. In parallel, a mobile robotic arm fetches bins containing the parts needed by the operator. To alleviate the complication and variability of the assembly task, the human assists by augmented reality technology that highlights which parts are needed and how to attach those parts. The model of this process is specified using the Business Process Model & Notation 2.0 (BPMN2.0) and shown in Figure 13. Video footage of the executed process with augmented reality and robotic support is available online (https://youtu.be/bqTDEZvOdVI).

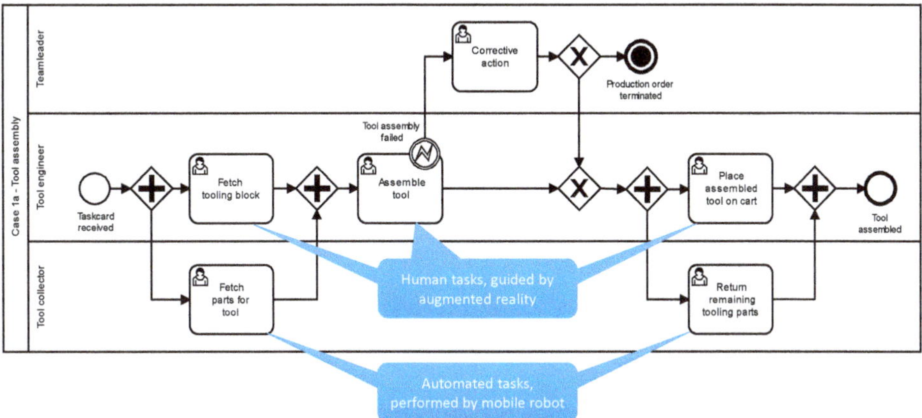

Figure 13. Process model used to enact the demonstration of Case 1a.

The model shown in Figure 13 is created using the Design Global subsystem of the HORSE System (see Section 3.3). Thereafter, the process model is parsed by the Global Execution to enact the process during the demonstration. Tasks, as specified in the process model, are globally instantiated and assigned to agents. The task instructions are sent to the Exec Local subsystem via the message bus. The Exec Local subsystem ensures synchronization between the robot and human and is guided by augmented reality.

The full stack of technology, as shown in Figure 12, is utilized to execute the tool assembly process. As a summary, the various technologies have the following roles in the process.

- Cloud-based process management to coordinate the activities of human and automated agents.
- IoT-enabled connectivity between cloud-based process management and multiple production agents.
- Augmented reality to guide the human agent through the tool assembly steps.
- Smart robotics to fetch and return tooling parts from the storage zone.

Pilot case 1 is a particularly compelling case for the HORSE Project because it is not a simple replacement of a human operator with a robot. Instead, the operations are granularized to determine which activities are more suited to human or robot execution. The cognitive and sensory abilities of the human are exploited for the complicated assembly task, but the monotonous fetching and returning tasks are allocated to a mobile robot. Therefore, instead of being replaced by automation, the human is rather supported by automation and allowed to focus on the task that requires deep concentration, which increases process throughput.

The primary limitation in this pilot case is related to the definition of augmented reality tasks. The images and instructions displayed during augmented reality supported tool assembly are currently manually programmed for a particular subset of tool configurations. This is perfectly adequate for demonstration purposes in a research and innovation project, but it must be simplified or at least streamlined for commercial purposes.

5.2. Pilot Case 2: Final Assembly of Automotive Parts

The second pilot site of the HORSE Project is a medium-sized factory in Spain that assembles highly customizable automotive parts. The case includes the final inspection and packaging of the assemblies before distribution to customers. The process involves three agents: (1) a robotic arm to pick up an assembly from the conveyor belt, present it for inspection, and, if accepted, place it in a box, (2) a sophisticated, bespoke camera system to inspect the assemblies, and (3) a human operator to evaluate assemblies flagged by the camera system to determine whether to discard or repair it. To deal with the complication of inspecting mass customized products, the human operator is assisted by an augmented reality system that indicates where and what the operator should inspect. Figure 14 shows the model used to enact the process of the second pilot case. Additional information and photos of pilot case 2 in operation are available online (http://www.horse-project.eu/Pilot-Experiment-1).

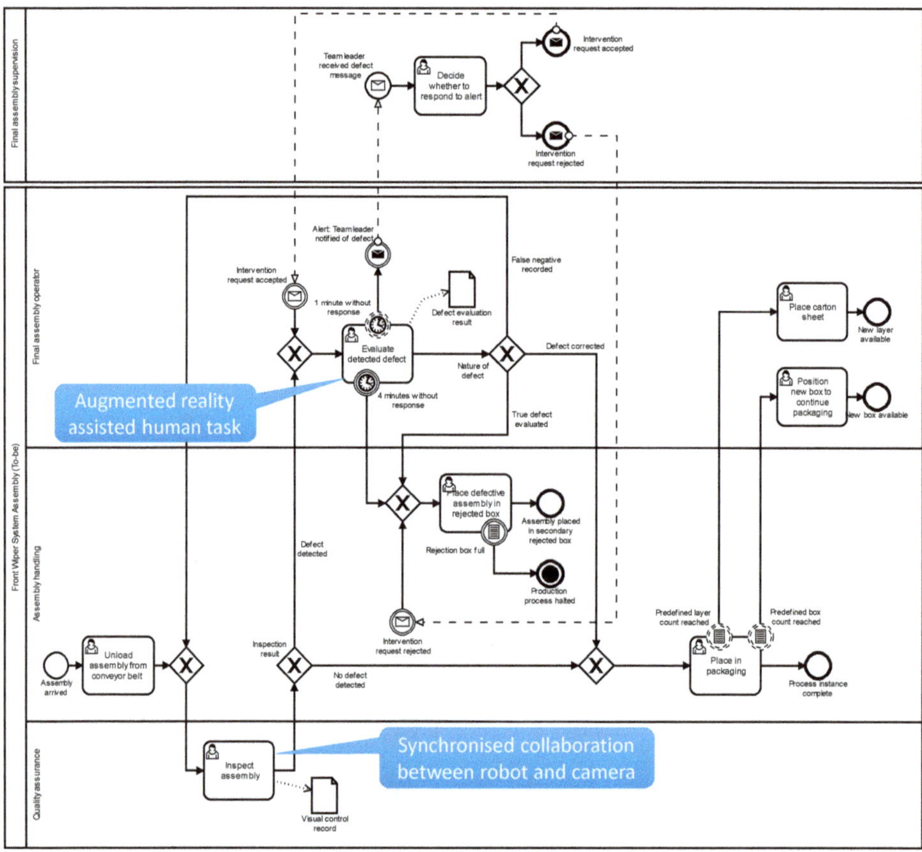

Figure 14. Assembly inspection and packaging process model of pilot case 2.

As can be seen from the lanes in Figure 14, the process involves four different roles, which are potentially performed by as many agents. The following technology is involved in this process.

- Cloud-based process management across multiple work cells.
- IoT-enabled connectivity between cloud-based process management and multiple production agents.
- Synchronized collaboration between smart robot and camera system to detect product defects.
- Augmented reality to guide the human agent with an inspection of highly customizable assemblies.

Pilot case 2 is a complex scenario involving four process participants and various eventualities that affect process execution. Communication from the factory floor is of utmost importance here to allow the HORSE System, track progress, and adjust accordingly. Communication is facilitated by handheld devices and sensors connected to the local network. The most compelling part of pilot case 2 is the human-robot collaborative task to inspect assemblies that are flagged as defective by the camera system. The Hybrid Task Supervisor module (see Section 3.5), in this case, is realized as a ROS application and uses state-machine models to synchronize the actions of the human and robot.

As with the first pilot case, pilot case 2 involves automation of some tasks in the process. Picking and placing is automated, but a human operator is still necessary to respond to detected defects and ensure that packaging material is available. This automation is beneficial but constraining. It is beneficial to the health of the human because the repetitive lifting and manipulation of heavy parts is eliminated. However, the robot can only handle one assembly at a time, which forces it to wait for the human if he or she is not immediately available. Therefore, the introduction of the HORSE System and associated smart technologies, in this case, improves operator health but not necessarily process performance.

5.3. Pilot Case 3: Separation of Castings

The third pilot site of the HORSE Project is a large foundry in Poland. The foundry uses sand casting to produce a large range of products for several industries including automotive, rail transport, and industrial equipment. Due to excess metal in the mold, four to eight castings are physically attached to each other after solidification. Those castings must be separated before further surface processing is performed. The separation can be automated. However, the high customizability of the product makes it infeasible to define cutting plans for all products. Each production run is ostensibly unique and, therefore, necessitates a new cutting plan. To handle such high product variability, a robot equipped with teaching-by-demonstration technology is deployed to lift some of the human operator. The operator physically moves the arm and the end-effector of the robot through the necessary cutting trajectories to teach the cutting plan instead of labor intensive approaches such as computer-aided manufacturing modeling or programming. Once the cutting plan is recorded, then the operator enables an execution mode to process the batch of products. Figure 15 shows the model used to enact the grape separation process, which highlights the teaching-by-demonstration task. Additional information and photos of pilot case 2 in operation are available online (http://www.horse-project.eu/Pilot-Experiment-2).

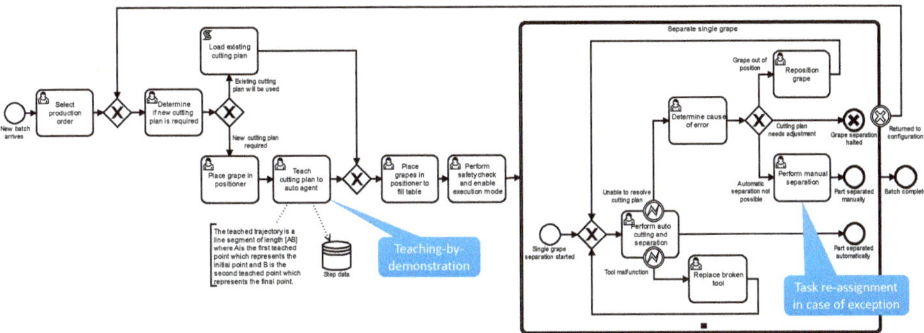

Figure 15. Grape separation process model of pilot case 3.

Although teaching-by-demonstration is the most noticeable feature of this process, several other technologies contribute to such a smart manufacturing process. The following technologies are involved in this process.

- Cloud-based process management to invoke the actions of humans, robots, and computer services.
- IoT-enabled connectivity between cloud-based process management and multiple production agents.
- Synchronized collaboration between human and robot during the teaching-by-demonstration task.
- Smart robotics with human intrusion detection to halt the execution in the case of safety risks.

Pilot case 3 is a clear demonstration of smart manufacturing with the utilization of teaching-by-demonstration and modern process management technology. More importantly, these technologies are demonstrated in an environment that is not particularly conducive to sophisticated and delicate computer systems. The factory floor is teaming with fine dust particles from the grinding operations. In this case, the cloud-based nature of the HORSE System proves valuable because only the robot is exposed to the dusty environment. However, this proves to be a limiting factor as well because the operator does not have access to any troubleshooting functionality. If something goes wrong, production management must be contacted to address the problem. Thus, Industry 4.0 technologies present many opportunities but also many new considerations when deploying smart technologies.

6. Inter-Organizational Process Perspective

Thus far, we explored how parts of the HORSE system can be deployed to achieve advantages in software management and computing infrastructure investments, i.e., advantages that do not influence the nature of the functionality offered. The focus of these developments is within one manufacturing enterprise. However, the technology employed in the HORSE System may offer improved functionality: cloud computing and the IoT can improve interoperability within manufacturing networks where a manufacturing process takes place across multiple sites or even autonomous enterprises.

Software modules that are used in a SaaS paradigm are typically of a standardized kind so that the same functionality can be used by multiple parties. Applying functional standardization to the modules at the global layer of the HORSE System in the context of multiple manufacturing enterprises that collaborate can improve interoperability between those enterprises. Consequently, it may be simpler to set up supply chains or supply networks with automated support for manufacturing processes (and related enterprise processes as discussed in Section 3.1) across enterprises. The current realization of the global layer of the HORSE System is run as a single application instance with multiple tenants for the three pilot cases. While the current pilot cases do not participate in the same supply chain, it does prove the feasibility of a single global layer serving multiple instances of the local layer across geographically separated sites. Figure 16 shows an expansion of the HORSE technology stack (see Figure 12) to include two sites with a different, illustrative set of things/devices.

Cross-organizational manufacturing is an ongoing research topic due to its numerous potential benefits [40,41]. The concept has been demonstrated in the CrossWork project [42,43], albeit without complete vertical integration down to the factory floor as in the HORSE project. In the CrossWork project, multiple manufacturing enterprises in the same supply chain network use a single, centralized process management system to synchronize their activities. These connected enterprises temporarily form an instant virtual enterprise, i.e., non-permanent collaborations with the sole purpose of producing a single product series. The CrossWork approach has been prototyped in the automotive industry, which provides a similarity with one of the pilot cases of the HORSE project. The CrossWork concept is illustrated in Figure 17 and it shows a global process that flows across four autonomous enterprises in a single supply chain network [42,43].

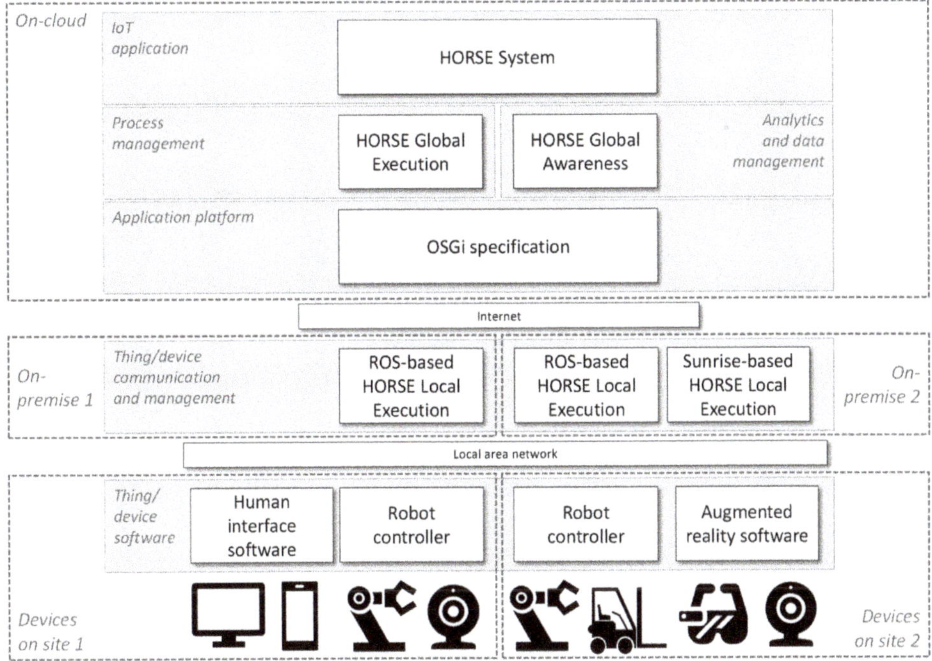

Figure 16. HORSE technology stack expanded to include multiple sites.

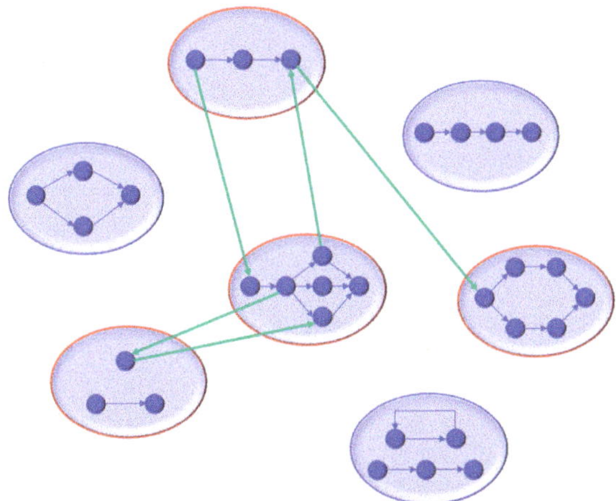

Figure 17. Networked manufacturing process crossing the boundaries of multiple organizations.

An alternative approach that allows more distribution of autonomy is the use of a multi-tenant SaaS process management solution that embodies the global execution module of HORSE for several enterprises in a chain and that also synchronizes the links between their manufacturing processes. Both alternatives lead to networked process management [44]. This is a relevant development since inter-organizational processes in manufacturing are receiving more attention in an Industry 4.0 setting [45,46]. This trend is required because of increasing product complexity, increasing producer specialization, and increasing mass customization of products [1].

7. Conclusions

This paper outlines the architecture design of the HORSE System. It is shown how structured, hierarchical design produces a modular architecture with clearly defined subsystems and interfaces. The system embodies the assimilation of traditional enterprise information systems (e.g., explicit process management) and advanced manufacturing technology (e.g., human-robot collaboration and the IoT). The HORSE System architecture is proposed as a reference architecture for a manufacturing operations management system for Industry 4.0. Thus, the architecture model of the HORSE System serves two purposes: (1) it can be used as a template or framework to position and develop a commercial-grade manufacturing operations management system for Industry 4.0 and (2) it helps frame the scientific inquiry into the management of manufacturing operations involving cloud computing, the IoT, and smart devices.

The proposed reference architecture model is the main contribution of this paper. Due to its complexity, it is presented and discussed in two views including the logical and physical views. The logical view details the functional operation of the system without any commitment to specific technologies. The logical view is elaborated at four levels of aggregation (two additional levels are covered in the full design report [29]). The physical view explores how the various technologies enabling Industry 4.0 are used to realize the HORSE System. Each module of the system is considered in order to determine whether it can be hosted in the cloud. To summarize, we show the leading drivers for cloud deployment for each of the four HORSE main sub-systems (as in Figure 6) in Table 3.

Table 3. Leading drivers for cloud deployment.

	Design-Time	Execution-Time
Global layer	Flexibility	Availability
Local layer	Configurability	Responsiveness

The HORSE System has three notable embedded characteristics that have potential advantages and disadvantages. First, cloud computing is emphasized as a fundamental enabling technology of the system. Cloud-based software lowers the barriers-to-entry for Industry 4.0. SMEs can focus on the use of smart devices to solve problems on the factory floor instead of being concerned with the installation and maintenance of software for those smart devices. However, cloud-based software especially with the SaaS model limits the flexibility available to SMEs. It is more time-consuming or expensive to make changes to a SaaS application than an application entirely under the control of the enterprise.

Second, the execution-time subsystem of the local layer of the system delivers real-time, sub-second synchronization between humans and robots. The time-critical nature of human-robot collaboration results in a clear separation between the cloud-based modules and modules for direct control of the things/devices on the factory floor. This separation contributes to the modular nature of the reference architecture and allows for technology heterogeneity on the factory floor. Multiple instances of local execution can be realized with different technology underpinnings. However, the separation places further importance on the reliability of Internet connectivity. If the cloud-based or Internet connectivity services are unavailable, all operations will be uncoordinated at best or suspended at worst.

Lastly, cross-functional, configurable manufacturing process management opens new ways for smart manufacturing by supporting flexible process definitions, dynamic allocation of tasks to human and robotic workers, and real-time coupling of work cell events for manufacturing processes. Such process management processes also hold promise beyond a single site or enterprise. Cloud-based process management supports improved interoperability in manufacturing chains and networks.

The HORSE Project is still ongoing and seeks to further enhance the system that bears its name. The design science research approach adopted in the project ensures practical relevance but it also increases the risk of overlooked problems or missed opportunities. The system design is informed by the real-world problems encountered by the three pilot cases and every attempt is made to extrapolate to more general problems. However, completely unrelated problems may exist in other factories, which may not be considered in the HORSE Project. Seven additional cases were subsequently added to the HORSE Project to evaluate the effectiveness of the HORSE System and to identify any shortcomings.

Lastly, as a research and innovation project, the HORSE Project did not create a commercial-grade manufacturing operations management system. Instead, a prototype was developed and implemented to demonstrate the feasibility of a system incorporating cloud computing, the Internet-of-things, and smart devices. Further complications will undoubtedly arise on the quest for a commercial-grade system, but, at least, the current system architecture can serve as a proven template for such a development.

Author Contributions: Conceptualization, J.E. and P.G. Funding acquisition, P.G. and I.V. Investigation, J.E., P.G., and I.V. Methodology, J.E. and P.G. Project administration, I.V. Software, J.E. and K.T. Validation, J.E. and K.T. Visualization, J.E. Writing—original draft, J.E. and P.G. Writing—review & editing, I.V. and K.T.

Funding: The HORSE project received funding from the European Union's Horizon 2020 Research and Innovation Program under Grant Agreement No. 680734.

Acknowledgments: All members of the HORSE Architecture Team are acknowledged for their contribution to the system design process.

Conflicts of Interest: The authors declare no conflict of interest.

References

1. Hofmann, E.; Rüsch, M. Industry 4.0 and the current status as well as future prospects on logistics. *Comput. Ind.* **2017**, *89*, 23–34. [CrossRef]
2. Lu, Y. Industry 4.0: A survey on technologies, applications and open research issues. *J. Ind. Inf. Integr.* **2017**, *6*, 1–10. [CrossRef]
3. Zhang, L.; Luo, Y.; Tao, F.; Li, B.H.; Ren, L.; Zhang, X.; Guo, H.; Cheng, Y.; Hu, A.; Liu, Y. Cloud manufacturing: A new manufacturing paradigm. *Enterp. Inf. Syst.* **2014**, *8*, 167–187. [CrossRef]
4. Tao, F.; Qi, Q. New IT Driven Service-Oriented Smart Manufacturing: Framework and Characteristics. *IEEE Trans. Syst. Man Cybern. Syst.* **2018**, 1–11. [CrossRef]
5. Qu, T.; Lei, S.P.; Wang, Z.Z.; Nie, D.X.; Chen, X.; Huang, G.Q. IoT-based real-time production logistics synchronization system under smart cloud manufacturing. *Int. J. Adv. Manuf. Technol.* **2016**, *84*, 147–164. [CrossRef]
6. Kang, H.S.; Lee, J.Y.; Choi, S.; Kim, H.; Park, J.H.; Son, J.Y.; Kim, B.H.; Noh, S.D. Smart manufacturing: Past research, present findings, and future directions. *Int. J. Precis. Eng. Manuf.-Green Technol.* **2016**, *3*, 111–128. [CrossRef]
7. Lucke, D.; Constantinescu, C.; Westkämper, E. Smart Factory—A Step towards the Next Generation of Manufacturing. In Proceedings of the Manufacturing Systems and Technologies for the New Frontier, 41st CIRP Conference on Manufacturing Systems, Tokyo, Japan, 26–28 May 2008.
8. Marvel, J.A.; Falco, J.; Marstio, I. Characterizing Task-Based Human–Robot Collaboration Safety in Manufacturing. *IEEE Trans. Syst. Man Cybern. Syst.* **2015**, *45*, 260–275. [CrossRef]
9. Dillmann, R.; Friedrich, H. Programming by demonstration: A machine learning approach to support skill acquisition for robots. In *Artificial Intelligence and Symbolic Mathematical Computation*; Calmet, J., Campbell, J.A., Pfalzgraf, J., Eds.; Lecture Notes in Computer Science; Springer: Berlin, Germany, 1996; Volume 1138, pp. 87–108.
10. Monostori, L. Cyber-physical Production Systems: Roots, Expectations and R&D Challenges. *Procedia CIRP* **2014**, *17*, 9–13. [CrossRef]
11. Tsarouchi, P.; Makris, S.; Chryssolouris, G. Human–robot interaction review and challenges on task planning and programming. *Int. J. Comput. Integr. Manuf.* **2016**, *29*, 916–931. [CrossRef]
12. Gerkey, B.P.; Matarić, M.J. A Formal Analysis and Taxonomy of Task Allocation in Multi-Robot Systems. *Int. J. Robot. Res.* **2004**, *23*, 939–954. [CrossRef]
13. Bauer, A.; Wollherr, D.; Buss, M. Human-robot collaboration: A survey. *Int. J. Hum. Robot.* **2008**, *05*, 47–66. [CrossRef]

14. Erasmus, J.; Vanderfeesten, I.; Traganos, K.; Grefen, P. The Case for Unified Process Management in Smart Manufacturing. In Proceedings of the 22nd International Enterprise Distributed Object Computing Conference (EDOC), Stockholm, Sweden, 16–19 October 2018; pp. 218–227.
15. *Worker at Volkswagen Plant Killed in Robot Accident*; Associated Press: New York, NY, USA, 2015.
16. Heyer, C. Human-robot interaction and future industrial robotics applications. In Proceedings of the IEEE/RSJ International Conference on Intelligent Robots and Systems, Taipei, Taiwan, 18–22 October 2010; pp. 4749–4754.
17. Vanderfeesten, I.; Erasmus, J.; Grefen, P. The HORSE Project: IoT and Cloud Solutions for Dynamic Manufacturing Processes. In Proceedings of the ESOCC 2016 Workshops, Vienna, Austria, 5–7 September 2016; Volume 9846, pp. 303–304.
18. Grefen, P.; Vanderfeesten, I.; Boultadakis, G. Supporting Hybrid Manufacturing: Bringing Process and Human/Robot Control to the Cloud (Short Paper). In Proceedings of the 5th IEEE International Conference on Cloud Networking (Cloudnet), Pisa, Italy, 3–5 October 2016; pp. 200–203.
19. van Aken, J.E. Management Research Based on the Paradigm of the Design Sciences: The Quest for Field-Tested and Grounded Technological Rules. *J. Manag. Stud.* **2004**, *41*, 219–246. [CrossRef]
20. Gregor, S.; Hevner, A.R. Positioning and Presenting Design Science Research for Maximum Impact. *MIS Q.* **2013**, *37*, 337–356. [CrossRef]
21. Hevner, A.R.; March, S.T.; Park, J.; Ram, S. Design Science in Information Systems Research. *MIS Q.* **2004**, *28*, 75–105. [CrossRef]
22. Kruchten, P.B. The 4 + 1 View Model of architecture. *IEEE Softw.* **1995**, *12*, 42–50. [CrossRef]
23. *Reference Architecture Model Industrie 4.0*, 1st ed.; DIN Deutsches Institut für Normung: Berlin, Germany, 2016.
24. Hankel, M.; Rexroth, B. *The Reference Architectural Model Industrie 4.0 (RAMI 4.0)*; German Electrical and Electronic Manufacturers' Association: Frankfurt am Main, Germany, 2015.
25. Chen, D. Enterprise-control system integration—An international standard. *Int. J. Prod. Res.* **2005**, *43*, 4335–4357. [CrossRef]
26. Kulkarni, V.; Reddy, S. Separation of concerns in model-driven development. *IEEE Softw.* **2003**, *20*, 64–69. [CrossRef]
27. Garcia, A.; Sant'Anna, C.; Chavez, C.; da Silva, V.T.; de Lucena, C.J.P.; von Staa, A. Separation of Concerns in Multi-agent Systems: An Empirical Study. In *Software Engineering for Multi-Agent Systems II*; Lucena, C., Garcia, A., Romanovsky, A., Castro, J., Alencar, P.S.C., Eds.; Lecture Notes in Computer Science; Springer: Berlin, Germany, 2004; Volume 2940, pp. 49–72.
28. Moreira, A.; Rashid, A.; Araujo, J. Multi-dimensional separation of concerns in requirements engineering. In Proceedings of the International Conference on Requirements Engineering, Paris, France, 29 August–2 September 2005; pp. 285–296.
29. Grefen, P.; Vanderfeesten, I.; Boultadakis, G. *D2.2a—HORSE Complete System Design—Public Version*; HORSE Consortium: Brussels, Belgium, 2017; p. 91.
30. Grefen, P.; Vanderfeesten, I.; Boultadakis, G. Developing a Cyber-Physical System for Hybrid Manufacturing in an Internet-of-Things Context. In *Protocols and Applications for the Industrial Internet of Things*; González García, C., García-Díaz, V., García-Bustelo, B.C.P., Lovelle, J.M.C., Eds.; Advances in Business Information Systems and Analytics; IGI Global: Hershey, PA, USA, 2018; ISBN 978-1-5225-3805-9.
31. Bass, L.; Clements, P.; Kazman, R. *Software Architecture in Practice*, 3rd ed.; SEI Series in Software Engineering; Addison-Wesley: Upper Saddle River, NJ, USA, 2013; ISBN 978-0-321-81573-6.
32. Erasmus, J.; Vanderfeesten, I.; Traganos, K.; Jie-A-Looi, X.; Kleingeld, A.; Grefen, P. A method to enable ability-based human resource allocation in business process management systems. In *Proceedings of PoEM 2018*; Lecture Notes in Business Information Processing; Springer: Vienna, Austria, 2018; Volume 335, pp. 37–52.
33. Polderdijk, M.; Vanderfeesten, I.; Erasmus, J.; Traganos, K.; Bosch, T.; van Rhijn, G.; Bennet, D. A Visualization of Human Physical Risks in Manufacturing Processes using BPMN. In Proceedings of the BPM 2017 Workshops, Barcelona, Spain, 10–15 September 2017.
34. Shawish, A.; Salama, M. Cloud Computing: Paradigms and Technologies. In *Inter-Cooperative Collective Intelligence: Techniques and Applications*; Xhafa, F., Bessis, N., Eds.; Studies in Computational Intelligence; Springer: Berlin, Germany, 2014; pp. 39–67, ISBN 978-3-642-35016-0.

35. Rimal, B.P.; Jukan, A.; Katsaros, D.; Goeleven, Y. Architectural Requirements for Cloud Computing Systems: An Enterprise Cloud Approach. *J. Grid Comput.* **2011**, *9*, 3–26. [CrossRef]
36. Chang, S.-C.; Yang, C.-L.; Cheng, H.-C.; Sheu, C. Manufacturing flexibility and business strategy: An empirical study of small and medium sized firms. *Int. J. Prod. Econ.* **2003**, *83*, 13–26. [CrossRef]
37. Mishra, R.; Pundir, A.K.; Ganapathy, L. Manufacturing Flexibility Research: A Review of Literature and Agenda for Future Research. *Glob. J. Flex. Syst. Manag.* **2014**, *15*, 101–112. [CrossRef]
38. Atzori, L.; Iera, A.; Morabito, G. The Internet of Things: A survey. *Comput. Netw.* **2010**, *54*, 2787–2805. [CrossRef]
39. Wortmann, F.; Flüchter, K. Internet of Things: Technology and Value Added. *Bus. Inf. Syst. Eng.* **2015**, *57*, 221–224. [CrossRef]
40. Lee, H.L.; Whang, S. Information sharing in a supply chain. *Int. J. Manuf. Technol. Manag.* **2000**, *1*, 79–93. [CrossRef]
41. Gerhard, J.F.; Rosen, D.; Allen, J.K.; Mistree, F. A Distributed Product Realization Environment for Design and Manufacturing. *J. Comput. Inf. Sci. Eng.* **2001**, *1*, 235–244. [CrossRef]
42. Grefen, P.; Mehandjiev, N.; Kouvas, G.; Weichhart, G.; Eshuis, R. Dynamic business network process management in instant virtual enterprises. *Comput. Ind.* **2009**, *60*, 86–103. [CrossRef]
43. Grefen, P.; Eshuis, R.; Mehandjiev, N.; Kouvas, G.; Weichhart, G. Internet-based support for process-oriented instant virtual enterprises. *IEEE Internet Comput.* **2009**, *13*, 65–73. [CrossRef]
44. Grefen, P. Networked Business Process Management. *Int. J. IT/Bus. Alignment Gov.* **2013**, *4*, 54–82. [CrossRef]
45. Weyer, S.; Schmitt, M.; Ohmer, M.; Gorecky, D. Towards Industry 4.0—Standardization as the crucial challenge for highly modular, multi-vendor production systems. *IFAC-PapersOnLine* **2015**, *48*, 579–584. [CrossRef]
46. Qin, J.; Liu, Y.; Grosvenor, R. A Categorical Framework of Manufacturing for Industry 4.0 and Beyond. *Procedia CIRP* **2016**, *52*, 173–178. [CrossRef]

Article

Transformation towards a Smart Maintenance Factory: The Case of a Vessel Maintenance Depot

Gwang Seok Kim * and Young Hoon Lee

Department of Industrial Engineering, Yonsei University, 50 Yonsei-ro, Seodaemun-gu, Seoul 03722, Korea; youngh@yonsei.ac.kr
* Correspondence: kimks747@yonsei.ac.kr

Abstract: The conceptualization and framework of smart factories have been intensively studied in previous studies, and the extension to various business areas has been suggested as a future research direction. This paper proposes a method for extending the smart factory concept in the ship building phase to the ship servicing phase through actual examples. In order to expand the study, we identified the differences between manufacturing and maintenance. We proposed a smart transformation procedure, framework, and architecture of a smart maintenance factory. The transformation was a large-scale operation for the entire factory beyond simply applying a single process or specific technology. The transformations were presented through a vessel maintenance depot case and the effects of improvements were discussed.

Keywords: smart factory; smart process transformation framework; smart maintenance architecture; smart maintenance factory

Citation: Kim, G.S.; Lee, Y.H. Transformation towards a Smart Maintenance Factory: The Case of a Vessel Maintenance Depot. *Machines* **2021**, *9*, 267. https://doi.org/10.3390/machines9110267

Academic Editors: Zhuming Bi, Li Da Xu and Puren Ouyang

Received: 31 August 2021
Accepted: 30 October 2021
Published: 2 November 2021

Publisher's Note: MDPI stays neutral with regard to jurisdictional claims in published maps and institutional affiliations.

1. Introduction

A smart factory is a production plant where the pillars of Industries 4.0 are implemented, including additive manufacturing (3D printing), augmented reality (AR), Internet of Things (IoT), big data analytics, autonomous robot, simulation, cyber-security, vertical and horizontal integration, and cloud computing [1]. The concept of a smart factory has become a keyword of manufacturing sites along with the technological development during the fourth industrial revolution.

Hyundai Heavy Industries, Daewoo Shipbuilding & Marine Engineering Co., Ltd., and Samsung Heavy Industries are the three major global ship manufacturers in Korea that lead smart ship and yard constructions by applying the new technologies of the fourth industrial revolution and utilizing IT systems [2]. In the report by SPAR Associates. Inc., which have been serving in the shipbuilding and repairing industry for over 45 years, about 23% of the total cost is ship acquisition cost, approximately 35% is labor cost, and the remaining 42% is repairing and maintenance in the Naval Ship Life Cycle Cost (LCC) Model [3]. Although the growth of a smart factory is limited to the ship building phase, shipbuilders are interested in further expanding the smart factory concept to the service phase in order to expand the scope of the maintenance, repair, and operations (MRO) business. It is necessary to expand the research on manufacturing-oriented smart factory research to cover the entire lifecycle.

This study takes focus on a practical case that expands the entire ship lifecycle from build phase to service phase, as shown in Figure 1.

The adaptability of the concept of a smart factory to a repair and maintenance factory (hereafter referred to as a maintenance factory) requires an understanding of the differences at work. Ship building is the process of assembling modules produced in factories and yards in accordance with the job schedule, whereas repair and maintenance tasks usually do not follow a fixed assembly process plan. Repair and maintenance work has the following variabilities:

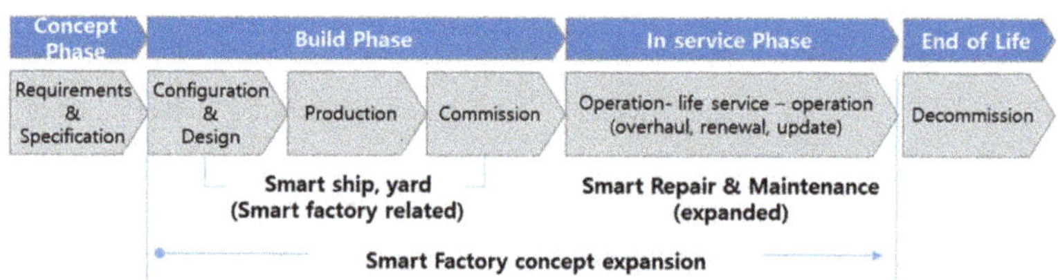

Figure 1. Ship lifecycle and smart factory expansion.

Variability in the plan: Ship building is progressed step by step following the working plan, and the supply of each relevant component is proceeded on a planned basis. In contrast, unplanned repair and maintenance requirements frequently occur since failures of ships are unpredictable. Moreover, repair and maintenance plans may vary due to the dock's idle state because ships must be drawn on the dry dock.

Variability in the process: The process of ship building is usually done at fixed workplaces and workstations. However, the process of repair and maintenance include: draw the ship to the dry dock—take away parts from the ship—taking parts out from the ship—moving to the factory—disassembling parts—cleaning—repairing—reassembling—testing performance—moving from the factory to the ship—taking in to the ship—assembling on the ship—trial run. The required jobs in the repair and maintenance process to repair/replace defective parts and improve performance are different from jobs in the ship building process.

Variability in the time: Planned demand and standard working hours are set for production and manufacturing. However, the work time of repair and maintenance varies depending on the degree of failure and repair requirements. Because the delivery date of each part is different (e.g., discontinued parts or parts that have a long delivery period), repair and maintenance are hard to complete in a timely manner.

Variability in the workplace: In the case of ship manufacturing, the parts are first assembled in the factory and then at the yard. On the other hand, repair and maintenance are not only proceeded in the factory; the process may be done on the ship or through remote maintenance system as needed.

The purpose of this paper is to present a method to extend the smart factory concept of the build phase to the service phase through empirical examples. Smart transformation procedure, the framework, and architecture were developed by repeating revisions and improvements in the process of establishing a smart transformation plan for about one year.

The four contributions of this study are as follows. First, this study provides the transformation procedure for a smart maintenance factory. Second, this study developed a smart process transformation framework for building a smart maintenance factory to improve highly volatile processes. Smart transformation occurs when site workers understand strategies and voluntarily draw actual changes. To this end, this study developed a practical template for site workers to participate in change and present their opinions. Third, this study proposed the architecture of a smart maintenance factory that shows the future look of a smart maintenance factory. This study suggested the value and technology for transforming an existing factory to a smart repair and maintenance factory while giving consideration to the characteristics of the process. Finally, smart transformation was implemented for the vessel maintenance depot. Transformation is a large-scale operation that changes the entire factory beyond simply applying a single process or specific technology.

The remainder of this paper is composed as follows. Section 2 reviews previous studies, Section 3 introduces the materials and methods, Section 4 presents the application cases, and in Section 5, we discuss the results, address the conclusions, and suggest directions for further research.

2. Literature Review

This study reviewed previous studies and categorized them into a practical application and research extension from ship building phase to ship servicing phase. Research of the building phase could be divided into smart manufacturing for the production process, smart management by information, and smart maintenance for 5M + 1E (Man, Machine, Material, Method, Measurement, and Environment), which supports production, as shown Figure 2.

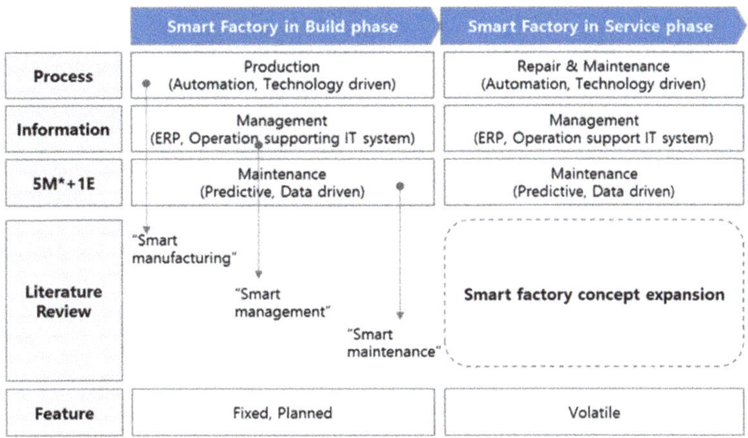

Figure 2. Research on smart factory and expansion.

Among numerous studies conducted in each area, the concept, the recent direction, and research on the implementation will mainly be analyzed.

Smart manufacturing is a term coined by several agencies such as the Department of Energy (DoE) and the National Institute of Standards and Technology (NIST) in the United States. It highlights the use of information and communication technology (ICT) and advanced data analytics to improve manufacturing operations on the shop floor [4,5], factory [6], and supply chain [7,8].

Smart manufacturing incorporates various technologies, including cyber-physical production systems (CPPS), IoT, robotics/automation, big data analytics, and cloud computing to realize a data-driven, connected supply network [9]. Intelligent manufacturing has often been used synonymously with smart manufacturing. Technologies and enabling factors associated with smart manufacturing were reviewed to compare the differences between smart manufacturing and intelligent manufacturing [4,10,11]. Studies on the features and availability of smart manufacturing and intelligent related technologies were carried out [12]. Compared to smart manufacturing, intelligent manufacturing focuses more on the technological aspect and less on the organizational aspect.

The opportunities and organizational issues to be considered during smart transformation were analyzed for SMEs [13]. A smart manufacturing performance measurement system was introduced based on exploratory and empirical research [14]. According to these studies, a guide for investment in smart manufacturing was presented to ensure the validity of investment. This study used a top-down approach to introduce the specific implementation scheme from top-level planning for industrial practice in smart manufacturing [15].

By referring to the studies of smart maintenance in the build phase, the applicability and implications for the service phase were examined. We analyzed the latest research on a smart maintenance concept, the changing trend of the existing IT system, recent intensive predictive maintenance studies, and maintenance architecture. Smart maintenance was defined as based on four main components through expert interviews and preliminary research on manufacturing plants. The four main elements are data-driven decision-making, human capital resources, internal integration, and external integration [16]

The Enterprise Resource Planning (ERP) system, which manages materials, budget, and planning, and the Manufacturing Execution System (MES), which manages field data, are the two representative factory management systems to support production. When a traditional factory is transformed into a smart factory, the existing ERP and MES systems would be advanced by applying cloud technology, big data-based data diagnostic, and AI technology. The existing information management systems of smart factory processes require more flexibility and a larger volume of data, which needs to connect to related systems, to analyze, and to visualize the activities inside the organization [14,17,18]. The introduction of AI systems that support decision making by making knowledge of the production process between each system and data collected from various smart devices is spreading [15,19].

In addition to academic research, the technology trends suggested by the operators in the maintenance industry were analyzed and reflected in this study. The five transformational Trends Reshaping Industrial Maintenance [20] are: (1) Additive (3D printing) Manufacturing in Maintenance, U.S. Department of Transportation to publish a notice which aims at raising awareness about the use of Additive Manufacturing in the maintenance, and preventive maintenance areas; (2) Internet of Things, Wireless Sensor Networks, and IoT-based automated data collection increases workers' productivity; (3) Augmented Reality (AR) for Training and Remote Maintenance. AR is enabling new paradigms for maintenance, including remote maintenance and maintenance customized to the workers' under-standing and skills; (4) Maintenance as a Service (MaaS) could become a game changer in industrial maintenance. It can motivate machine vendors to provide the best service while also providing versatile, reliable, and functional equipment; (5) Supply Chain Collaboration. Streamlining the supply chain management information has been proven to be extremely beneficial for industrial maintenance as well, as it reduces the delivery times for parts. At the same time, supply chain operators benefit from maintenance insights, such as predictive maintenance.

The other operator of Manufacturer [21] stated that ERP and MES should be data-driven management solutions to support Industry 4.0 as follows: (1) Modern ERP systems must be built fundamentally different from the ground up; (2) The ERP system should also be architected to interact with external systems with application programming interfaces (APIs) available for any and all entities of the system; (3) IoT should be able to make almost any product a smart, connected product. IoT operations support traditional MES for shop floor automation and control adding the flexible communication and data collection protocols. The research theme of smart maintenance is changing from Time Based Management (TBM) to Condition-Based Maintenance (CBM) [22] for predictive maintenance and research on optimizing preventive intervention through CBM [23]. Studies on collaborative-based architecture [24], web platform [25], decision support [26], sensor and data analysis for predictive maintenance were discussed for CBM [27].

A study on the basics for designing the maintenance process of Industry 4.0 [28] and a study on performance and KPI design are also presented [29]. SMEs do not exploit all the resources for implementing Industry 4.0 and often limit themselves to the adoption of Cloud Computing and the IoT [30,31]. These limitations will be experienced not only in this study but also in most organizations.

In order to expand the smart factory concept to the service phase, practical application research, and applications were emphasized [6,7,32]. The difficulties in applying the concept of a production-oriented smart factory to a smart factory in the maintenance area were discussed by site workers. The staff of vessel maintenance depot stated the limitations in applying smart factory to the field of repair and maintenance sector. "Despite the rapid growth of technology due to the fourth industrial revolution, smart factories are applied in the maintenance field, not the manufacturing field, but in reality, there are technical limitations. Since nobody has ever done it, no one has been able to strongly suggest to what extent it should be specified or what is needed to realize the future of the smart vessel maintenance depot. [33], (p. 30)."

We reviewed the smart factories for the ship building phase, including the smart manufacturing, management, and maintenance of a smart factory and the need for research expansion in the service phase. The previous studies on the procedure, modeling, and architecture to build a smart maintenance factory were also reviewed.

Reference [34] presented a step-by-step methodology for the efficient planning of a smart factory from the initial idea to the final realization in the real environment. The construction process from design, smart machining, smart monitoring, smart control, smart scheduling, industrial applications, and the phases of data utilization were introduced [35]. The importance of stepwise starting from a small scale were argued [36,37]. The new technology application in a limited area were demonstrated by building a smart factory [38]

Existing studies introduced different models for the smart factory model. Reference [39] proposed a human-centered model, [40] suggested IoT-based, [41] proposed IoT and cloud computing, and [35] proposed a cloud-based control system as smart control systems. Operation values are defined based on the environment and work of the maintenance factory. Reference [42] provided an overview of these principles in terms of the general scope of Industry 4.0, and [7] investigated and analyzed the principles of a smart factory and proposed modularity, interoperability, decentralization, virtualization, service orientation, and real-time capability. Reference [36] suggested features of connectivity, optimization, transparency, proactivity, and agility for smart factory construction.

Different studies on smart factory architecture were presented. Reference [24] introduced a data-based smart maintenance architecture and reference model to perform predictive equipment maintenance in a factory. Reference [43] described the three major criteria of the general system architecture, including mechatronic changeability, individualized mass production, and internal/external networking. Product, production layer, supply layer, integration layer, and IT are defined as the five layers. Reference [44] emphasized the integration of industrialization and informatization as the core of China's smart manufacturing implementation strategy and proposed the standards framework and reference architecture of a smart factory. Reference [45] presented a technical architecture and argued that the interoperability of the systems or components of the architecture at every level is imminent. References [46–48] proposed the hierarchical architecture of a smart factory including four layers, namely the physical resource layer, network layer, data application layer, and terminal layer.

3. Materials and Methods

As suggested in previous studies, starting on a small scale [36,37] or limited area [38] seems to be an effective way to minimize risk, but there may be situations in which it is stopped or scaled down due to various obstacles in the process. Procedures for smart conversion of specific tasks are limited in their application of transforming the entire plant. Few studies suggested a method for the procedure for building a smart maintenance factory in the field. We developed a hybrid procedure that combines the top-down procedure that presents the future image and goals of a smart maintenance factory and the bottom-up procedure that reflects on the site conditions and requirements for changes. After approximately one year of revision and improvement, this procedure was eventually developed as shown in Figure 3

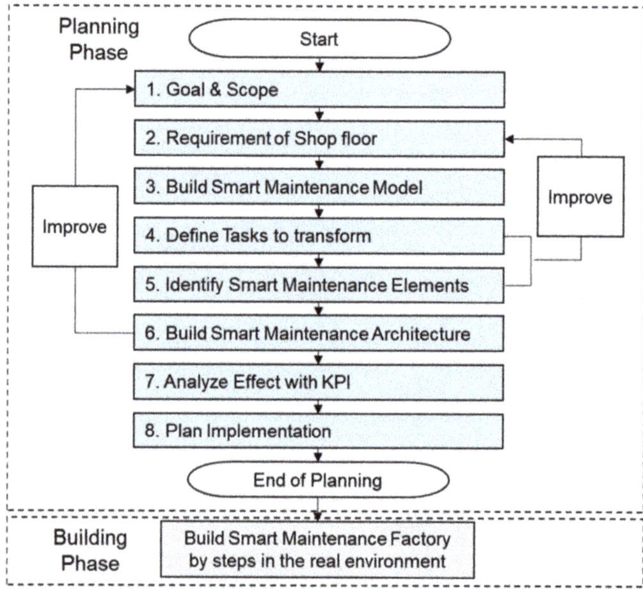

Figure 3. Implementation procedure.

(1) Goal and Scope

It is desirable to set the goal of a smart maintenance factory based on the organization's missions or policies, instead of the goal of an individual task or technology. In general, the primary goal of maintenance factory management is to minimize variability, which is a characteristic of maintenance, in order to complete the maintenance on time. The safety of workers, jobs on the site, and the quality of maintenance could be ensured. Another important goal is to increase the availability of ships. As a large-scale investment, increasing availability is compared to increasing the value of investment.

(2) Requirements of Shop Floor

The requirements of transforming each workplace of the maintenance factory to a smart factory are analyzed for selecting tasks to be improved. Such transformation requirements are handled systematically based on the smart process transformation framework developed. The characteristic of the smart process transformation framework presented in this study is that it leads to changes in the repair and maintenance process through the following three elements of smart transformation: data, system, and automation (technology).

While the traditional approach derived the necessary elements of data, system, and automation through the direction of process change, the smart process transformation framework in this study intends to lead process change through actual technologies of the fourth industrial revolution. In addition, based on long-term field analysis and opinions from site workers, this study identified that data, IT systems, and automation (technology) take the lead in major changes in smart maintenance plants and thus reflected them in the framework.

As illustrated in Figure 4, the smart process transformation framework is composed of data, systems, and automation (technology) that induce process change. The smart transformation strategy should be defined with KPI and the infrastructure and communication network should be selected to support the smart transformation. The organizational capabilities required for the smart transformation operation should also be planned.

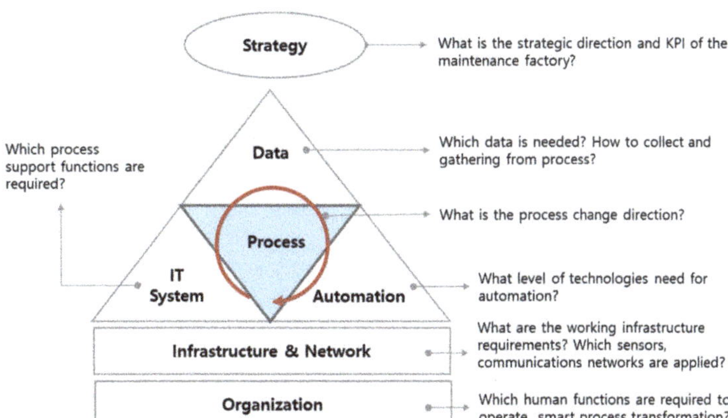

Figure 4. Smart process transformation framework.

Ongoing effort and time are required to obtain the improvement needed from field managers through the smart process transformation framework. Getting new opinions is not easy since most workers and managers are immersed in existing work methods. Thus, it is necessary to provide a specific and convenient template that shows the transformation direction to figure out the challenges to change the viewpoints of site workers and managers. The template effectively helps derive the requirements of the site and encourage participation of the working site. Figure 5 illustrates the template for concretization of changes on the site, which reflects the elements of the smart process transformation framework.

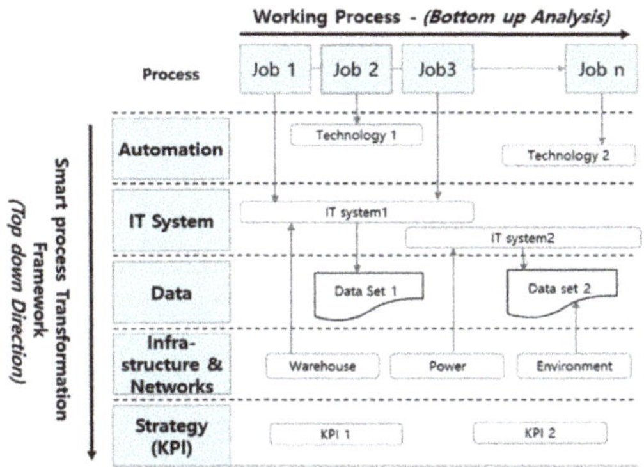

Figure 5. Template for analysis of on-site requirements.

The template consists of the working process on the horizontal axis and the smart transformation framework shown in Figure 4 on the vertical axis. The requirements for jobs constituting the process are classified and organized into automation, IT system, data, operating infrastructure and communication network, and strategy. Automation describes smart technologies and equipment needed for smart transformation. In the IT system, the requirements and functions of the operation support system that support the job are described. Data describes the required data set and format. Support infrastructure requirements such as warehouse and power system are described as infrastructure, and requirements for sensors and communication networks are described in network. Strategies are described in connection with improvement effects in each job and KPIs. In addition, deletion, and integration of jobs in the process are also described.

Workers on the site are sensitive to organizational changes and may consider the transformation towards a smart maintenance factory a negative effect on job stability [30] It is a reaction that can be easily felt in the field. Thus, it is recommended not to ask workers on the site for opinions on changes in organization and manpower.

(3)   Building Smart Maintenance Model

The model of a smart maintenance factory is according to the requirements on changes to be made in strategic direction and operational values. Studies have introduced different models such as a human-centered model [39], IoT-based model [40], and cloud computing [35,41] for a smart factory. Principles were suggested, such as modularity, interoperability, decentralization, virtualization, service orientation, real-time capability, connectivity, optimization, transparency, proactivity, and agility for smart factory construction [7,36].

The smart maintenance factory model consists of missions, a target model, operational values, and the technology enabling the technologies of the fourth industry revolution for process transformation. The target model of a smart maintenance factory can be defined in various shapes by integrating the as-is state and strategic direction. When defining the smart maintenance model, it is important to consider how to respond to variability, which is a characteristic of maintenance work. Agile response, predictive, and readiness should be defined as core operating values in order to respond to changes in the work plan, process, time, and workplace of repair and maintenance work.

As an enabling technology for realizing operational values, it enables agile response to changes on site through central control and mobile network. For predictive maintenance, operational data of the vessel are collected and analyzed by a big data system. In addition, inventory and work readiness are promoted through on-site operational support IT systems and smart warehouses. These operating values and required enablers can be derived through on-site interviews.

(4)   Define Tasks to Transform

Tasks define the activities required for smart factories to achieve high performance, from policy, process improvement, data improvement, system construction, infrastructure improvement, and organizational change. Defining tasks is necessary in connection with the workload, schedule, budget, and expected effects to be pursued in the future, so an organizational consensus process is required through several meetings. After defining the task, the selection of new technologies and targeting levels to implement the task should be followed.

(5)   Identify Smart Maintenance Elements

It is a step to define specific technological enablers to implement smart transformation with the selected tasks. Enablers consist of changes in maintenance policies, work standards, and the selection of appropriate technologies. The technology enablers consist of ICT systems for managing the working process and necessary 5M + 1E, automation, data collection, data system for management and analysis, a network that connects data to the entire maintenance factory, a central control system that manages it in real time, and a smart factory infrastructure.

Selecting the right technologies is one of the main tasks for the smart maintenance element. Reference [49] reviewed research on technology selection over the past 20 years to find out various new technologies needed for manufacturing and various methods for technology selection due to the emergence of the fourth industrial revolution. In selecting Industry 4.0 technology, it is recommended that long-term development or selection of unstable technology be avoided. The hasty or wrong technology selection will weaken the overall work productivity of the plant. It is recommended to apply validated and reliable innovative technologies to the field at the initial stage, and to choose a business operator that has sustainable development and maintenance capabilities. As an advanced work, a written confirmation on the time and degree of the technology application should be submitted for applying a particular technology that is required.

The required technologies and the purpose of utilizing technologies for smart maintenance are referred from smart manufacturing. Reference [50] insisted on end-to-end ICT-based integration between the manufacturing technologies of smart machines, warehousing systems, and production facilities that have developed digitally and feature end-to-end ICT-based integration. Reference [34] mentioned equipment, a cloud-based control system, communication network for real-time data collection and control, and the importance of power monitoring to secure stable operation.

(6)   Build Smart Maintenance Architecture

By configuring the architecture based on the operational values, the future shape of the smart maintenance factory and its implementation technology are identified. The architecture of a smart maintenance factory presents the direction of technical realization for constructing the smart maintenance factory. The architecture may be presented differently by point-in-time, by factory, or by composition direction.

Studies have introduced the data-based smart maintenance architecture [24], the internal/external networking architecture [43], the integration of industrialization and informatization architecture [44], the interoperability technical architecture [45], and the hierarchical architecture of smart factory [46–48].

This study set one offsite layer and six onsite layers to classify the values and required technological elements. The feature of the smart maintenance factory architecture in this study contains seven layers according to the implementation characteristics, and maintenance was divided into factory maintenance and remote maintenance of vessel according to the required services and technologies. It also expressed the connection between business management and field management as shown in Figure 6.

Layers		Values	Systems and Technological Enablers		
Business O&M		Business Value	ERP and Management System		Off-Site
Control Layer			Deport (Factory) Control System	Remote Control System	
Intelligent Layer			On-site Operating Supporting System	Remote Maintenance Supporting System	
Data Layer		Operational Values of On-Site	Data Diagnostic System	Data Monitoring System	On-Site
Automation Layer			Industry 4.0 Technology Application	Industry 4.0 Technology Application	
Network Layer			IoT & Communication Network	IoT & Communication Network	
Shop Flow	Process Layer		Working process	Part Repairing Manual	
	Infrastructure Layer		Common Working Environments	Machines and Materials on Board	

Figure 6. Architecture of a smart maintenance factory.

The first layer, Business O&M, is composed of ERP components that emphasize management efficiency. It is composed of systems that share information through the organization, such as scheduling, human resource, budgeting, procurement, etc. The other six layers are field-oriented functions. Operational values such as visibility, stability, and speed should be set to guide the design direction of each layer.

The control, intelligence, data, and network layers transform the visualized and data-driven maintenance factory through close interlocking. The control layer manages risks and fluctuations in the schedule and environment through the central control screen. In the intelligent layer, the field workers want to secure the process visibility of their jobs through the operation support system. In the data layer, the maintenance process is identified by sharing specific data on-site. To this end, it consists of technical elements of data standardization, collection, and sharing. For remote maintenance, the operational conditions are monitored through remote diagnostic and operational supporting system.

In the automation layer, productivity and safety are focused. It is composed of industry 4.0 technologies applicable to each workplace to secure safety and productivity. The network layer emphasizes connectivity for collecting data and sharing information of workers and materials through wireless and wire. The remote maintenance is connected to the vessel via satellite.

The shop flow emphasizes speed, quality, and safety. The shop flow layer consists of the actual work processes and operational infrastructures. For remote maintenance, the vessel is considered as the shop floor.

(7)   Analyze Effect with KPIs

KPIs and effect analysis are the high priority parts of executing Industry 4.0 technologies in tasks and securing budgets. Reference [51] pointed out the direction, the contents of the role, and the importance of key performance indicators of application Industry 4.0 technologies. Reference [52] demonstrated the expected effect of building a smart factory and analyzed the effect of smart factory adoption with an empirical analysis based on a sample representing local manufacturing units.

It is important to define KPIs based on the missions and operational value of the maintenance factory. Although it is appropriate to present KPIs and the improvements of a single task if a single task-oriented process is applied, it is essential to select the expected effects and KPIs for the maintenance factory if the entire factory is transformed step by step, as presented in this study. Linking KPIs with the expected effects is an operation optimization plan that is done after the establishment of a smart maintenance factory by aligning strategy and implementation.

(8)   Plan Implementation

There are two ways to construct a smart maintenance factory. The first way is to prioritize technology development to evaluate performance and expected effects, and then apply them to the workplace on a technical basis. It is to expand according to the technology development stage. The other way is to select a specific workplace as a pilot, apply all technologies that are under consideration, and then evaluate the performance and deploy the technologies horizontally to other work processes. It is to expand by the organization or process unit.

While the first way has the disadvantage that it requires a long transformation time, failure of technology may also cause failure of the smart transformation. The second way is preferred if the smart maintenance factory needs to constructed in the short term, because members of the organization are guided to participate in the processes, and improvements or difficulties can be identified immediately in the field. Showing step-by-step outputs will be an important factor that accelerates the implementation of smart transformation to ensure enough budgets throughout the process. In the implementation phase, the priority of transformation tasks will be determined according to budget size and urgency.

4. Application

The vessel maintenance depot was operated by the owner of a vessel established about 70 years ago, and it is in the process of smart transformation from mechanical and hydraulic maintenance to 3D printers, articulated lifts, and pilot-level remote maintenance. Currently, the smart factory is evaluated as Level 2.0 of mechanization and computerization.

In the vessel maintenance depot, vessels are towed to the dry dock and the broken parts are separated from the vessel to be repaired in the factory. Vessels are sent to the maintenance depot for regular maintenance and emergency repair due to failure. This study derived the requirements for building a smart maintenance factory from more than 20 maintenance processes. The field issues to be improved through smart maintenance factory construction are shown in Table 1 below.

Table 1. Issues to be transformed in the vessel maintenance depot.

Classification	Issues On Site	Improvement Direction
Process	Need to improve chronic delay in repair parts.	Predictive Maintenance
	Need to improve poor linkage between schedules	
	Need fine-grained management of worker and working hours	
	Avoid concentration on maintenance work at a specific time	
	Neutralization of planned schedules by sudden maintenance	
Data	Need information system to share and collect data on site	Data Diagnostic Management
	Data-based analysis and managing on-site data are required for predictive maintenance	
	Need to identify maintenance history on the site	
System	Maintenance knowledge should be secured	Onsite Operation Supporting /Monitoring
	Insufficient support for the on-site maintenance process by the existing system	
	Integrated control and work monitoring system are needed on site	
	Real-time monitoring system is required for safety management	
	Need to check equipment status information for timely maintenance	
	Supporting of remote maintenance are increasing	
Automation	Need automation for work safety and reducing work lord	Automation (safety and productivity)
	Need to improvement of workability, convenience, and safety	
	Insufficient quality control and deterioration of quality control ability	
	Technology required to perform precise tasks such as positioning a ship	
	Precise measurement system required for direct production and quality assurance of manufactured products	
	Technology and equipment required to support remote maintenance	
Network	Need for mobile device, sensors, and network for data entry and automatic data collection in the field	IoT & Mobile
Infrastructure	Emphasis on additional and emergency power supply system	Safety and Health
	Warehouse needed for immediate supply of inventory	
	Removal of hazardous substances such as hazardous gas, waste oil, and dust during maintenance.	

The transformation of a smart maintenance factory is to establish a predictive maintenance process that is to respond to changes in demands and plans above all else. To this end, the current task-oriented management would be transformed to data-oriented management. The central operation supporting system of the management level would be transited to the on-site supporting system including the monitoring system. The automation was expected to reduce the burden of moving heavy parts and to improve work quality from poor precision. The entire maintenance depot would be connected and transformed into a mobile workplace. A safe and healthy work environment was also required in order to respond to environmental regulations. The research in this paper belongs to the design stage, not the implementation stage, and the simulation of the expected result is presented as a goal of improvement.

A diagnosis of the current factory at the smart factory level was conducted through years of learning, and the transformation from the existing factory to a smart maintenance factory was confirmed. Nonetheless, the managers of the vessel maintenance depot intended to carry forward the smart transformation of the existing factory, but they are facing practical difficulties in planning to build the smart maintenance factory [33].

(1)   Goal and Scope

This goal has been formulated as a unique mission of the maintenance depot. The goal of the maintenance depot is to maintain and improve performance during the life cycle of a vessel, and to improve the productivity of maintenance factory.

(2)   Requirement of Shop floor

For each process, field workers were able to present automation, system, data, infrastructure, and expected effects according to the provided template. In addition, some processes were able to be integrate in response to changes. Figure 7 illustrated the practical requirements that were derived by using templates of the smart process transformation framework.

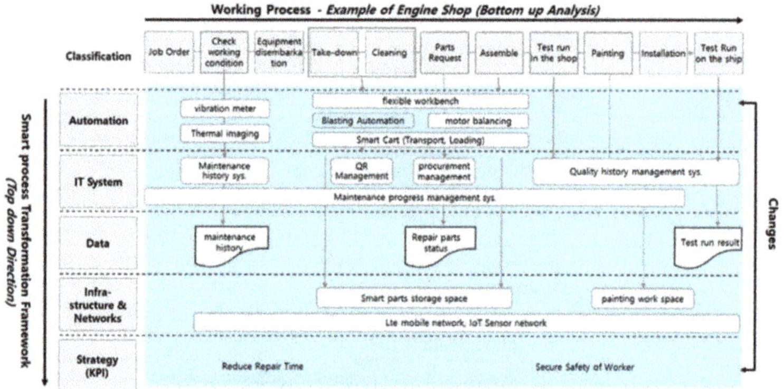

Figure 7. Example of requirements analysis for smart transformation.

(3)   Build Smart Maintenance Model

The maintenance depot requests an on-site smart maintenance factory model that can support and manage tasks at the site. The on-site smart maintenance factory is no longer equipment-oriented. Instead, a data-and technology-centered construction direction was born after repeated discussions and strategic presentations. With consideration of the requirements and strategic direction of the derived smart transformation, the strategic model for constructing a smart maintenance factory focuses on two directions.

First, the goals of a data-oriented smart maintenance factory include visualization of the maintenance process and predictive maintenance. Second, the technology-oriented smart maintenance factory is to adopt Industry 4.0 technologies considering job specialty of maintenance.

(4)   Define Tasks to Transform

Tasks to build a smart maintenance factory of data driven and technology driven:

Task 1: data diagnostic management;
Task 2: on-site operation support system;
Task 3: semi-automation;
Task 4: remote maintenance system;
Task 5: IoT and mobile workplace;

Task 6: operational and environmental infrastructure.

The six tasks were determined in consideration of the analyzed issues in the field and the required direction of resolution. In order to implement the six tasks, about a dozen detailed tasks have been embodied. The tasks were finalized through collaboration and considerable time and consultation with managers and the project team.

(5)   Identify Smart Maintenance Elements

Regarding automation and cutting-edge technologies for the smart maintenance factory, Industry 4.0 technologies that suit the specificities of the tasks in a maintenance factory were decided to be applied. Among the various Industry 4.0 technologies, performance-validated technologies and currently applicable technologies that do not require long development and preparation time were prioritized.

In order to build a data diagnostic management of task 1, the data set must be defined in advance. Since the existing IT system focuses on data storage and loss prevention, the data management for utilization is insufficient. There is a lot of data stored in the system, but usually the data is improper or the data format is incorrect for factual use. It is necessary to establish governance rules for data standardization and collection in the field. The big data system for analyzing a large amount of collected data and Quick Response (QR) for sharing data in the fields are major systems for a data-driven smart factory. The QR system was applied to allow verification of the maintenance history and objects at the site.

In order to implement task 2, the on-site support system is connected to the central management system to share ERP information. It also plays roles of data collection, job history identification, and job processing status on the site. The information of workers, job status, job schedules, working inventories, and dangerous environments on the site can be visually managed through the central control room with a huge screen. The control room ensures comprehensive and rapid response to the various changes on the site.

The cloud-based smart control requires standardized maintenance works and procedures. Standardization is difficult due to variability in the maintenance area. Thus, unlike manufacturing, establishing a separate on-site support system and linking it with the central information and data sharing system may be a method to construct smart maintenance effectively in a short period of time.

For task 3, semi-automation is more preferable to full-automation considering the changeable working conditions. Full automation is recommendable on the basis of a stable work environment, planned work schedules, and repetitive work procedures. Nevertheless, it is not proper to apply full automation to a maintenance factory where works are unstable and highly changeable.

The specific technology for semi-automation was selected in consideration of the working conditions and technology availability. 3D scanning technology for accurately locating a vessel and smart vehicle/cart technology for the convenient movement of workers are selected. For smart moving between processes, smart vehicles were selected as the optimal technology rather than Automated Guided Vehicle (AGV) due to environmental variations, advanced works, and excessive costs. In order to replace discontinued parts, 3D scanning and printing technologies were applied to produce the parts internally without design works. Co-robots are brought onto the site to assemble/disassemble heavy components that need to be repaired, promoting convenience of work. In addition, wearable equipment was introduced to reduce health and safety risks because workers no longer need to manually move heavy weights frequently. CPS (Cyber-Physical Systems) requires digitization of all landmarks of the factory, which takes a considerable amount of time and expense. In this case, a lower-level or simple CPS technologies should be selected. It is not necessary to apply the most advanced Industry 4.0 technologies to build a smart maintenance factory.

In order to implement task 4, an AR system was implemented to support remote maintenance. Workers and managers had a preconceived notion that VR technology shown in the media could change the way of work. However, in order to apply VR technology, it could only be applied to preliminary preparation work and repetitive and limited work environments. In addition, there have been studies [5,15,26,53] on the limitations of the application of AR technology to maintenance work in advance. Simple AR technology that can give immediate instructions while sharing a maintenance site through a remote screen was finally selected.

In order to implement task 5: the Internet of Things (IoT) and a mobile workplace, a mobile network based on IoT and Long Term Evolution (LTE) was built as a communication network. The operation condition data of engines and the major equipment of vessels is collected and transmitted through a satellite network in real-time. The data collected from the vessel is diagnosed in the big data system for predictive maintenance. The collected data are aggregated and operated on the IoT platforms and the on-site support system, and are connected to the big data analysis system for predictive maintenance.

To implement task 6, operational and environmental infrastructure, facilities for hazardous substances and the environment from maintenance were built as separate collective facilities. The Energy Saving System (ESS) is equipped for a stable power supply for the increase in power consumption due to the smart maintenance factory.

(6)   Build Smart Maintenance Architecture

Figure 8 illustrates the technical architecture for the smart maintenance factory.

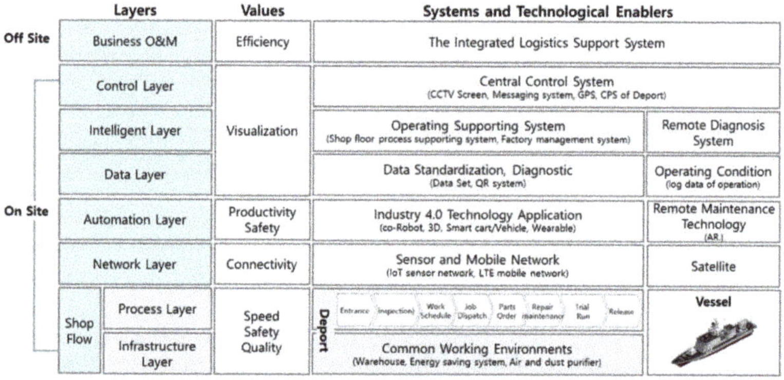

Figure 8. Technology architecture of a vessel smart maintenance factory.

The operating system of a smart maintenance factory promoted the integrated logistics support system of Business O&M (Operation and Management), for information connectivity between the upper operating system and the on-site operating system. The central control system consists of technologies similar to CCTV technology, such as big screen, location identification, emergency call, remote conference, and messaging technologies. The operating support system (OSS) includes a factory management system and a task management system to support works at the site. The data sets are identified and standardized in type and collection period considering usage from the site. The QR system was applied extensively to most works and materials at the maintenance site. Automation is mainly composed of Industry 4.0 technologies. In this case, semi-automation and ease of movement were focused due the characteristics of the works.

Three types of networks were applied: near distance communication by using IoT, long-range communication via LTE, and remote communication via satellite. The wireless and sensor networks allowed workers to access information conveniently and assured that data created at the field could be collected immediately. Changes that occurred in each work process due to smart data and technology should be carried out in future construction and implementation stages. Acquiring data and technical experts to operate smart factories is another task to be prepared to increase organizational competency in the future.

The operating infrastructure for building a smart maintenance factory is the power supply system. Compared to a traditional machine-centered maintenance factory, a smart maintenance factory emphasizes the importance of a stable power supply. Thus, it is essential to build an ESS system to assure an affordable and stable power supply.

Moreover, environmental pollutants such as noise, dusts, gases, and hazardous oil generated during the maintenance process were to be carried out in a separate and safe workplace. Furthermore, a separate smart warehouse was built to store inventory and material for maintenance in order to minimize wait time. For the purpose of promoting a worker use environment with work safety and convenient information utilization, such as PDA terminals, smart helmets, and wearables, the smart maintenance factory was constructed to associate with each technology and system.

(7)   Analyze Effect with KPIs

It is proper to define KPIs based on the missions and operational values of a smart maintenance factory. In this study, the analysis of an expected value requires an analysis of the factory unit. In the case of a single operation, in order to improve the expected effect, an excessive load or performance degradation of related operations may result, and a view of managing the whole is necessary. Aligning KPIs with the expected effects is a plan for operation optimization after the establishment of a smart maintenance factory by linking strategy and implementation.

In connection with the goals of the vessel maintenance depot, the maintenance factory used mean time to failure (MTTF) and mean time to repair (MTTR) as KPIs to increase the availability rate of vessels and improve the productivity of the maintenance factory. Based on the two KPIs, the expected effect was calculated by improving the vessel availability rate and shortening the maintenance time through the establishment of a smart maintenance factory. Through this, the alignment among the Goal-KPI-Expected effect was secured.

While constructing a smart maintenance factory, the question about the benefits of smart transformation has arisen. The Korean Ministry of SMEs and Startups analyzed the effect of smart factory introduction for 5003 SMEs from 2014 to 2017, and as a result, productivity increased by 30% [54]. The report also provided the results of an application system by companies. In this study, we set improvement goals for each stage and system, based on the empirical productivity improvement effect and the MTTR and MTTF reference values managed internally.

Since this project was in the stage of designing smart transformation, the goal to be obtained in the future was set rather than the actual application result. The goals for each stage are shown in Table 2.

Table 2. Goals of improvement.

Transformation	Contribution Weight	MTTR Improvement					MTTF Improvement				
		Year0	Year1	Year2	Year3	Year4	Year0	Year1	Year2	Year3	Year4
Big data	10%	0.0%	0.6%	1.5%	2.4%	3.0%	0.0%	0.2%	0.6%	1.0%	1.2%
IoT & Mobile	10%	0.0%	0.6%	1.5%	2.4%	3.0%	0.0%	0.2%	0.6%	1.0%	1.2%
OSS	25%	0.0%	1.5%	3.8%	6.0%	7.5%	0.0%	0.6%	1.5%	2.4%	3.0%
Remote Diagnostic	15%	0.0%	0.9%	2.3%	3.6%	4.5%	0.0%	0.4%	0.9%	1.4%	1.8%
Automation	20%	0.0%	1.2%	3.0%	4.8%	6.0%	0.0%	0.5%	1.2%	1.9%	2.4%
Infrastructure	20%	0.0%	1.2%	3.0%	4.8%	6.0%	0.0%	0.5%	1.2%	1.9%	2.4%
Total Improvement	100%	0%	6%	15%	24%	30%	0%	2%	6%	10%	12%

In this project, when operating the smart maintenance factory for five years, the expected effect was estimated to be up to 30% for productivity improvement and up to 12% for vessel operation rate improvement. The total cost of transforming the entire smart maintenance factory was estimated to be about 30 million USD and the expected payback period was three to five years (excluding equipment costs for infrastructure and automation).

(8) Plan Implementation

It took about one year to analyze the current status of the existing maintenance factories and establish a construction plan for smart transformation on factory basis, rather than on the particular job or technology basis.

In this study, we transformed by resolving the issues presented in Table 1. The deport of mechanical and hydraulic maintenance feature was transformed to a data-driven and technology-driven maintenance factory. The first change was to establish a data-driven management with data set definition, gathering, analysis, and standardization on the basis of big data governance. Second, QR, IoT (Bluetooth), and LTE-based networks enabled data communication for the entire smart maintenance depot. Third, an on-site operation support IT system was established to support workers, process, and field managers with information and technical support. In particular, in order to secure the visibility of the site, depot, 5M, and environmental status and information were centralized on the central control system. Fourth, a remote data collection system and AR technology were applied for remote maintenance.

Fifth, semi-automation was focused on 3D printing/scanning, Auto Guided Vehicle, and co-Robot to ensure the accuracy of work and the convenience of moving heavy equipment. Finally, a smart warehouse, Energy Saving System, and a workshop for hazardous environments were constructed separately for health and safety. Figure 9 illustrates the changed image of the smart maintenance depot.

The design and developed jobs of the proposed construction plan of the smart maintenance factory will be promoted to ensure that the smart maintenance factory is to be implemented step by step. The pilot working process (shop) was selected and deployed all technologies that were under consideration and will evaluate the performance and deploy the technologies to other work processes over the next five years.

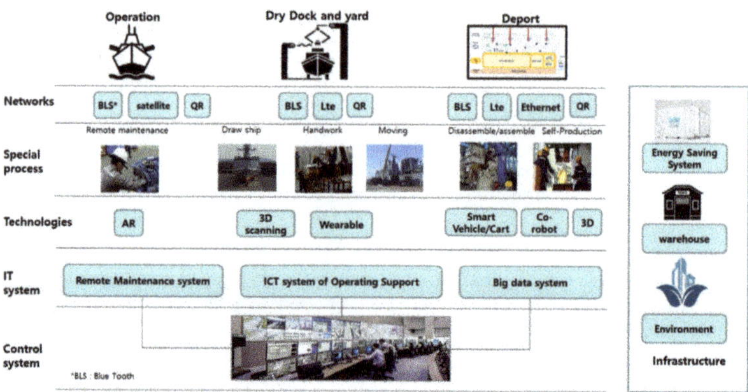

Figure 9. Future image of a transformed smart depot.

5. Discussion and Conclusions

In terms of the entire lifecycle, smart factories have been mainly studied in the building phase, and the need to expand research to the service phase has been suggested [6,7,32]. This study emphasized the service phase of the smart factory for a vessel maintenance depot. Due to the variability of repair and maintenance work, there was a limit to applying the smart factory concept in the manufacturing field as it is. The procedures and methods were proposed for the smart transformation of maintenance factory by applying Industrial 4.0 technology.

In order to respond to the variability of maintenance works, methods for securing on-site agility and predicting maintenance were presented. For an agile response to the field, we mainly set the direction of visualization on-site to control the processes in line through the visual management and to secure the linkage between each job through the on-site operation support system. For predictive maintenance, data of the status of the currently operating vessels was collected in real-time and analyzed by the big data system. The entire maintenance depot was built as a mobile workplace with IoT and LTE enabling data-based factory management.

In data-driven factories, data standardization should be implemented through data governance. Although the goal is to achieve efficient management with big data analysis, the reality is that factories neither collect/manage data nor have a modeling concept for the data. Since the IT system has stored data without considering the data format or type, it has been difficult to utilize the stored data.

The technology-oriented automation was implemented rather than equipment-oriented automation. For automation, semi-automation was proposed rather than full automation considering the work characteristics. Instead of the investment-oriented levels of a smart factory, effect-driven automation should be given the top priority in consideration of the difficulty of learning for operators and the work environment of the factory.

While IT innovation improves the process by applying the necessary technologies and systems, smart transformation changes existing processes through the application of the fourth industrial technologies of data, systems, automation, and infrastructure. In other words, the approach of smart transformation is the opposite perspective to the legacy IT approach.

The concept of a smart factory in the ship building phase needed to be changed for the application of the smart maintenance factory, but most of the technologies including 3D, AR, big data, IoT, and mobile could be accommodated. The applied enablers in the study were IoT system, wireless system, operation support IT system, big data system, remote support system, semi-automation, and infrastructure improvement. Among them, the contribution to improvement was evaluated to be the highest through the operating support system that supports the connection and visualization on-site.

Through the linkage between the mission, KPIs, and expected effect, the goals of smart transformation for the plant rather than the individual work could be carried out. This linkage is considered as a basic strategic direction, enabling the performance-oriented smart maintenance factory operation.

In summary, we conducted a study to expand the smart factory concept, which was limited to the ship building phase and to the ship servicing phase. The smart transformation procedure, framework, and architecture for a smart maintenance factory were proposed with a practical case. The designed practical case was embodied as a data-driven and technology-driven smart factory.

The results of this paper will be a reference model for building a smart maintenance factory in the service area. Future studies should verify the actual increases of MTTFs and reductions of MTTR through smart maintenance factory operation. The expansion and improvement of the approach to business areas other than vessel maintenance could also be studied further.

Author Contributions: Conceptualization, methodology, G.S.K.; supervision, Y.H.L. All authors have read and agreed to the published version of the manuscript.

Funding: This research received no external funding.

Institutional Review Board Statement: Not applicable.

Informed Consent Statement: Not applicable.

Data Availability Statement: Not applicable.

Conflicts of Interest: The authors declare no conflict of interest.

References

1. Rüßmann, M.; Lorenz, M.; Gerbert, P.; Waldner, M.; Justus, J.; Engel, P.; Harnisch, M. Industry 4.0 the Future of Productivity And Growth In Manufacturing Industries. *BCG* **2015**, *9*, 54–89. Available online: https://www.bcg.com (accessed on 29 August 2021).
2. Kim, D.W.; Lee, H.N. "Smart Ships Are Floating". Shipbuilders in Korea Taking the Lead in Manufacturing Smart Ships. Available online: http://www.aitimes.com/news/articleView.html?idxno=137618 (accessed on 30 October 2021).
3. *Naval Ship Life Cycle Cost (LCC) Model*; SPAR Associates, Inc.: New Orleans, MD, USA, 2011.
4. Thoben, K.D.; Wiesner, S.; Wuest, T. "Industrie 4.0" and Smart Manufacturing—A Review of Research Issues and Application Examples. *Int. J. Autom. Technol.* **2017**, *11*, 4–16. [CrossRef]
5. Martin, B.; McMahon, M.E.; Riposo, J.; Kallimani, J.G.; Bohman, A.; Ramos, A.; Schendt, A. *A Strategic Assessment of the Future of U.S. Navy Ship Maintenance: Challenges and Opportunities*; RAND: Santa Monica, CA, USA, 2007.
6. Osterrieder, P.; Budde, L.; Friedli, T. The smart factory as a key construct of industry 4.0: A systematic literature review. *Int. J. Prod. Econ.* **2020**, *221*, 107476. [CrossRef]
7. Mabkhot, M.M.; Al-Ahmari, A.M.; Salah, B.; Alkhalefah, H. Requirements of the smart factory system: A survey and perspective. *Machines* **2018**, *6*, 23. [CrossRef]
8. James, M.; Jimmy, I. Big Data Analytics for Smart Manufacturing: Case Studies in Semiconductor Manufacturing. *Process* **2017**, *5*, 39. [CrossRef]
9. Sjödin, D.R.; Parida, V.; Leksell, M.; Petrovic, A. Smart Factory Implementation and Process Innovation: A Preliminary Maturity Model for Leveraging Digitalization in Manufacturing Moving to smart factories presents specific challenges that can be addressed through a structured approach focused on people, processes, and technologies. *Res. Technol. Manag.* **2018**, *61*, 22–31. [CrossRef]
10. Wang, B.; Tao, F.; Fang, X.; Liu, C.; Liu, Y.; Freiheit, T. Smart Manufacturing and Intelligent Manufacturing: A Comparative Review. *Engineering* **2021**, *7*, 738–757. [CrossRef]
11. Mittal, S.; Khan, M.A.; Romero, D.; Wuest, T. Smart manufacturing: Characteristics, technologies and enabling factors. *Proc. Inst. Mech. Eng. Part B J. Eng. Manuf.* **2019**, *233*, 1342–1361. [CrossRef]
12. Zhong, R.Y.; Xu, X.; Klotz, E.; Newman, S.T. Intelligent Manufacturing in the Context of Industry 4.0: A Review. *Engineering* **2017**, *3*, 616–630. [CrossRef]
13. Kerin, M.; Pham, D.T. A review of emerging industry 4.0 technologies in remanufacturing. *J. Clean. Prod.* **2019**, *237*. [CrossRef]
14. Vogel-Heuser, B.; Bauernhansl, T.; ten Hompel, M. (Eds.) *Handbuch Industrie 4.0 Bd.4, Allgemeine Grundlagen*, 2nd ed.; Springer: Berlin/Heidelberg, Germany, 2017; ISBN 978-3-662-53254-6.
15. Dornhöfer, M.; Sack, S.; Zenkert, J.; Fathi, M. Simulation of Smart Factory Processes Applying Multi-Agent-Systems—A Knowledge Management Perspective. *J. Manuf. Mater. Process.* **2020**, *4*, 89. [CrossRef]
16. Bokrantz, J.; Skoogh, A.; Berlin, C.; Wuest, T.; Stahre, J. Smart Maintenance: An empirically grounded conceptualization. *Int. J. Prod. Econ.* **2020**, *223*, 107534. [CrossRef]
17. Capestro, M.; Kinkel, S. *Industry 4.0 and Knowledge Management: A Review of Empirical Studies*; Springer International Publishing: Berlin/Heidelberg, Germany, 2020; ISBN 978-3-030-43589-9_2.
18. Mantravadi, S.; Møller, C.; Li, C.; Schnyder, R. Design choices for next-generation IIoT-connected MES/MOM: An empirical study on smart factories. *Robot. Comput.-Integr. Manuf.* **2021**, *73*, 102225. [CrossRef]
19. Lang, V.; Weingarten, S.; Wiemer, H.; Scheithauer, U.; Glausch, F.; Johne, R.; Michaelis, A.; Ihlenfeldt, S. Process Data-Based Knowledge Discovery in Additive Manufacturing of Ceramic Materials by Multi-Material Jetting (CerAM MMJ). *J. Manuf. Mater. Process* **2020**, *4*, 74. [CrossRef]
20. Prometheus Group Homepage. Available online: https://www.prometheusgroup.com/posts/5-transformational-trends-reshaping-industrial-maintenance (accessed on 10 October 2021).
21. The Manufacturer Homepage. Available online: https://www.themanufacturer.com/press-releases/erp-solutions-support-industry-4-0 (accessed on 10 October 2021).
22. Noman, M.A.; Nasr, E.S.A.; Al-Shayea, A.; Kaid, H. Overview of predictive condition based maintenance research using bibliometric indicators. *J. King Saud Univ.-Eng. Sci.* **2019**, *31*, 355–367. [CrossRef]
23. Goti, A.; Oyarbide-Zubillaga, A.; Alberdi, E.; Sanchez, A.; Garcia-Bringas, P. Optimal Maintenance Thresholds to Perform Preventive Actions by Using Multi-Objective Evolutionary Algorithms. *Appl. Sci.* **2019**, *9*, 3068. [CrossRef]
24. Balogh, Z.; Gatial, E.; Barbosa, J.; Leitao, P.; Matejka, T. Reference Architecture for a Collaborative Predictive Platform for Smart Maintenance in Manufacturing. In Proceedings of the IEEE 22nd International Conference on Intelligent Engineering Systems (INES), Las Palmas de Gran Canaria, Spain, 21–23 June 2018. [CrossRef]
25. Aksa, K.; Aitouche, S.; Bentoumi, H.; Sersa, I. Developing a Web Platform for the Management of the Predictive Maintenance in Smart Factories. *Wirel. Pers. Commun.* **2021**, *119*, 1–29. [CrossRef]
26. Gopalakrishnan, M.; Skoogh, A.; Salonen, A.; Asp, M. Machine criticality assessment for productivity improvement. *Int. J. Product. Perform. Manag.* **2019**, *68*, 858–878. [CrossRef]
27. Pech, M.; Vrchota, J.; Bednář, J. Predictive Maintenance and Intelligent Sensors in Smart Factory: Review. *Sensors* **2021**, *21*, 1470. [CrossRef] [PubMed]
28. Fusko, M.; Rakyta, M.; Krajcovic, M.; Dulina, L.; Gaso, M.; Grznar, P. Basics of Designing Maintenance Processes in Industry 4.0. *MM Sci. J.* **2018**, *1*, 2252–2259. [CrossRef]

29. Biedermann, H.; Kinz, A. Lean Smart Maintenance—Value Adding, Flexible, and Intelligent Asset Management. *BHM Berg-Und Hüttenmännische Mon.* **2019**, *164*, 13–18. [CrossRef]

30. Moeuf, A.; Pellerin, R.; Lamouri, S.; Tamayo-Giraldo, S.; Barbaray, R. The industrial management of SMEs in the era of Industry 4.0. *Int. J. Prod. Res.* **2018**, *56*, 1118–1136. [CrossRef]

31. Masood, T.; Sonntag, P. Industry 4.0: Adoption challenges and benefits for SMEs. *Comput. Ind.* **2020**, *121*, 103261. [CrossRef]

32. Strozzi, F.; Colicchia, C.; Creazza, A.; Noè, C. Literature review on the 'smart factory' concept using bibliometric tools. *Int. J. Prod. Res.* **2017**, *55*, 6572–6591. [CrossRef]

33. Shin, S.M.; Chang, C.W.; Kown, S.B.; Kim, B.K. Smart Factory of Naval Maintenance Depot Construction concept and development direction. *J. Soc. Nav. Archit. Korea* **2020**, *57*, 27–30.

34. Resman, M.; Turk, M.; Herakovic, N. Methodology for planning smart factory. *Procedia CIRP* **2020**, *97*, 401–406. [CrossRef]

35. Zheng, P.; Wang, H.; Sang, Z.; Zhong, R.Y.; Liu, Y.; Liu, C.; Mubarok, K.; Yu, S.; Xu, X. Smart manufacturing systems for Industry 4.0: Conceptual framework, scenarios, and future perspectives. *Front. Mech. Eng.* **2018**, *13*, 137–150. [CrossRef]

36. Burke, R.; Laaper, S.; Hartigan, M.; Mussomeli, A.; Sniderman, B. *The Smart Factory Responsive, Adaptive, Connected Manufacturing A Deloitte Series on Industry 4.0, Digital Manufacturing Enterprises, and Digital Supply Networks*; (n.d.); Deloitte University Press: Westlake, TX, USA, 2017.

37. Automation World Home Page. Three Steps to Start the Smart Manufacturing Journey. Available online: https://www.automationworld.com/products/software/blog/13316170/three-steps-to-start-the-smart-manufacturing-journey (accessed on 17 October 2016).

38. Yoon, S.C.; Um, J.; Suh, S.H.; Stroud, I.; Yoon, J.S. Smart Factory Information Service Bus (SIBUS) for manufacturing application: Requirement, architecture and implementation. *J. Intell. Manuf.* **2019**, *30*, 363–382. [CrossRef]

39. Longo, F.; Nicoletti, L.; Padovano, A. Smart operators in industry 4.0: A human-centered approach to enhance operators' capabilities and competencies within the new smart factory context. *Comput. Ind. Eng.* **2017**, *113*, 144–159. [CrossRef]

40. Shariatzadeh, N.; Lundholm, T.; Lindberg, L.; Sivard, G. Integration of Digital Factory with Smart Factory Based on Internet of Things. *Procedia CIRP* **2016**, *50*, 512–551. [CrossRef]

41. Erasmus, J.; Grefen, P.; Vanderfeesten, I.; Traganos, K. Smart hybrid manufacturing control using cloud computing and the internet-of-things. *Machines* **2018**, *6*, 62. [CrossRef]

42. Hermann, M.; Pentek, T.; Otto, B. Design Principles for Industrie 4.0 Scenarios. In Proceedings of the 2016 49th Hawaii International Conference on System Sciences (HICSS), Koloa, HI, USA, 5–8 January 2016. [CrossRef]

43. Gorecky, D.; Weyer, S.; Hennecke, A.; Zühlke, D. Design and Instantiation of a Modular System Architecture for Smart Factories. *IFAC-Papers* **2016**, *49*, 79–84. [CrossRef]

44. Li, Q.; Tang, Q.; Chan, I.; Wei, H.; Pu, Y.; Jiang, H.; Li, J.; Zhou, J. Smart manufacturing standardization: Architectures, reference models and standards framework. *Comput. Ind.* **2018**, *101*, 91–106. [CrossRef]

45. Zeid, A.; Sundaram, S.; Moghaddam, M.; Kamarthi, S.; Marion, T. Interoperability in smart manufacturing: Research challenges. *Machines* **2019**, *7*, 21. [CrossRef]

46. Zhang, D.; Wan, J.; Hsu, C.H.; Rayes, A. Industrial technologies and applications for the Internet of Things. *Comput. Netw.* **2016**, *101*, 1–4. [CrossRef]

47. Shu, Z.; Wan, J.; Zhang, D.; Li, D. Cloud-Integrated Cyber-Physical Systems for Complex Industrial Applications. *Mob. Netw. Appl.* **2016**, *21*, 865–878. [CrossRef]

48. Chen, B.; Wan, J.; Shu, L.; Li, P.; Mukherjee, M.; Yin, B. Smart Factory of Industry 4.0: Key Technologies, Application Case, and Challenges. *IEEE Access* **2018**, *6*, 6505–6519. [CrossRef]

49. Hamzeh, R.; Xu, X. Technology selection methods and applications in manufacturing: A review from 1990 to 2017. *Comput. Ind. Eng.* **2019**, *138*, 106123. [CrossRef]

50. Kagermann, H.; Walhster, W.; Helbig, J. *Recommendations for Implementing the Strategic Initiative INDUSTRIE 4.0: Securing the Future of German Manufacturing Industry*; Final Report of the Industrie 4.0 Working Group; Forschungsunion Deutsche Wissenschaft, 2013. Available online: https://www.din.de/blob/76902/e8cac883f42bf28536e7e8165993f1fd/recommendations-for-implementing-industry-4-0-data.pdf (accessed on 10 October 2021).

51. Žižek, S.Š.; Nedelko, Z.; Mulej, M.; Čič, Ž.V. Key Performance Indicators and Industry 4.0—A Socially Responsible Perspective. *Our Econ.* **2020**, *66*, 22–35. [CrossRef]

52. Büchi, G.; Cugno, M.; Castagnoli, R. Smart factory performance and Industry 4.0. *Technol. Forecast. Soc. Chang.* **2020**, *150*, 119790. [CrossRef]

53. Wikitude Home Page. Wikitude Augmented Reality: The World's Leading Cross-Platform AR SDK. Available online: https://www.wikitude.com (accessed on 10 October 2021).

54. MSS Home Page. Available online: https://www.mss.go.kr/site/smba/ex/bbs/View.do?cbIdx=86&bcIdx=1011893&parentSeq=1011893, (accessed on 10 October 2021).

Article

A Product Conceptual Design Method Based on Evolutionary Game

Yun-Liang Huo [1], Xiao-Bing Hu [1,*], Bo-Yang Chen [1] and Ru-Gu Fan [2]

[1] School of Science and Engineering, Sichuan University, Sichuan 610065, China; huo_yunlliang@163.com (Y.-L.H.); shanbenyunque8018@163.com (B.-Y.C.)
[2] Sinohydro Jiajiang Hydraulic Machinery Company Limited, Sichuan 614000, China; liofrg@163.com
* Correspondence: huxb@scu.edu.cn

Received: 3 December 2018; Accepted: 4 March 2019; Published: 5 March 2019

Abstract: In this paper, an intelligent-design method to deal with conceptual optimization is proposed for the decisive impact of the concept on the product-development cycle cost and performance. On the basis of matter-element analysis, an effective functional-structure combination model satisfying multiple constraints is first established, which maps the product characteristics obtained by expert research and customer-requirements analysis of the function and structure domain. Then, the Evolutionary Game Algorithm (EGA) was utilized to solve the model, in which a strategy-combination space is mapped to the solution-search space of the conceptual-solution problem, and the game-utility function is mapped to the objective functions of concept evaluation. Constant disturbance and Best-Response Correspondence were applied cross-repeatedly until the optimal equilibrium Pareto state corresponding to the global optimal solution was obtained. Finally, the method was simulated on MATLAB 8.3 and applied to the design for fixed winch hoist, which greatly shortens its design cycle.

Keywords: conceptual design; intelligent design; evolutionary game; domain mapping

1. Introduction

Research on product conceptual design is booming with regard to the direct influence of the concept on the quality of the final product, and the vast majority of researchers agree that how to scientifically evaluate candidate concepts and how to express a product concept with an accurate model are two vital tasks in conceptual design [1]. Hence, advanced models and effective evaluation systems have been intensively addressed by researchers worldwide. Danni et al. [2] presented an evaluation and selection method composed of three modules: data mining, concept reconstruction, and decision support, to improve the efficiency of concept review and evaluation. Sun et al. [3] established an effective conceptual model for new-product concept development from two theoretical backgrounds about organizational learning, and the model was applied to the design of a large scramjet with satisfying results. Wang et al. [4] proposed an optimization decision model for product conceptual design to help enterprises select key technical characteristics under the condition that cost and time maximally meet customer requirements. Christoph F. et al. [5] presented methodology integration with a knowledge model for conceptual design in accordance with model-driven engineering, and the work extended Gero's Function-Behavior-Structure model. Based on Bunge's Scientific Ontology, Chen et al. [6] developed an explicit and complete conceptual foundation for the establishment of a new conceptual design model. Varun Tiwari et al. [7] proposed a novel way of performing design-concept evaluations, where instead of considering the cost and benefit characteristics of the design criteria, the work identifies the best concept that satisfies constraints imposed by the team of designers, as well as fulfilling as many of a customer's preferences as possible. To obtain the best comprehensive performance of mechanical products, Wang et al. [8] established an evaluation model for product conceptual design based on the principle of maximum-entropy value, and solved the model by constructing a Lagrange function.

The above work mainly focuses on product-model expression and product conceptual evaluation. However, there could be many generated concepts through its combination nature, and the evaluation of a larger number of concepts, one by one, is a very difficult work, although many novel and effective methods of concept evaluation have been proposed [6–8]. As a result, the best design concept cannot easily be obtained, and the internalization of the conceptual-design process becomes critical.

Computational intelligence, which consists of an evolutionary neural network and fuzzy logic, is a novel technology aiming to bring intelligence into computation [9]. Attempts have been made in recent years for the application of computational intelligence. Manu Augustin [10] proposed a framework that uses a fuzzy inference process for evaluating each initial concept against identified decision criteria, to select and/or evolve improved concepts. Integrated with ACO, Ma et al. [11] presented a mathematical programming model to quantitatively predict change-propagation impact, and improved the intelligence of change-propagation prediction during the design process. Ming-Chyuan et al. [12] proposed an integrated procedure that involves neural-network training and genetic-algorithm simulations within the Taguchi quality-design process to aid in searching for an optimal solution with more precise design-parameter values for improving product development. Oliviu Matei et al. [13] addressed the automated product-design problem with two distinct evolutionary approaches: genetic algorithms and evolutionary ontology. S.H. Ling [14] developed intelligent particle-swarm optimization (iPSO), where a fuzzy-logic system, developed based on human knowledge, is proposed to determine the inertia weight for the swarm movement of the PSO and the control parameter of a newly introduced cross-mutated operation.

Although the above methods greatly contribute to the process of conceptual-design intelligence, the main focus is to study the commonality of various problem models [4,6,13,14]. While the model can be solved to obtain a feasible solution, they ignore the personality of the problem. If we choose or design a specific algorithm to solve a specific problem, the efficiency and accuracy of the solution is improved [8]. In view of this, we explored the establishment of a constraint model for product design, focusing on the functional variables and constraints of the model, and the optimal or approximately optimal solution of the functional variable combination of the Evolutionary Game Algorithm (EGA) search model was completed in this paper. In order to accurately express information during conceptual design, product characteristics are extracted at first via customer-requirement analysis and the application of expert knowledge, and the Analytic Network Process (ANP) is used to assess their importance. Then, a model of product conceptual design is established by means of mapping product characteristics to the functional and structural domains while comprehensively taking all constraints of product conceptual design into account. Finally, to quickly solve the model, intelligent algorithm EGA, with fast convergence speed, was used [15], and the optimal solution was obtained after multiple evolutions.

The paper is organized as follows. Section 1 introduces the process of how a product-optimization design model is established. Section 2 briefly introduces EGA. Section 3 provides a practical example to illustrate how the method performs. Section 4 concludes the paper.

2. Modeling for Product Conceptual Design

2.1. Matter-Element Description

The matter-element model is a representation of objects for computer storage, recognition, and operation, which is widely employed in product design and reliability assessment. Yue et al. [16] applied matter-element theory to ecological-risk assessment, and successfully evaluated the Gannan Plateau. To solve the formal description in the modular design of mechanical products, Huang et al. [17] introduced extension theory into Reconfiguration Design Technology (RDT), and built the matter-element model. Based on the model, the selection, matching, and transformation of a mechanical product and its modules were researched. Liu [18] proposed an assessment approach by combining extension and ensemble empirical-mode decomposition (EEMD) to describe the bearing performance-degradation (BPD) process that was denoted by the matter-element model. Lv et al. [19] presented a new method for equipment-criticality evaluation based on a fuzzy matter-element model.

In this paper, a model of product conceptual design based on matter-element analysis was constructed. Firstly, the function tree and structure tree could be obtained by mapping product characteristics to functional and structural domains, before which the product characteristics and their importance must be obtained through expert investigation and customer-requirement analysis. Then, in order to obtain the utility function of a product, various constraints in conceptual design are comprehensively considered, and the utility vector of the product characteristics is given to each substructure with the knowledge of the expert team. Finally, a matter-element model of product conceptual design is established with both utility attributes and design-constraint attributes invested to express product information.

The above model can be described as *Pro* = (*S_Attrib*, *U_Attrib*, *C_Attrib*, *Cl_Attrib*), where *U_Attrib* denotes the model-evaluation information of each product concept; *Cl_Attrib* denotes the hierarchical information of the matter element; *C_Attrib* denotes the constraint information including functional, structural, and relational constraints during product conceptual design; and *S_Attrib* denotes product-feature information. In order to express it more clearly, the matter-element model of product conceptual optimization design is expressed as follows:

$$
\begin{pmatrix}
Pro, & U\_Attrib, & v_1 \\
 & Cl\_Attrib, & v_2 \\
 & C\_Attrib, & v_3 \\
 & S\_Attrib, & v_4
\end{pmatrix}
\tag{1}
$$

where v_i (i = 1, 2, 3, 4) is the value of an attribute belonging to a matter element, the larger the v_1, the better the concept; v_2 denotes the hierarchical information of the matter element; v_3 indicates whether the solution is a feasible solution, for example, v_3 = 0 means that the solution is feasible without breaking any constraint; and v_4 is the combined information of the substructures for achieving a functional unit. The detailed information of each attribute is expressed via its submatter elements, and the process of finding the optimal solution is transformed into the process of searching for a matter element of a product concept with a maximum v_1 under constraint conditions v_3 via combination of submatter elements.

2.2. Matter-Element Description

Product-characteristics set *PC* is obtained through the brainstorming of experts and technicians involved in all phases of the product life cycle, with customer requirements being taken into consideration (the *i*th element in *PC* is denoted by PC_i). The original *PC* should be processed to obtain the new one, as their relationships may be inclusion, cross, and independence. Generally speaking, there are mutual relations between elements in *PC*, customer-requirement set *CR* (the *j*th element in *CR* is denoted by CR_j) and PC_i, which should all be taken into account when synthetically analyzing the importance of *PC*. The Analytic Network Process (ANP) method is a widely used decision-making algorithm, mainly to determine the relative importance of a group with inter-related elements in a multiobjective decision-making problem; therefore, it is adopted to analyze a *PC* and calculate its importance.

1. Analyzing the importance of a *PC* driven by *CR*

    Assume that each PC_i is independent from the others. Importance vector $w_s = (w_1, w_2, \ldots w_m)$ is obtained according the customers' preference for each requirement. For each PC_i, relative importance matrix R_i between *CR* and PC_i is evaluated by an expert team; for element $r_{ij} \in [0,9]$ in R_k, which indicates the importance of RC_i for RC_j when pursuing PC_k, if $r_{ij} \neq 0$, then $r_{ji} = 1/r_{ij}$; else, $r_{ij} = r_{ji} = 0$.

In addition, the Analytic Hierarchy Process (AHP) [20] is used to obtain relative-importance vector w_i = $(w_{1i}, w_{2i} \ldots w_{ii}, w_{mi})$, where $\sum_{j=1}^{m} w_{ji} = 1$, and importance matrix $W_{cr\text{-}pc}$ of a PC driven by CR can finally be obtained.

$$R_k = \begin{pmatrix} r_{11} & r_{12} & \cdots & r_{1i} & \cdots & r_{1m} \\ r_{21} & r_{22} & \cdots & r_{2i} & \cdots & r_{2m} \\ \vdots & \vdots & \vdots & \vdots & \cdots & \vdots \\ r_{i1} & r_{i2} & \cdots & r_{ii} & \cdots & r_{im} \\ \vdots & \vdots & \vdots & \vdots & \vdots & \vdots \\ r_{m1} & r_{m2} & \cdots & r_{mi} & \cdots & r_{mm} \end{pmatrix} \quad W_{cr-pc} = \begin{pmatrix} w_{11} & w_{12} & \cdots & w_{1i} & \cdots & w_{1n} \\ w_{21} & w_{22} & \cdots & w_{2i} & \cdots & w_{2n} \\ \vdots & \vdots & \vdots & \vdots & \cdots & \vdots \\ w_{i1} & w_{i2} & \cdots & w_{ii} & \cdots & w_{in} \\ \vdots & \vdots & \vdots & \vdots & \vdots & \vdots \\ w_{m1} & w_{m2} & \cdots & w_{mi} & \cdots & w_{mn} \end{pmatrix} \tag{2}$$

where w_{ij} denotes the impact degree of CR_i on PC_j, and vector $w^{(1)} = w_s \times w_{cr\text{-}pc}$ denotes the importance of a PC driven by CR.

2. Gaining mutual importance among elements of PC

Relative-importance degree matrix R'_i that is similar to R_i is obtained when considering the correlations between PC_i and the others. r_{ij} in R'i indicates the importance of PC_i for PC_j when pursuing PC_k, and importance vector $w^{(2)} = (w_1^{(2)}, w_2^{(2)}, \ldots \ldots w_n^{(2)})$ is also obtained by AHP, where $\sum_{j=1}^{n} w_i^{(2)} = 1$.

$$R_k' = \begin{pmatrix} r_{11} & r_{12} & \cdots & r_{1i} & \cdots & r_{1n} \\ r_{21} & r_{22} & \cdots & r_{2i} & \cdots & r_{2n} \\ \vdots & \vdots & \vdots & \vdots & \cdots & \vdots \\ r_{i1} & r_{i2} & \cdots & r_{ii} & \cdots & r_{in} \\ \vdots & \vdots & \vdots & \vdots & \vdots & \vdots \\ r_{n1} & r_{n2} & \cdots & r_{ni} & \cdots & r_{nn} \end{pmatrix} \quad W_{pc} = \begin{pmatrix} w_{11} & w_{12} & \cdots & w_{1i} & \cdots & w_{1n} \\ w_{21} & w_{22} & \cdots & w_{2i} & \cdots & w_{2n} \\ \vdots & \vdots & \vdots & \vdots & \cdots & \vdots \\ w_{i1} & w_{i2} & \cdots & w_{ii} & \cdots & w_{in} \\ \vdots & \vdots & \vdots & \vdots & \vdots & \vdots \\ w_{n1} & w_{n2} & \cdots & w_{ni} & \cdots & w_{nn} \end{pmatrix} \tag{3}$$

$$w_i^{(2)} = \frac{\sum_{j=1}^{n} w_{ji}}{n} \tag{4}$$

3. Gaining the importance of PC

The importance of PC_i is shown in Equation (5) by comprehensively considering the two relationships mentioned above.

$$w_i = \frac{w_i^{(1)} \times w_i^{(2)}}{\sum_{k=1}^{n} w_k^{(1)} \times w_k^{(2)}} \tag{5}$$

2.3. Modeling Process

2.3.1. Multidomain PC mapping

With the fuzzy, complex, and tedious relationships between PC and the product structure, inaccuracy of information mapping and loss of information occur if we directly map the PC to the product-structure domain. Therefore, considering the correspondence between product function and structure in axiomatic design [21], the functional domain is introduced as an intermediate medium between PC and product structure, guiding mapping the PC to the product domain, and completing the product-structure design of the specific PC_i.

2.3.2. Function Decomposition

The process of *PC* multidomain mapping is shown in Figure 1, where the product-function tree is obtained by progressively decomposing product function to the tiniest independent functional units; the structure tree corresponding to the function tree is obtained by an expert team that enumerates the component structure corresponding to each function in the product-design database; the cell located at the bottom of the structure tree is called a substructure.

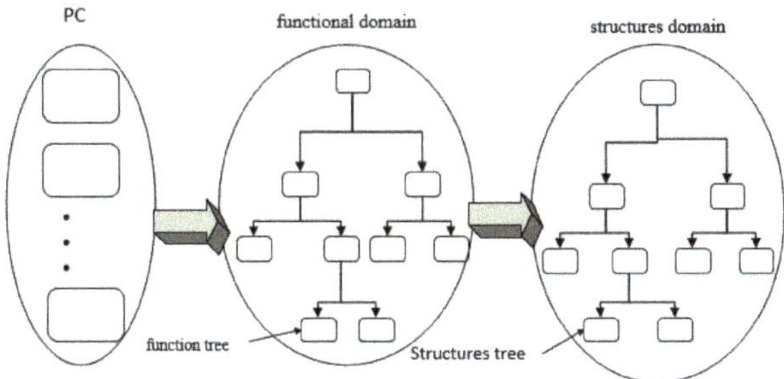

Figure 1. Product-characteristics set (PC) multidomain-mapping diagram.

2.3.3. Concept Modeling

Both functional units and substructures are denoted by the matter element after finishing the multidomain mapping of *PC*. The optimal-design concept is obtained by solving the model through the EGA via mapping product features to game players. A mechanical-product concept is expressed in Figure 2; it has *n* functional units, and *i*th functional units have *k* substructures.

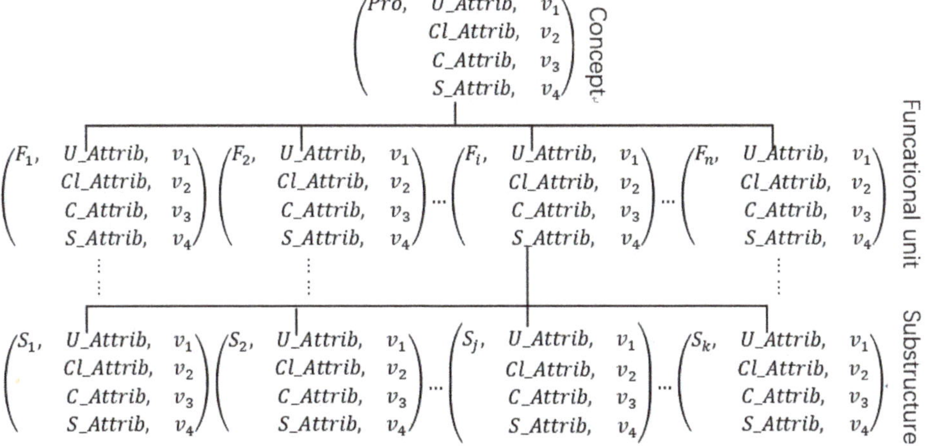

Figure 2. Matter-element model of a mechanical product.

where the information of the entire product concept is denoted by the matter-element model, for example, the specific information of the *i*th functional unit of the product, which includes structural information *S_Attrib*, constraint information *C_Attrib*, utility information *U_Attrib*, and hierarchical information *Cl_Attrib*, is denoted by the second-level matter element. The substructures to achieve a functional unit are denoted by third-level matter elements. It should be noted that the third-substructure-layer matter elements are alternative substructures, which are optional strategies of the game player, since the effectiveness of structural combinations has not been judged; therefore, no constraint information is required.

2.3.4. Values of Obtained Matter-Element Attributes

Linguistic terms such as 'very unimportant' and 'medium' are usually used to assess an attribute's importance, as they are always fuzzy during product design. Some linguistic terms should be transferred to crisp numbers for accurate analysis and calculations.

- Strategy variables and utility vectors are obtained

For m substructures s_{ij} ($j = 1, 2 \ldots m$) corresponding to a functional unit f_i ($i = 1, 2 \ldots n$), one of them must be chosen to achieve f_i during conceptual design, and the choice information of f_i for m substructures can be donated by the value of S_Attrib. For example, if $m = 8$ and the fourth substructure is chosen, then the value of S_Attrib of f_i is $v_4 = 00010000$, and the utility vector for the PC of substructure s_{i4} is used to calculate the utility value of the product concept. A typical mapping relationship between product features and matter-element attribute values is expressed in Figure 3.

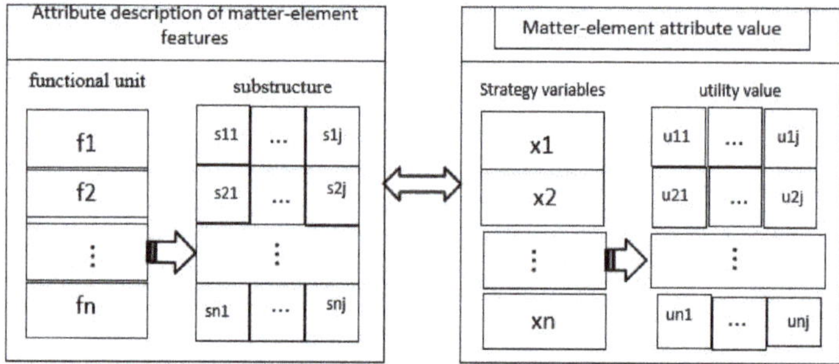

Figure 3. Schematic of property values.

where u_{ij} is a utility vector of a jth substructure of an ith functional unit provided by experts and designers based on a nine-point scale [22], which denotes the utility index of s_{ij}; x_i is the strategy variable of functional unit f_i, which denotes the choice information of alternative substructures. In the matter-element model proposed above, x_i is the value of S_Attrib for a game player. The frequently used nine-point scale is shown in Figure 4.

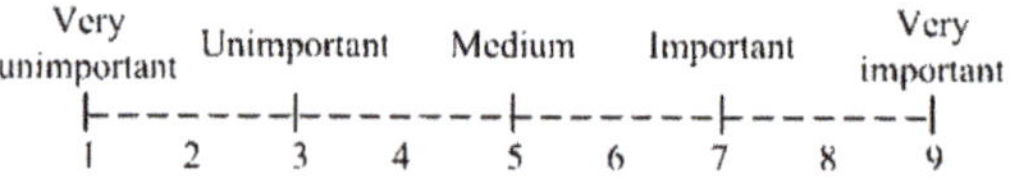

Figure 4. Nine-point scale.

- C_Attrib value is obtained

Product-design constraints, including functional, structural, and related constraints are ultimately embodied in the portfolio optimization of product substructures. In the optimization model proposed in this paper, a uniform expression C was used to specify dependency constraints that can denote the multiple constraint forms, and a dual constraint is taken as an example, shown in Equation (4).

$$C(x_i, x_j) = \{(u_{ik}, u_{jp})\} \tag{6}$$

where x_i and x_j are variables denoting the constraint relationship between functional units i and j, and the ranges of x_i and x_j are expressed as u_{ik} and u_{jp}, respectively. C indicates that j must choose the pth substructure if i chooses the kth. The constraint between a fixed winch hoist coupling and its service brake is used as an example.

$$C\,(b,\,c) = \{(\text{Wheel break, Wheel coupling}),\,(\text{Disc break, Disc coupling})\} \tag{7}$$

This shows that the wheel brake must be matched with the wheel coupling; otherwise, the number of constraints on the current composition strategy increases. If the number of constraints in the current combination strategy is i, then the value of $C\_Attrib\ v_3$, which is used to decrease the utility value of a concept in an evolutionary game, is i.

- *Cl_Attrib* value is obtained

The *Cl_Attrib* attribute in the model mainly denotes the hierarchical information of the matter element. As shown in Figure 2, $v_2 = 1$ indicates it is just a matter element of the product concept rather than a component.

2.4. Benefits

•By focusing on functional variables and constraints of the model, the obtained solution is the optimal solution that satisfies the constraint.

•Comprehensively considering PCs and CRs makes products perform well in terms of performance and personalization.

•A modular product functions as a player in the EGA that performs well on combinatorial optimization problems, and quickly obtains the optimal solution.

3. Introduction of Evolutionary-Game Algorithm

Considering that product conceptual design is actually a combinatorial optimization problem, EGA was employed to solve the above optimization model as it is effective in solving combinatorial optimization problems [23]. The optimal solution is obtained through the game for functional-unit-layer matter elements, a combination of substructure-layer matter elements, and comparison between matter elements in the conceptual layer.

The EGA is a novel kind of intelligent computation algorithm based on economic game theory and dynamic evolution calculation, which takes maximum utility as its optimization objective and searches the whole solution space by combining the strategies of game players, and simultaneously considers local and global performances. Compared with the selection process of a stochastic genetic algorithm, the EGA converges to a global optimal solution with probability 1, and is more certain in evolution [24].

3.1. Key Issues

3.1.1. Fundamental Theorems

A basic game consists of game player i, strategy set S, and utility u; the two fundamental theorems for EGA are shown as follows.

- If strategy combination S^* satisfies Equation (8) for any strategy $s_i \in S_i$ of any game player i, then it is called an S^* Nash equilibrium, and S_i is the strategy set of i. The specific form of Equation (8) is as follows:

$$u_i\left(s_i^*, S_{-i}^*\right) \geq u_i(s_i, S_{-i}) \tag{8}$$

where S_{-i} is the strategy combination of players without i, $S_{-i}{}^*$ is the Nash equilibrium of strategy combinations of players without i, and s_i^* is the optimal strategy for i in a Nash equilibrium. It is called a strict Nash equilibrium when

$$u_i\left(s_i^*, S_{-i}^*\right) = u_i(s_i, S_{-i}) \tag{9}$$

- Assuming that $S_{-i} = \prod S_k$, where $k = 1,2 \ldots n$ and $k \neq i$. If Equation (10), established as follows, is satisfied, then B_i is called the Best-Response Correspondence for player i.

$$B_i(s_{-i}) = \{s_i^* \in S_i : u_i(s_i^*, S_{-i}^*) \geq u_i(s_i, S_{-i}), \forall s^{(i)} \in S_i\} \tag{10}$$

Underlying the meaning of the Best-Response Correspondence is a process where i chooses the strategy with the maximum utility in the current situation. The dynamic process that all game players complete a Best-Response Correspondence in turn is called the Optimal-Response Dynamic.

3.1.2. EGA Expression

The specific form of the evolutionary-game algorithm is expressed as EGA= $\{G, S_0, \alpha, \beta, \tau\}$, and each member of the EGA is described in detail as follows.

- Game structure G

The game structure is described as $G = [I, S, U]$, where I, S, and U denote the information of the game players, the current situation, and utility, respectively. For the model described in this work, game-player set I is obtained by mapping functional units to the strategy variables, and k substructures for realizing a functional unit are mapped to the strategy set of the player.

The mathematical description is $s_{ij} \in \{0,1\}$, where $i \in I$ ($1 \leq i \leq n$) and $1 \leq j \leq k$; for example, if $k = 7$ and the third substructure is selected when he functional unit i generates a strategy, according to Section 2.3.3, the strategy of player i transfers to binary code 0010000. Then, the strategy combination of n players constitutes a solution S (also called a situation) in the above model. Equation 11 is the form of utility function $f(S)$ that is used to calculate utility value U of the current situation.

$$U_i = \begin{cases} f(S) & if\ satisfy\ the\ constraint \\ f(S) - f_{max} & else \end{cases} \quad i \in I$$

$$f(S) = \sum_{i=1}^{n} G_i W_i \tag{11}$$

$$G_i = \sum_{k=1}^{m} u_{ij} * W_{PC}$$

where u_{ij} is the utility vector of substructure s_{ij}, m is the element number of PC; W_{PC} is the importance degree of m PC_i calculated by Equation (5); W_i is the importance degree of game player i in product conceptual design, where $\sum_{i=1}^{n} W_i = 1$; n is the number of game players; and f_{max} is the maximum-utility value of the current evolutionary generation. Compared with penalty functions in other algorithms that are difficult to determine forms, f_{max} can be directly calculated. It should be noted that game player i is only bound by the constraint rules associated with itself during the game.

- Initial situation S_0

The EGA starts with S_0, which is initialized by a randomization method.
- Optimization operator α

Game theory is based on the assumption that all game players are economic, and in the process of evolution, each game player pursues maximum-utility values. Hence, the Best-Response Correspondence is called the optimization operator for the maximum-utility value of a game player made by it.
- Equilibrium perturbation operator β

In order to ensure that the solution obtained by the EGA is globally optimal, equalization perturbation operator β is employed to break the current Nash equilibrium state reached after several iterations; then, a new Nash equilibrium state is obtained by performing the Best-Response

Correspondence of each player after the balance state is sequentially broken. The specific calculation form of β is shown in Equation (12):

$$\beta(s_i) = \left\{ \begin{array}{ll} s_i & if\, X_i \geq p_i \\ Z_i & else \end{array} \right. \tag{12}$$

where p_i is the perturbation probability assigned according to the importance degree of each functional unit, and the functional units with much contribution to the utility value of a solution are easier to deviate the system from the original state. Therefore, higher disturbance probability should be attached to them, and the functional units with less contribution to the utility value should be given lower probability. X_i is a decimal randomly generated from 0 to 1; si indicates that the disturbance operator changed nothing and the former strategy is maintained; Z_i is the disturbance operator that means a strategy is randomly selected from the strategy set of player i to replace the current one.

- Termination condition τ

In a given situation, the process of Optimal-Response Dynamic is called one round, and the Nash equilibrium state of the situation is reached after two rounds. Two rounds achieving a Nash equilibrium state are defined as a generation. Setting the iteration termination condition as $\tau \geq T$, and T is the preset iteration generation.

3.2. EGA Process

The specific process of EGA is shown in Figure 5.

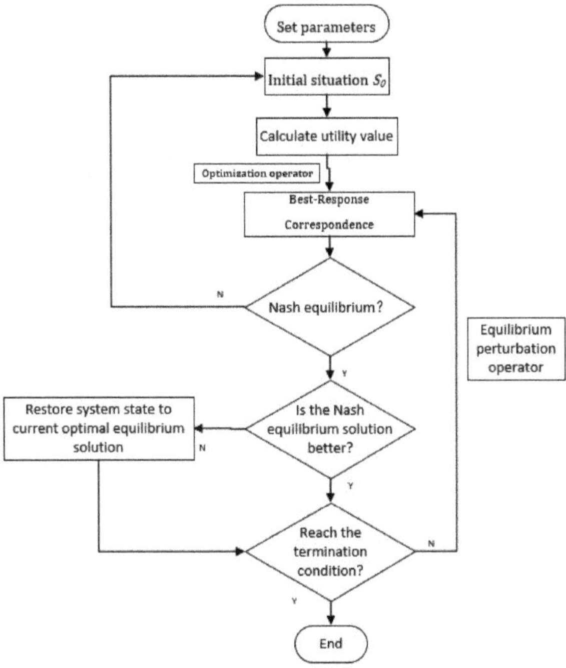

Figure 5. Evolutionary Game Algorithm (EGA) flowchart.

Step 1: Set parameters.
First, maximum iteration number T and disturbance probability pi are set.
Step 2: Algorithm initialization.

Update game structure to $G = G_0$ with the strategy randomly initialized; then, initial situation S_0 is generated, and the system starts to evolve from S_0 when $\tau = 0$.

Step 3: Calculate current-situation utility value.

Calculate the U of the current situation based on $f(S)$.

Step 4: Application of optimization operator α.

α is first used to estimate updated player utility, and then to update the strategy combination of game players from S_j to S_{j+1} when the updated one is better than the before; otherwise, keep S_j unchanged.

Step 5: Stability of the situation.

If the situation at timer $\tau = \tau(i)$ satisfies $U_{i+1} = U_i$, then strategy S_i is stable and its corresponding solution is a Nash equilibrium solution.

Step 6: Application of equalization perturbation operator β.

The new situation is achieved by applying β to the current situation; then, update situation S_j to S_{j+1} and calculate the utility value of S_{j+1}. Finally, estimate whether it is a stable evolution strategy again.

Step 7: Estimation of termination condition.

The algorithm terminates when $\tau \geq T$ is satisfied; otherwise, it returns to Step 6.

The EGA steps can be regarded as a stochastic process in a Nash equilibrium solution space that continuously updates the current stable solution with a better Nash equilibrium until the optimal situation equilibrium is reached. Since the main operation of EGA is only to compare the utility value between different strategy combinations, the global optimal solution can always be obtained by reasonably setting the number of iterations, because the utility of the global optimal solution is greater than other feasible solutions, and the utility of all feasible solutions is greater than infeasible solutions. Compared with frequently used evolutionary algorithms, such as Genetic Algorithm, Ant-Colony Algorithm, and Artificial Neural Networks, which involve complex mutation operations, path calculation, and network learning, respectively, the speed and efficiency of EGA are obvious advantages.

4. Case Study

A 3600 KN fixed winch hoist that was supported by the Sinohydro hydraulic machinery company was taken as an example to validate the method mentioned above. *PC* set F for the fixed winch hoist was obtained by product investigation, customer-requirement analysis, technical, economic, and social environments, and, finally, an expert team. Given F = {① low complexity, ② manufacturability, ③ assembly ability, ④ reliability, ⑤ mechanical strength ⑥ environment-friendly, ⑦ brake, ⑧ low noise, ⑨ Lifting stability, ⑩ low cost, ⑪ synchronicity, ⑫ high energy-conversion efficiency, and ⑬ lightweight}. How to express and implement the element-matter model of a fixed winch hoist is introduced below.

4.1. Modeling of the Fixed Winch Hoist

4.1.1. Design Knowledge

A fixed winch hoist is a heavy-tonnage lifting machine that works in the water-conservancy and hydropower industries, and it is composed of 10 components. As shown in Figure 6, only the structure of a movable pulley is related to the lifting force, and the structure of the other components could have different structures according to different *PCs*. Hence, the process of conceptual design according to a *PC* is transformed into the process of selecting the optimal structure of each component based on product characteristics.

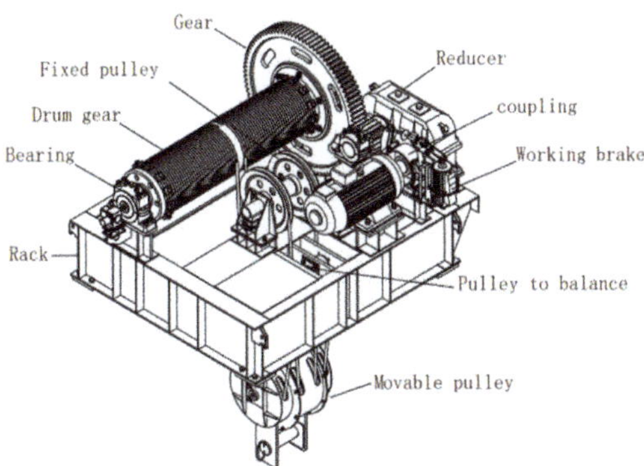

Figure 6. Schematic diagram of fixed winch hoist.

In order to solve the problem of fixed-winch-hoist conceptual design from the perspective of product characteristics, the function tree was first obtained by an engineer through functional decomposition, with the substructure set for each functional unit enumerated as shown in Table 1. Then, based on knowledge and customer-requirement constraints, they were obtained as shown in Figure 7. Finally, a model for product conceptual design was established, as shown in Figure 8.

Table 1. Function units and their alternative substructure.

Functional Unit	Structure	Alternative Substructure
Lifting	S_1 Drum gear	s_{11} single helix with intermediate rope, s_{12} single helix with sides rope, s_{13} Single fold with center rope, s_{14} single fold with sides rope s_{15} double fold with intermediate rope, s_{16} double fold with sides rope, s_{17} double helix with sides rope, s_{18} double helix with intermediate rope.
Balance	S_2 Pulley to balance	s_{21} balanced pulley suspension, s_{22} balanced pulley placement,
Stabilization	S_3 Fixed pulley	s_{31} fixed pulley is placed vertically, s_{32} fixed pulley is hung vertically, s_{33} fixed pulley is arranged in parallel, s_{34} No fixed pulley
Working brake	S_4 Working brake	s_{41} wheel brake, s_{42} disc brake
Reducer	S_5 Reducer	s_{51} horizontal speed reducer, s_{52} suspension reducer
Support	S_6 Bearing	s_{61} Antifriction bearing, s_{62} sliding bearing, s_{63} hybrid bearing
Power transmission	S_7 Gear and coupling	s_{71} wheel coupling with gear, s_{72} disc coupling with gear, s_{73} wheel coupling, s_{74} disc coupling
Safety brake	S_8 Safety brake	s_{81} safety brake, s_{82} no safety brake
Master support	S_9 Rack	s_{91} motor fixed pulley same side, s_{92} motor fixed pulley different side

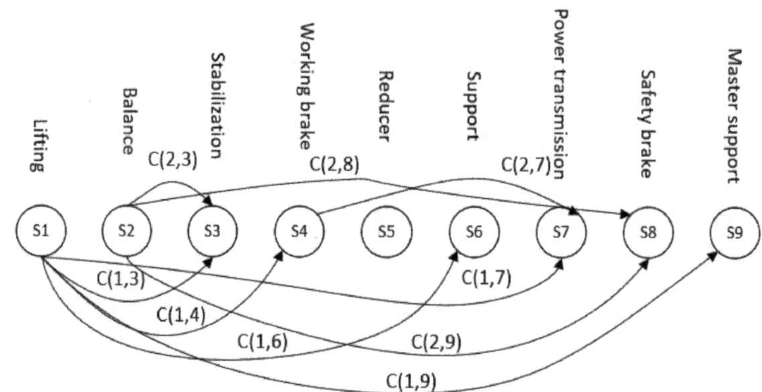

Figure 7. Constraint expression of hoist conceptual design.

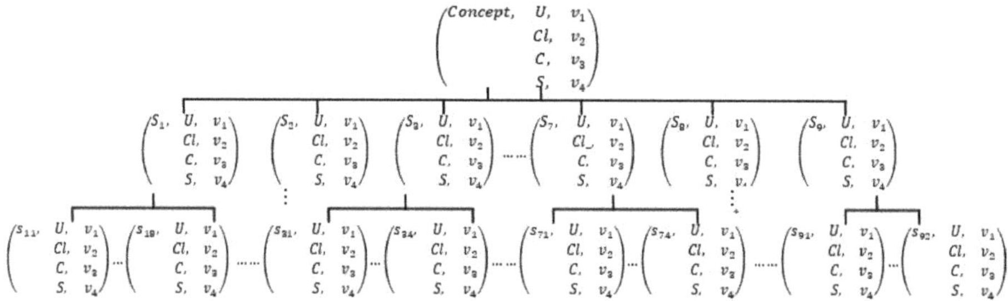

Figure 8. Matter-element model of fixed winch hoist.

4.1.2. Acquiring *PC* importance

Analyzing the importance of PC driven by CR

According to customer preferences, the weight vector for five customer requirements $CR = \{$① maintainability, ② long service life, ③ work stability and reliability, ④ energy utilization rate, ⑤ environment-friendly$\}$, $w_s = (0.29, 0.30, 0.31, 0.09, 0.05)$ appeared, and *PC* importance relationship matrix $W_{cr\text{-}PC}$ driven by customer requirement is obtained using the Analytic Hierarchy Process. Relative importance matrix R_i between *CR* and PC_i is evaluated by an expert team, and R_1 was taken as an example

$$
CR_1\ CR_2\ CR_3\ CR_4\ CR_5 R1 =
\begin{array}{c}
CR_1 \\ CR_2 \\ CR_3 \\ CR_4 \\ CR_5
\end{array}
\begin{pmatrix}
1 & \frac{1}{5} & \frac{1}{3} & 0 & \frac{1}{3} \\
5 & 1 & \frac{1}{4} & 0 & \frac{1}{5} \\
3 & 4 & 1 & 0 & 0 \\
0 & 0 & 0 & 1 & \frac{1}{5} \\
3 & 5 & 0 & 5 & 1
\end{pmatrix}
$$

According to AHP, $W_{cr\_PC}(1,j) = \dfrac{\sum_{j=1}^{5} R_{11j}}{\sum_{i=1}^{5} \dfrac{\sum_{j=1}^{5} R_{11j}}{5}}$

$$
W_{cr\_PC}
=
\begin{pmatrix}
0.06 & 0.21 & 0.36 & 0.11 & 0.10 & 0 & 0.13 & 0.07 & 0.09 & 0.39 & 0.03 & 0 & 0.07 \\
0.20 & 0.10 & 0.07 & 0.13 & 0.21 & 0 & 0.13 & 0.14 & 0.13 & 0.23 & 0.07 & 0 & 0.08 \\
0.25 & 0.14 & 0.10 & 0.49 & 0.49 & 0 & 0.47 & 0.09 & 0.49 & 0.14 & 0.53 & 0 & 0.17 \\
0.05 & 0.12 & 0.29 & 0.12 & 0.07 & 0.1 & 0.21 & 0.18 & 0.15 & 0.17 & 0.31 & 0.7 & 0.19 \\
0.44 & 0.43 & 0.18 & 0.14 & 0.13 & 0.9 & 0.06 & 0.52 & 0.14 & 0.05 & 0.06 & 0.3 & 0.49
\end{pmatrix}
$$

$w^{(1)} = w_s \times W_{cr\text{-}PC} = (0.3380, 0.1628, 0.1909, 0.2353, 0.2469, 0.0550, 0.2399, 0.1272, 0.2325, 0.2335, 0.2245, 0.0850, 0.1365)$.

Gaining Mutual Importance among PC Elements

Using the Analytic Hierarchy Process to obtain mutual importance vector $w^{(2)}$ among elements in PC. Relative importance matrix R'_i among PC_i is evaluated by the expert team, and R'_1 between PC_1 and the others was taken as an example.

$$R'_1 = \begin{array}{c} \\ PC_1 \\ PC_2 \\ PC_3 \\ PC_4 \\ PC_5 \\ PC_6 \\ PC_7 \\ PC_8 \\ PC_9 \\ PC_{10} \\ PC_{11} \\ PC_{12} \\ PC_{13} \end{array} \begin{array}{ccccccccccccc} PC_1 & PC_2 & PC_3 & PC_4 & PC_5 & PC_6 & PC_7 & PC_8 & PC_9 & PC_{10} & PC_{11} & PC_{12} & PC_{13} \\ \left(1 \right. & \tfrac{1}{5} & \tfrac{1}{2} & \tfrac{1}{5} & \tfrac{1}{3} & 0 & 0 & 0 & \tfrac{1}{3} & \tfrac{1}{6} & 0 & \tfrac{1}{6} & \left. \tfrac{1}{6} \right) \\ 5 & 1 & \tfrac{1}{4} & \tfrac{1}{6} & \tfrac{1}{3} & \tfrac{1}{6} & 0 & \tfrac{1}{2} & 0 & \tfrac{1}{7} & 0 & \tfrac{1}{6} & 1 \\ 2 & 4 & 1 & \tfrac{1}{2} & 0 & \tfrac{1}{2} & 0 & 0 & 0 & \tfrac{1}{5} & 0 & 1 & \tfrac{1}{2} \\ 5 & 6 & 2 & 1 & \tfrac{1}{7} & \tfrac{1}{2} & 0 & \tfrac{1}{2} & \tfrac{1}{7} & \tfrac{1}{3} & \tfrac{1}{7} & 1 & \tfrac{1}{4} \\ 3 & 3 & 0 & 7 & 1 & 0 & \tfrac{1}{5} & 0 & \tfrac{1}{2} & \tfrac{1}{6} & 0 & \tfrac{1}{3} & \tfrac{1}{7} \\ 0 & 6 & 2 & 2 & 0 & 1 & 0 & \tfrac{1}{7} & \tfrac{1}{2} & \tfrac{1}{5} & 0 & 0 & \tfrac{1}{7} \\ 0 & 0 & 0 & 0 & 5 & 0 & 1 & 0 & \tfrac{1}{5} & \tfrac{1}{3} & 0 & \tfrac{1}{9} & \tfrac{1}{3} \\ 0 & 2 & 0 & 2 & 0 & 7 & 0 & 1 & \tfrac{1}{2} & \tfrac{1}{3} & 0 & 0 & \tfrac{1}{3} \\ 3 & 0 & 0 & 7 & 2 & 2 & 5 & 2 & 1 & \tfrac{1}{3} & \tfrac{1}{5} & 0 & \tfrac{1}{2} \\ 6 & 7 & 5 & 3 & 6 & 5 & 3 & 3 & 3 & 1 & \tfrac{1}{3} & \tfrac{1}{6} & \tfrac{1}{7} \\ 0 & 0 & 0 & 7 & 0 & 0 & 0 & 0 & 5 & 3 & 1 & \tfrac{1}{5} & 0 \\ 6 & 6 & 1 & 3 & 0 & 9 & 0 & 0 & 0 & 6 & 5 & 1 & 0 \\ 6 & 1 & 2 & 4 & 7 & 7 & 3 & 3 & 2 & 7 & 0 & 0 & 1 \end{array}$$

Similar to the calculation method of $W_{cr\text{-}PC}$, the first column vector of W_{pc} was obtained: $W_{pc}(:,1) = (0.2359, 0.6712, 0.7462, 1.2086, 1.4282, 0.9495, 0.5368, 1.1111, 3.2800, 1.7718, 1.2462, 2.8462, 3.3077)$ according to R_1. Finally, similar to the calculation method of $w^{(1)}$, $w^{(2)} = (0.1110, 0.1321, 0.0852, 0.1231, 0.0742, 0.0173, 0.0952, 0.0952, 0.0903, 0.1123, 0.0548, 0.0100, 0.0300)$ was finally obtained.

Gaining PC Importance

Calculating the importance of PC according to Equation (5), $w = (0.1681, 0.0965, 0.0730, 0.1299, 0.0822\ 0.0043, 0.1025, 0.0543, 0.0942, 0.1176, 0.0552, 0.0038, 0.0184)$.

4.1.3. Acquiring Functional-Unit = Importance

W_i denotes the importance of functional unit i in the whole product concept, where $\sum_{i=1}^{n} W_i = 1$, and as for nine functional units of the hoist, ① lifting, ② balance, ③ stabilization, ④ working brake, ⑤ reducer, ⑥ support, ⑦ power transmission, ⑧ safety brake, and ⑨ master support, importance vector $W_i = (0.17, 0.11, 0.11, 0.10, 0.13, 0.09, 0.13, 0.07, 0.09)$ was obtained based on knowledge.

4.1.4. Obtaining PC Substructure Utility Vector

According to expert analysis, the impact of each candidate substructure on the PC was quantified by 0 to 9. As shown in Table 2, the larger the number is, the greater the impact. Particularly, 0 indicates that the substructure had no effect on this index.

Table 2. Substructure utility vector for a *PC*.

	Substructure Utility Vector																	
PC	s11	s12	s13	s14	s15	s16	s17	s18	s21	s22	s31	s32	s33	s34	S41	…	s91	s92
F1	7	7	8	8	9	9	9	9	2	2	5	7	9	1	5	…	2	2
F2	9	9	7	7	9	9	9	9	2	2	5	7	9	1	0	…	2	2
F3	9	9	9	9	7	7	7	7	5	3	5	9	7	1	3	…	2	2
F4	1	1	1	1	1	1	1	1	4	4	5	5	7	3	7	…	2	2
F5	5	5	5	5	7	7	7	7	4	4	5	5	7	3	7	…	2	2
F6	7	7	5	5	9	9	7	7	3	3	3	5	7	1	5	…	2	2
F7	0	0	0	0	0	0	0	0	0	0	0	0	0	0	9	…	2	2
F8	0	0	1	1	1	1	0	0	0	0	0	0	0	0	5	…	0	0
F9	7	7	5	5	7	7	9	9	6	6	7	7	7	3	0	…	0	0
F10	4	4	3	3	5	5	6	6	4	4	5	7	9	0	5	…	0	0
F11	0	0	0	0	0	0	0	0	0	0	0	0	0	0	0	…	0	0
F12	0	0	0	0	0	0	0	0	0	0	0	0	0	0	5	…	7	9
F13	3	3	3	3	3	3	3	3	3	3	3	3	3	7	7	…	7	9

4.2. Model Solution Process

In order to solve the above model with the EGA, the functional units were mapped to the game players; the substructures were mapped to the strategy set for the player; and the constraints condition was mapped to the game rules. Perturbation probability pi of each functional unit was given according to the importance degree of the functional units for the entire hoist design, where p_i = (0. 781, 0.547, 0.452, 0.343, 0.433, 0.536, 0.412, 0.246, 0.435), and maximum evolution generation T = 250 was set according to multiple experiment simulations, as shown in Figure 9.

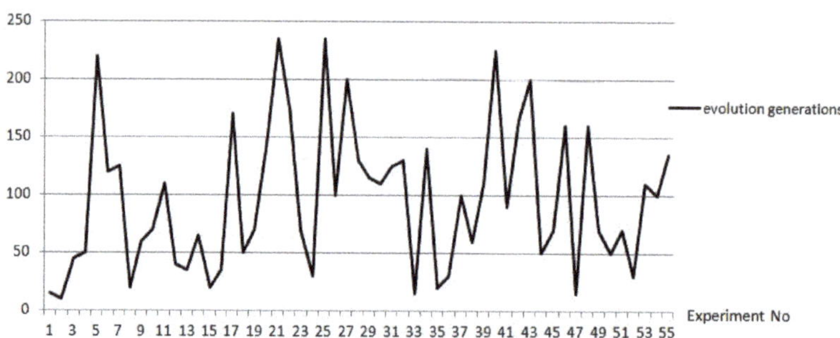

Figure 9. Nash equilibrium generation distribution.

Based on the above conditions, the evolutionary game was carried out in software environment MATLAB 8.3, in which the result of the conceptual design was shown in Figure 10. The optimal product design that meets the relevant *PC* was finally achieved as a single fold with center rope; fixed pulley placed vertically; horizontal speed reducer; balanced pulley placement; sliding bearing; wheel coupling; safety brake; and motor fixed pulley different side. The original data in this work were provided by a water-conservancy and hydroelectric-machinery company.

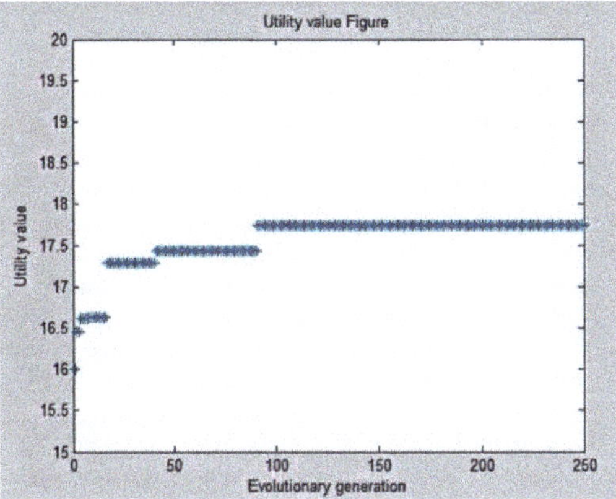

Figure 10. Solution optimization process.

4.3. Results and Discussion

Under the same conditions, the hoist concept was designed by the company's designers using an empirical design system of the company. As shown in Figure 11, the result was: single fold with center rope; balanced pulley placement; sliding bearing; wheel coupling; safety brake; and motor fixed pulley different side.

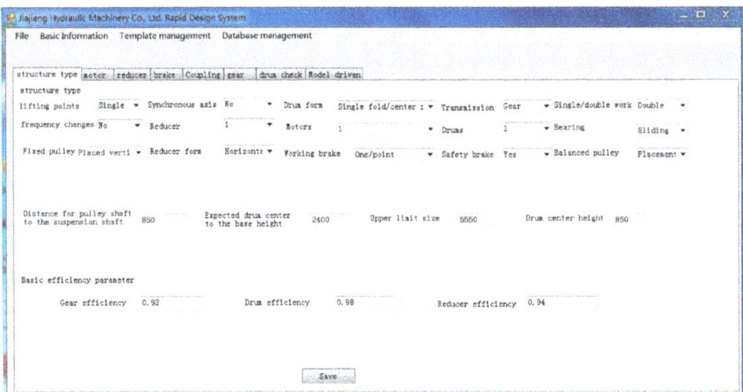

Figure 11. Current design-method result.

Comparing the concept designed by the company engineers with that achieved by the method proposed in this paper, the main difference was that engineers think that a motor fixed pulley different side makes the structure more compact, while there is neither CR nor PC related to compactness. From this point of view, the proposed design method in this paper is less advanced in the application of expert knowledge. A comprehensive comparison was made from occupation, design cycle, experiential knowledge, result reliability, and economy. Results are shown in Table 3.

Obviously, this method needs to be improved in the acquisition and learning of empirical knowledge, but performs well in other aspects.

Table 3. Performance comparison.

Project	Occupation	Design Cycle	Experiential Knowledge	Result Reliability	Economy	Total
Current Design Method	•Empirical design system •Experienced designer	1–2 days	•Design experience •Knowledge base	Reliable	Poor	General
Method of this Paper	Computer	0.5 h	Improved knowledge base	Reliable	Well	great
Improved	Greatly	Greatly	Little	Little	Good	Greatly

5. Conclusions

Product conceptual design was investigated in this paper. Based on the achieved experiment results, the following conclusions are derived:

It was found that a problem could be effectively solved by the method proposed in this paper. Using this process, the design cycle was reduced to 0.5 hour, and occupation and economy greatly improved.

The method does not perform well in empirical-knowledge application. Hence, our future work will focus on how to more accurately acquire design knowledge and objective–subjective expert knowledge.

Author Contributions: Conceptualization, Y.-L.H. and X.-B.H.; methodology, Y.-L.H.; software, Y.-L.H.; validation, X.-B.H., B.-Y.C. and R.-G.F.; formal analysis, X.-B.H.; investigation, Y.-L.H. and R.-G.F.; resources, R.-G.F.; data curation, R.-G.F. and Y.-L.H.; writing—original draft preparation, Y.-L.H. and B.-Y.C.; writing—review and editing, Y.-L.H.; visualization, Y.-L.H.; supervision, X.-B.H.; project administration, X.-B.H.; funding acquisition, X.-B.H.

Funding: This research was funded by Sichuan Province Science and Technology Support Program, grant number (2017GZ0146, 2018GZ0125); Made in China 2025-Sichuan Province program, grant number 2018ZZ011. The APC was funded by Sichuan Province Science and Technology Support Program.

Conflicts of Interest: The authors declare no conflict of interest.

References

1. Tan, J.-R.; Liu, Z.-Y. *Intelligent Manufacturing*; China Machine Press: Beijing, China, 2017; pp. 154–196.
2. Chang, D.; Chen, C.-H. Product concept evaluation and selection using data mining and domain ontology in a crowd sourcing environment. *Adv. Eng. Inform.* **2015**, *29*, 759–774. [CrossRef]
3. Sun, M.-B.; Zhao, Y.-X.; Zhao, G.-Y.; Liu, Y. A conceptual design of shock-eliminating clover combustor for large scale scramjet engin. *Acta Astronaut.* **2017**, *130*, 34–42. [CrossRef]
4. Wang, P.-J.; Gong, Y.-D.; Xie, H.-L. Applying CBR to Machine Tool Product Configuration Design Orientedto Customer Requirements. *Chin. J. Mech. Eng.* **2016**, *30*, 60–73. [CrossRef]
5. Christophe, F.; Bernard, A.; Coatanéa, É. RFBS: A model for knowledge representation of conceptual design. *CIRP Ann. Manuf. Technol.* **2010**, *59*, 155–158. [CrossRef]
6. Chen, Y.; Zhao, M.; Xie, Y.; Zhang, Z. A new model of conceptual design based on Scientific Ontology and intentionality theory. Part II: The process mode. *Des. Stud.* **2015**, *38*. [CrossRef]
7. Tiwari, V.; Jain, P.K.; Tandon, P. Product design concept evaluation using rough sets and VIKOR method. *Adv. Eng. Inform.* **2016**, *30*, 16–25. [CrossRef]
8. Wang, H.-W.; Liu, G.; Yang, Z.-D. Multi-criteria comprehensive evaluation method of mechanical product design scheme. *J. Harbin Inst. Technol.* **2014**, *46*, 99–103.
9. Huang, H.-Z.; Bo, R.-F.; Chen, W. An integrated computational intelligence approach to product schematic generation and evaluation. *Mech. Mach. Theory* **2006**, *41*, 567–583. [CrossRef]
10. Augustine, M.; Yadav, O.P.; Jain, R.; Rathore, A.P.S. Concept convergence process: A framework for improving product concepts. *Comput. Ind. Eng.* **2010**, *59*, 367–377. [CrossRef]

11. Ma, S.; Jiang, Z.; Liu, W.; Huang, C. Design Property Network-Based Change Propagation Prediction Approach for Mechanical Product Development. *Chin. J. Mech. Eng.* **2017**, *30*, 676–688. [CrossRef]

12. Lin, M.-C.; Lin, Y.-H.; Lin, C.-C.; Chen, M.-S.; Hung, Y.-C. An integrated neuro-genetic approach incorporating the Taguchi method for product design. *Adv. Eng. Inform.* **2015**, *29*, 47–58. [CrossRef]

13. Matei, O.; Contras, D.; Pop, P.; Vălean, H. Design and comparison of two evolutionary approaches for automated product design. *Soft Comput.* **2016**, *20*, 4257–4269. [CrossRef]

14. Ling, S.H.; San, P.P.; Chan, K.Y.; Leung, F.H.F.; Liu, Y. An intelligent swarm based-wavelet neural network for affective mobile phone design. *Neurocomputing* **2014**, *142*, 30–38. [CrossRef]

15. Lin, X.-H.; Feng, Y.-X.; Tan, J.-R. Constraint model and calculating methold by evolutionary game algorithm for product conceptual design. *J. Zhejiang Univ. (Eng. Ed.)* **2012**, *46*, 533–540.

16. Yue, D.-X.; Zeng, J.-J.; Yang, C.; Zou, M.-L.; Li, K.; Chen, G.-G.; Guo, J.-J.; Xu, X.-F.; Meng, X.-M. Ecological risk assessment of the Gannan Plateau, northeastern Tibetan Platea. *J. Mount. Sci.* **2018**, *15*, 1254–1267. [CrossRef]

17. Huang, H.; Jin, L.; Rui, Z.-Y.; Guo, W.-D. Reconfigurable design of complex mechanical product design based on matter-element model. *Comput. Integr. Manuf. Syst.* **2013**, *19*, 2686–2696.

18. Liu, Y.-M.; Zhao, C.-C.; Xiong, M.-Y.; Zhao, Y.-H.; Qiao, N.-G.; Tian, G.-D. Assessment of bearing performance degradation via extension and EEMD combined approach. *J. Cent. South Univ.* **2017**, *24*, 1155–1163. [CrossRef]

19. Lv, F.; Yang, Y.-S.; Guo, C.-Q. New approach of criticality analysis of equipment based on improved fuzzy matter-element model. *J. Jilin Univ. (Eng. Technol. Ed.)* **2014**, *44*, 111–116.

20. Xiang, X.; Luo, Y.; Cheng, H.; Sheng, Y.; Wang, Y.; Zhang, Y. Biogas engineering technology screening based on analytic hierarchy process and fuzzy comprehensive evaluation. *Trans. Chin. Soc. Agric. Eng. (Trans. CSAE)* **2014**, *30*, 205–212.

21. Wang, T.-C.; Chen, B.-F.; Bu, L.-F. Extension Configuration Model of Product Scheme Design Based on Axiomatic Design. *China Mech. Eng.* **2012**, *23*, 2269–2276.

22. Chan, L.K.; Kao, H.P.; Wu, M.L. Rating the importance of customer needs in quality function deployment by fuzzy and entropy methods. *Int. J. Prod. Res.* **1999**, *37*, 2499–2518. [CrossRef]

23. Xu, M. *Research on Game Theory Idea Based Optimization Algorithm*; University of Science and Technology of China: Hefei, China, 2006.

24. Ye, J. *Optimization by Evolution Game*; Huazhong University of Science and Technology: Wuhan, China, 2003.

Article

Automatic Test and Sorting System for the Slide Valve Body of Oil Control Valve Based on Cartesian Coordinate Robot

Pingping Liu *, Gangjun Li, Rui Su and Guang Wen

College of Mechanical Engineering, Chengdu Technological University, No. 1, Zhongxin Avenue, Pidu Distric, Chengdu 611730, China; lgjun1@cdtu.edu.cn (G.L.); suruictu@cdtu.edu.cn (R.S.); wg66360336@cdtu.edu.cn (G.W.)
* Correspondence: lpping1@cdtu.edu.cn

Received: 21 October 2018; Accepted: 11 December 2018; Published: 13 December 2018

Abstract: Current industrial robotics technology is often not well integrated with the enterprise's on-site environment and actual working conditions and small and medium-sized enterprises are unable to achieve product automation due to production cost constraints. In order to meet the medium and small scale production of the slide valve body of OCV (Oil Control Valve) of a certain enterprise and its special process requirements, the automatic test system and sorting system based on the production environment of the enterprise are studied and designed. Firstly, according to the production conditions and process requirements of the enterprise, the overall design scheme of the automatic production line is put forward based on the existing automatic assembly system. Secondly, the test description is further improved by analysing and interpreting the test requirements of the products in detail and the automatic test system and test process are designed. Finally, according to the sorting process requirements, a Cartesian coordinate robot sorting system with two-terminal manipulators parallel operation is designed and its sorting motion scheme is optimized. The automatic test system and sorting system are seamlessly connected with the automatic assembly system, which can efficiently complete the automatic test and sorting of products and meet the production cycle time.

Keywords: slide valve body of OCV; automatic production line; test system; sorting system; Cartesian coordinate robot

---

1. Introduction

The production and testing of OCV valve are very difficult owing to its complexity and high precision. Although some international manufacturers have relatively complete production and testing systems but the technology is closed. The OCV valve test system produced by Schaeffer Group in Germany has powerful functions. It can simulate the control of the OCV valve by ECU (Electronic Control Unit), adjust the oil temperature and pressure in a wide range and monitor the parameters of the OCV valve. The system has high test accuracy and repeatability. The production of OCV valves and actuators in most countries is still in the exploratory stage and few enterprises have a complete production and testing system. An OCV valve performance test bench with higher automation and better test accuracy was studied and designed by Tang and Deng [1]. An OCV valve flow experimental platform was built by Xie to test the dynamic response characteristics of the OCV valve and verify the reliability of the experiment [2]. An OCV displacement test bench was designed and developed by Yao, Li and a Ningbo company [3,4]. The above research only provides guidance for OCV valve testing but fails to provide specific test methods and reference test indicators, which cannot meet the production needs.

Automatic production lines are usually composed of the assembly system, the test system and the sorting system, which can automatically complete all or part of the manufacturing process of the product. Adopting automation line in mass production can improve labour productivity, improve and stabilize product quality, improve labour conditions, shorten production cycle and have remarkable economic benefits [5–7]. Cartesian coordinate robots have been widely used in automatic production lines of various products because of their advantages such as strong load capacity, easy combination and expansion, low cost and so on [8]. They are usually used for handling and sorting. However, the mature, stable and reliable industrial robot technology on the market is often unable to combine well with the on-site environment and the actual working conditions of enterprises; in addition, many enterprises have realized automatic assembly of small and medium-sized products but the testing and sorting of products are still manual operation due to the limitation of production cost, which is not conducive to the improvement of product quality and stability. Moreover, poorly designed production lines will not only reduce the efficiency of equipment but also may make the production lines too complex. Therefore, according to the product type, production scale and actual working conditions of the enterprise, it is of practical significance to study and design the industrial robot production line suitable for the enterprise.

Because the production of OCV valve is not common, this paper takes the production of OCV slide valve body as the background, aiming at efficient testing and sorting. The main purpose of the study is to provide the actual test indicators as reference for test analysis and test description and to propose a design scheme. At the same time, the design of a new sorting system and sorting movement is emphasized and a novel application case of rectangular coordinate robot in a specific production line is provided. According to the actual situation of small and medium-sized enterprises, the sorting system of production line is reasonably designed, the cycle time is allocated and the sorting movement is optimized, so as to reduce the cost and improve the sorting efficiency.

2. Design of the Automatic Production Line

2.1. Presentation of Problems

OCV is an important part of the traditional VVT (Variable Valve Timing) system and it is the oil control valve that controls the timing of the engine intake valve. Most automotive engines use the structure of extrapolated OCV to achieve the best performance [9–11]. OCV is essentially an electromagnetic sliding valve which controls the direction and flow of engine oil [12], including the magnet part and the slide valve body part. The slide valve body is also an important part of a new type of intermediate VVT system.

The structure of the slide valve body of OCV is mainly composed of a centre bolt, a valve sleeve, a valve core, a reset spring and a one-way valve ball and so forth, as shown in Figure 1. The oil inlet and the oil outlet are arranged at both ends of the centre bolt and the oil inlet end is screwed into the intermediate locking phase device through the thread to connect with the oil circuit of the intake camshaft; the valve core of the oil outlet end is connected with the magnet part of OCV; four pairs of holes are evenly opened on the circumferential surface of the centre bolt and the four holes A and four holes B are respectively connected with the A cavity and the B cavity of the intermediate locking phase device [13].

The automatic production line of the slide valve body of OCV should consist of an automatic assembly system, a test system and a sorting system. The automatic assembly system mainly completes the automatic assembly of the product parts; the test system is responsible for the automatic test of the basic indicators of the product; the sorting system, also known as the loading and unloading robot, on the one hand completes the automatic loading of the test products: loading the assembled products to the test system; on the one hand, completes the automatic unloading and sorting of the tested products: unloading the qualified products to the qualified storehouse and the unqualified products to the unqualified storehouse after testing.

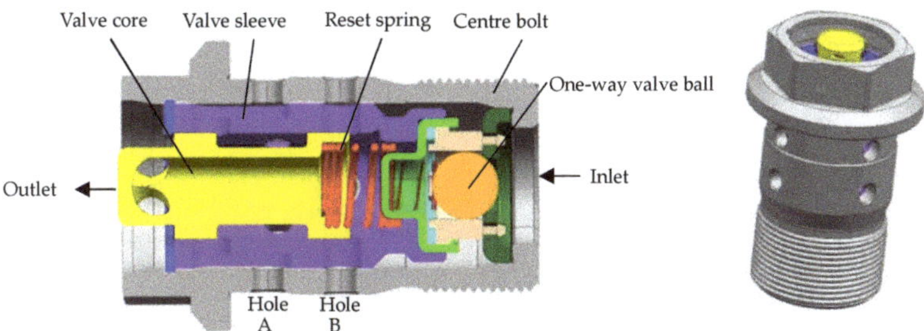

Figure 1. Structure of the slide valve body of OCV.

The automatic assembly system of the slide valve body of OCV already exists in the enterprise, as shown in Figure 2. The assembly system consists of five single-row assembly stations and a multi-station rotary automatic feeding system is set beside each station. The rotary automatic feeding system is assisted by manual feeding and the manipulators automatically feed products to the assembly line and perform the assembly work. The products are conveyed in sequence by the clamping conveying system in the direction of the arrow shown in the diagram and the products are gradually assembled from station 1 to station 5.

The automatic test system and sorting system matching with the existing assembly system of the enterprise are designed according to the production cycle time, test requirements, sorting process requirements and so on. Detailed test requirements and process requirements are shown in Section 2.3 of this article.

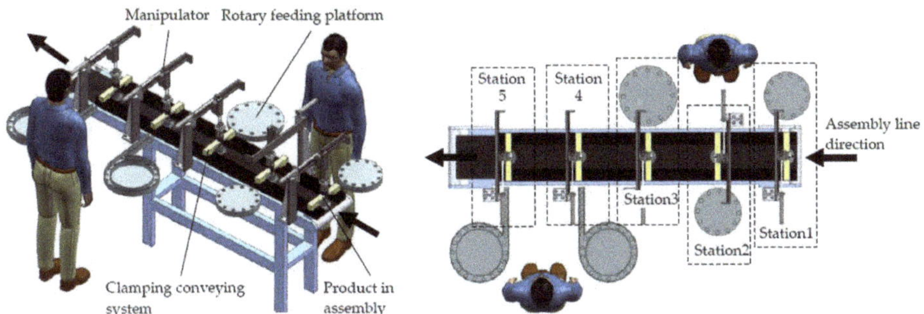

Figure 2. The automatic assembly system.

2.2. General Design of the Automatic Production Line

The whole production line designed in this paper based on the production scale, production environment and process requirements of the slide valve body of OCV is shown in Figure 3. It is composed of an automatic assembly system, a test system and a sorting system. It is 1.2 m wide and 1.5 m high. The automatic test system and sorting system are arranged at the exit of the assembly line; the test system is located on the ground and the sorting system is arranged above the test system and supported by it. The overall workflow of the automatic production line is as follows:

(1)　The products are automatically assembled and sent to the waiting test station of the automatic test system.

(2)　The product to be tested at the waiting test station is loaded to the test station of the test system and is tightened by the sorting system.

(3)　The automatic test system starts testing products and reads data;

(4)　The tested products are unscrewed and unloaded by the sorting system.

(5) The qualified products are unloaded to the qualified storehouse and the unqualified products to the unqualified storehouse by the sorting system.

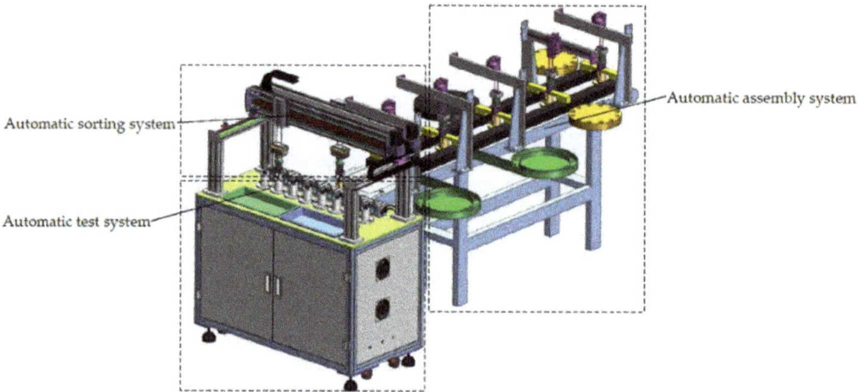

Figure 3. General structure of the automatic production line.

2.3. Design of the Automatic Test System

2.3.1. Test Requirements Analysis

The automatic test system tests the force, flow, displacement and other parameters of the slide valve body of OCV delivered by the sorting system. The main test requirements are shown in Table 1.

Table 1. Test requirements.

NO.	Category	Requirements	Remarks
1	Test media	Dry air	
2	Functional tests	1. Exercise (2 cycles minimum) to check centre bolt assembly integrity & friction. Test for First Contact (initial point) and Spool Stroke. 2. Spring rate test (three points force tests: F1, F2, F3) 3. Midpoint force 4. Holding point force 5. Holding point leakage 6. Mid-width force 7. Hysteresis force 8. A point force 9. B point force 10. Initial (start) point flow 11. Full travel flow/valve core travel	See Figures 4 and 5 and Table 2 for detailed test description. Air test parameters are to be determined after the fluid-to-air correlation study. Flow shows air flow and related oil flow in the same time.

In addition, in order to design the test system, the relevant data in Figures 4 and 5 and Table 2 is added in this paper to further improve the test description by analysing and interpreting the test requirements.

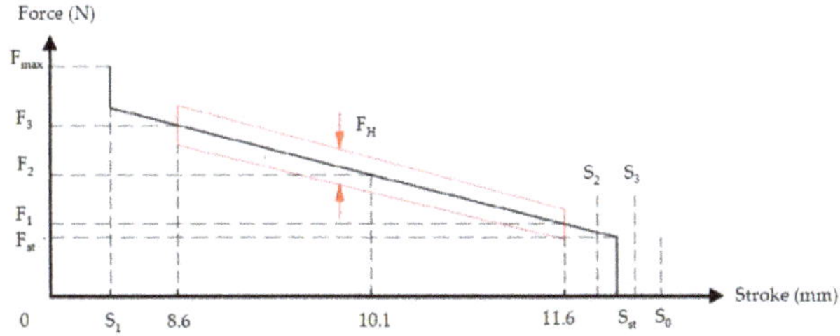

Figure 4. Stroke and force Curve.

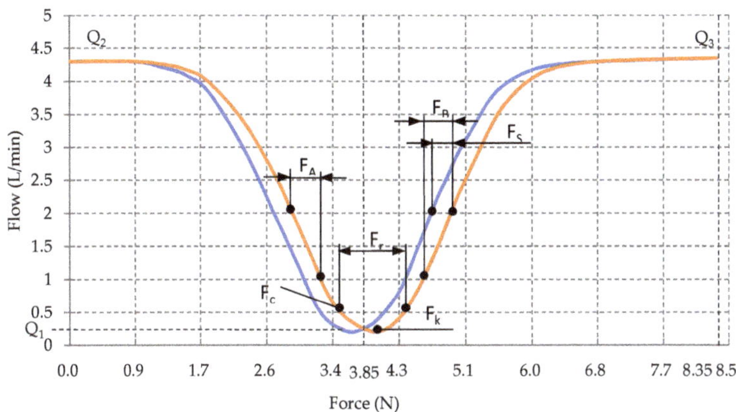

Figure 5. Force and flow characteristic curve.

Table 2. Test description and test parameters.

NO.	Test Description	Test Parameters
1	Exercise (2 cycles minimum): Tests: First contact Maximum travel/Spool stroke	First contact: TBD mm Max travel: TBD mm (maximum)
2	Spring rate test (3 points force tests): (F_1), (F_2), (F_3)	$(F_1) = 1.4 \pm 0.8$ N @ 11.6 mm $(F_2) = 4.45 \pm 0.95$ N @ 10.1 mm, $(F_3) = 7.5 \pm 1.1$ N @ 8.6 mm
3	Midpoint force (F_c)	$(F_c) = 3.2 \pm 0.8$ N at TBD mm
4	Holding point force (F_k)	$(F_k) = 4 \pm 0.8$ N at TBD mm
5	Holding point leakage (Q_1)	$(Q_1) < 0.2$ L/min
6	Mid-width force (F_r)	$(F_r) = 1.6 \pm 0.6$ N
7	Hysteresis force (F_s), Hysteresis band width (F_H)	$(F_s) < 0.1$ N, $(F_H) < 2.2$ N
8	Point A force (F_a)	$(F_a) = 0.2 \pm 0.1$ N
9	Point B force (F_b)	$(F_b) = 0.2 \pm 0.1$ N
10	Start point flow (S_{st}), (Q_2)	$(Q_2) > 4$ L/min Start Point: $(S_2) < (S_{st}) < (S_3)$ Force $(F_{st}) = 0.9 \pm 0.4$ N Lower Travel $(S_2) = 11.8$ mm Upper Travel Limit $(S_3) = 12.1$ mm
11	Full travel flow and Valve core travel Full travel (F_{max}), Flow (Q_3), Valve core travel (S)	$(Q_3) > 4$ L/min, $(S) = 3.5 \pm 0.2$ mm, $(F_{max}) > 10$ N, $(S1) = $ TBD mm

2.3.2. Scheme Design of the Automatic Test System

According to the test requirements and description of the slide valve body of OCV, the test scheme of the test system is presented, as shown in Figure 6. And the automatic test system of the slide valve body of OCV is designed based on this scheme.

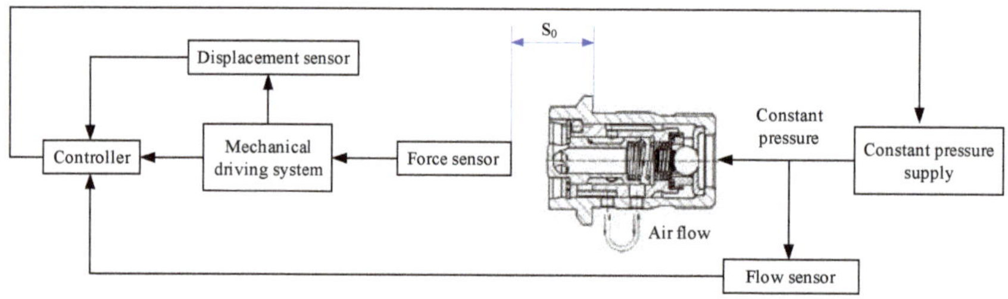

Figure 6. Testing scheme for the test system.

The inlet of the slide valve body of OCV is connected with a constant pressure source to provide a constant pressure air input; the flow sensor and controller are connected with the inlet to detect the input air flow and read data; and the valve core at the outlet is connected with a force sensor, a mechanical driver, a displacement sensor and a controller.

The automatic test system of the slide valve body of OCV is designed based on the above test scheme, as shown in Figure 7. It is mainly composed of a control cabinet and a test platform.

The control cabinet uses aluminium profile as its frame and the overall size is 1200 mm (length) × 700 mm (width) × 870 mm (height). The constant pressure source and its regulating device, power supply, main control board and valve group are arranged in the control cabinet.

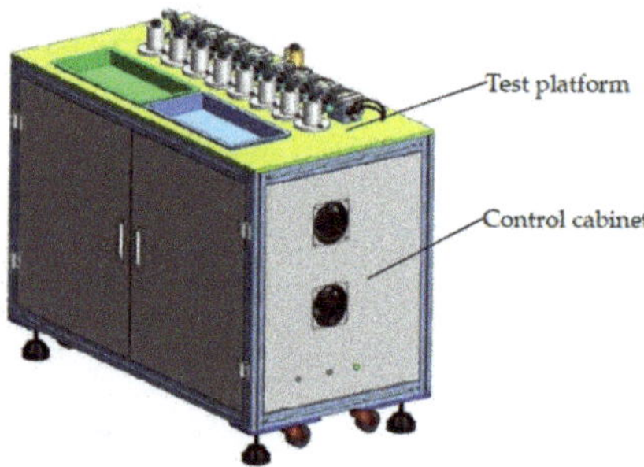

Figure 7. Composing of the automatic test system.

The test platform is composed of a panel, several test fixtures, airflow and force detection devices, two finished product storehouses and so forth, as shown in Figure 8. The panel is 1200 mm (length) × 620 mm (width) × 10 mm (thickness), made of aluminium alloy plate, fixed on the top of the control cabinet of the system, which supports the whole test platform. The test platform can be divided into a waiting test station, a test station and a finished product station according to the product flow direction. The waiting test station consists of a product to be tested and its fixture, which is used to receive the finished product sent by the automatic assembly system. Eight identical test items and their test fixtures are arranged side by side at the test station for receiving the products sent from the waiting test station. The test fixture is made of nylon material and designed according to the configuration and installation requirements of the centre bolt; the test fixture has a test interface on it and is fixed on the panel. The airflow and force detection device is designed as a whole with compact structure and the coupling and separation of the flow test port and the force value test port are completed simultaneously by the expansion of the light cylinder piston. Each test product corresponds to a set of detection devices. The finished product station adopts two plastic containers embedded under the panel and it is divided into a qualified product storehouse and an unqualified product storehouse.

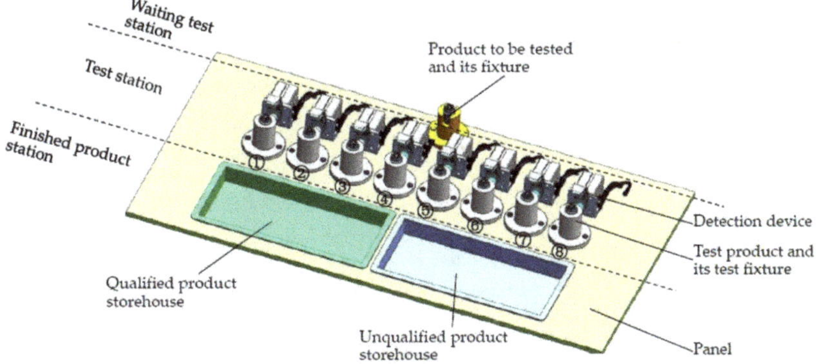

Figure 8. Composing of the test platform.

According to the actual working conditions and the requirement of production cycle time, the test process is designed as follows:

(1) Feed the product to be tested into the test fixture of the test station and tighten it.
(2) Connect the airflow and force value detection device;
(3) Ventilation, power up, start testing and reading data;
(4) Disconnect the airflow and force detection device;
(5) Unscrew and remove the tested product.

2.4. Design of the Automatic Sorting System

2.4.1. General Design Requirements of the Sorting System

On the one hand, the automatic sorting system completes the automatic loading work of the test system: loading the assembled products to the test station from the waiting test station of the test system; on the other hand, completes the automatic unloading and sorting work of the test system: unloading the tested products from the test station to the finished product station, unloading the qualified products to the qualified product storehouse, the unqualified products to the unqualified product storehouse. The whole working rhythm should meet the production cycle time of the whole production line, the design requirements of the automatic sorting system are shown in Table 3.

Table 3. Design requirements of the automatic sorting system.

NO.	Category	Requirements
1	Part Handling	Automatic load. Automatic unload accept to an accept conveyor. Automatic unload reject to the reject conveyor.
2	Cycle Time	Estimated 16 s/part

Combined with the automatic test system and testing process designed in this paper, sorting actions can be refined as follows:

(1) Grab, unscrew and vertically lift the product to be tested at the waiting test station;

(2) Move the product to be tested to the test station;

(3) Put the product to be tested vertically into the test fixture and tighten it;

(4) Grab, unscrew and lift the tested product vertically;

(5) Move the tested product to the corresponding product storehouse.

2.4.2. Type Selection of the Sorting Robot

A Cartesian coordinate robot is generally composed of the linear motion part, the control part, the drive part and the terminal manipulator. The most basic component of the Cartesian coordinate robot is the linear motion unit and various combinations of linear motion units can be used to form one dimensional, two-dimensional and three-dimensional robot [14–16]. Cartesian coordinate robots are generally supported at both ends and have higher rigid strength under the premise of fixed stroke and given structure size; three joints are independent and the kinematics solution is simple without coupling or singular state; at the same time, the accuracy and position resolution of Cartesian coordinate robots are not easily affected by external factors and it is easy to achieve very high working accuracy [17,18]. The Cartesian coordinate robot has great bearing capacity, flexible assembly and convenient operation and maintenance. It can be seen from the above that the sorting action of this production line is not complicated and the selection of the Cartesian coordinate robot can completely meet the requirements.

2.4.3. General Design of the Sorting Robot

The sorting robot designed in this paper is made up of the robot body and the support bracket, as shown in Figure 9. The support bracket is made up of four 40 mm × 80 mm aluminium profiles and is fixed on the panel of the test platform to support the robot body. The robot body includes X, Y, Z axis linear motion units and terminal manipulators, which can realize the linear motion of the terminal manipulators along X axis, Y axis and Z axis in space, as well as the rotation (R axis) around Z axis and the grabbing action. The Y axis of robot body is supported by two sides and driven unilaterally. A Y axis linear motion unit is set on the right side of the robot body and a Y axis supporting slide rail is set on the left side, which are respectively fixed at the top of the support bracket. The X axis of the robot body has two sets of parallel linear motion units, which are fixed back to back on the Y axis connecting plates. The Z axis has two sets of linear motion units, which are respectively fixed on the connecting plates of the two sets of X axis linear motion units and two sets of terminal manipulators are respectively fixed on the ends of the Z axis.

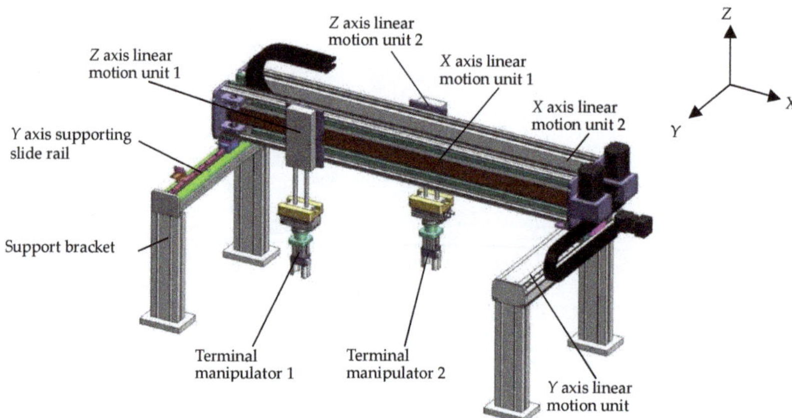

Figure 9. General configuration of the sorting robot.

2.4.4. Scheme Design of Sorting Motion

The sorting system has two sets of terminal manipulators completing the sorting work in parallel. The linear motions along X axis, the linear motions along Z axis, the rotations around their respective Z axis and the grabbing motions of the two terminal manipulators are independent of each other, while their linear motions along Y axis are synchronous. This configuration can combine (1) and (4) of the sorting actions described in 2.4.1 and combine (2), (3) and (5) into following actions:

(1) The system detects the number of the product (assuming NO. 2 represented by ② in Figure 8 or Figure 10) that was first completed at the test station and the corresponding detection device is disconnected. The terminal manipulator 1 moves to the No. 2 test position along X axis, drops along Z axis, grabs, unscrews and lifts up the No. 2 tested product. At the same time, The terminal manipulator 2 aligns with the product to be tested at the waiting test station, then drops along Z axis, grabs, unscrews and lifts the product to be tested and moves the object along X axis to the position corresponding to the test position No. 2 according to the system instructions. As shown in Figure 10.

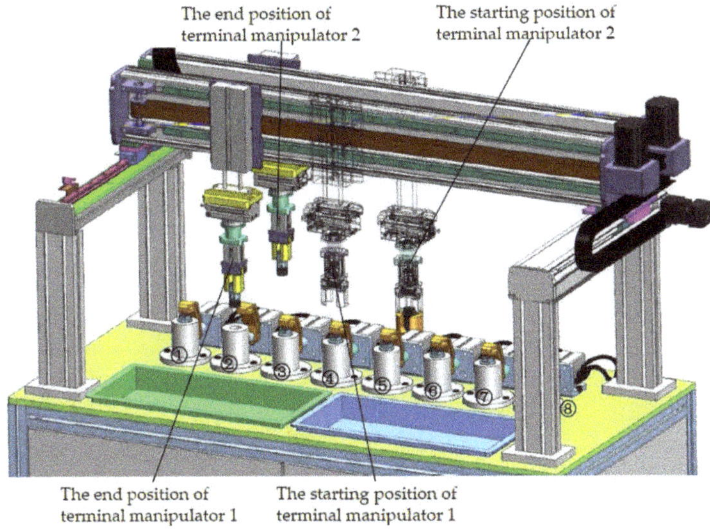

Figure 10. Sketch map of sorting action 1.

(2) The system judges whether the tested product No. 2 is qualified or not. The terminal manipulator 1 performs two-axis simultaneous motion according to the system instructions, that is, moving along X axis and aligning the tested product NO. 2 to the centre of the corresponding product storehouse while moving along Y axis. The paw is released when the terminal manipulator 1 reaches the top of the product storehouse and drops the product into the product storehouse. At the same time, the terminal manipulator 2 moves to the No. 2 test station along Y axis synchronously with the terminal manipulator 1 and drops the product to be tested down into the No. 2 test fixture along Z axis and screws it.

At this point, a work cycle is completed, as shown in Figure 11. You can compare Figure 11 with Figure 10 to see the position changes of the terminal manipulator 1 and the terminal manipulator 2.

The sorting system has two sets of terminal manipulators working in parallel. Two sets of X axis linear motion units are fixed back to back on the Y axis connecting plates. This design can eliminate the deflection torque of X axis horizontal motion in the Y direction and improve the stability of the system. More importantly, it can further improve the production efficiency and meet the production cycle time.

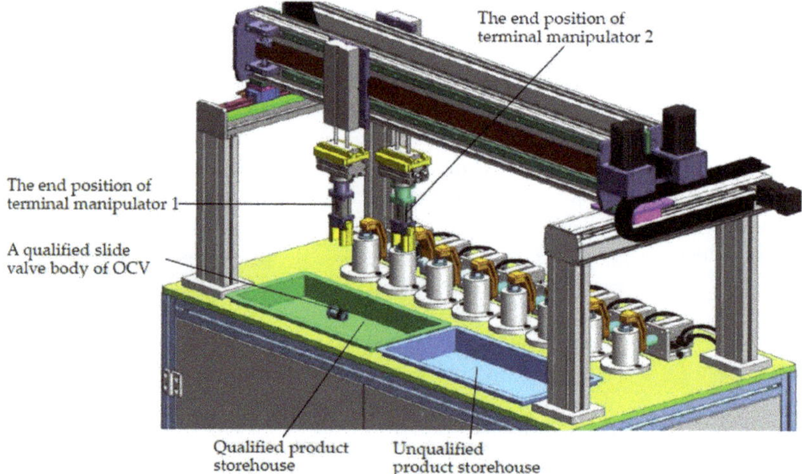

Figure 11. Sketch map of sorting action 2.

2.4.5. Design of Robot Transmission Form and Detailed Parameters

The linear motion unit is generally supported by the aluminium profile and the inner part is the transmission device and the guide rail. The linear motion is accomplished by moving the slider. The internal drive forms of linear motion unit mainly include the ball screw, the synchronous tooth belt and the gear rack, which have different characteristics [19]. The linear motion unit based on synchronous toothed belt is composed of the servo motor, the pulley bracket bearing, the pulley, the synchronous belt, the coupling and so forth. It is commonly used in high speed, high acceleration, heavy load equipment and with low cost.

(1) The X axis

The X axis is mainly used to complete the X direction movement of products. Its maximum load includes a Z axis linear motion unit, a terminal manipulator, a slide valve body of OCV and so forth. The total load of X axis is about 3.5 kg. The maximum running speed of X axis is 1.2 m/s and the maximum acceleration is 2.4 m/s^2. According to the size of the test platform, the effective travel of X axis is 770 mm and the span is small. The two linear motion units of X axis adopt synchronous belt linear modules and their motors are installed in opposite directions. Detailed parameters are shown in Table 4.

Table 4. Parameters of the robot body.

Object	Transmission Form	Type	Effective Travel	Maximum Speed	Maximum Acceleration	Load Capacity
X axis	synchronous belt linear module	/	1000 mm	2.7 m/s	3 m/s^2	30 kg
Y axis	synchronous belt linear module	/	500 mm	2.7 m/s	3 m/s^2	30 kg
Z axis	double shaft linear cylinder	TN20×80-S	80 mm	0.5 m/s	/	20 kg (@0.5 MPa)
Terminal manipulator	Rotary cylinder	MSQB10-A	190°	0.2~1 s/90°	/	7.4 kg (radial); 0.89 N.m (@0.5 MPa)
	Finger cylinder	MHZ2-16D	6 mm	/	/	4.5 kg

(2) The Y axis

The Y-axis is mainly used to complete the Y direction movement of products. The maximum load of the Y axis includes two X axis linear motion units, two Z axis linear motion units, two terminal manipulators, two slide valve bodies and so forth. Because the Y axis is supported by two sides, a single linear motion unit only takes half of the total load, 18 kg. the maximum running speed of Y axis is 1.2 m/s and the maximum acceleration is 2.4 m/s^2. According to the size of the test platform, the effective travel of Y axis is 212 mm. The synchronous belt linear module is selected and the detailed parameters are shown in Table 4.

(3) The Z axis

The Z axis is mainly used to complete the vertical lifting and dropping of products. The load of Z axis consists of a terminal manipulator and a slide valve body of OCV with a total load of about 2 kg. The maximum running speed of Z axis is 0.4 m/s and the effective travel is 60 mm, which can be realized by using the double shaft linear cylinder. The detailed parameters are shown in Table 4.

(4) The terminal manipulator

The robot terminal manipulator can be divided into clamp and suction type according to the clamping principle [20]. The terminal manipulator of this sorting system needs to complete the grabbing and rotating action of the slide valve body of OCV. The pneumatic manipulator is adopted [21,22], the rotation of the product is realized by a rotary cylinder and the product is grabbed by a finger cylinder with parallel air claws. The rotary cylinder and the finger cylinder are selected from standard type products, as shown in Table 4. At the same time, to adjust the air claw size and improve the grabbing stability, the structure of the air claw is redesigned according to the shape of the slide valve body of OCV, as shown in Figure 12. The rotary cylinder is connected with the cylinder telescopic rod of Z axis through the mounting plate of rotary cylinder and the finger cylinder is connected with the rotary cylinder through the mounting plate of finger cylinder.

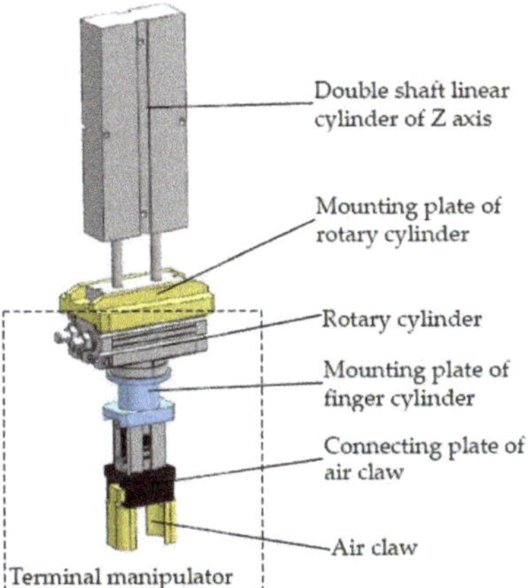

Figure 12. Structure of Z axis and the terminal manipulator.

3. Conclusions

In this paper, the automatic production line of the slide valve body of OCV is designed based on the existing production conditions of an enterprise. The automatic test system is designed based on the further improved test description and the sorting system is a Cartesian coordinate robot with two terminal manipulators working parallel, which can realize more reasonable and efficient testing workflow and sorting motion. The automatic test system and sorting system are seamlessly connected with the automatic assembly system and the automatic test and sorting of the products can be completed efficiently on the basis of the specified production cycle time to meet the needs of medium and small scale production. The automatic production line designed in this paper has the advantages of low cost, practicability and effectiveness and can solve the problems of low production efficiency, high labour cost and unstable product quality. In addition, the scheme design of the automatic production line can be applied to medium and small scale automatic production equipment of other valve products, especially the design of the automatic test and the sorting system.

Author Contributions: Formal investigation and analysis, P.L., G.L. and G.W.; analysing and improving the test description, P.L. and R.S.; methodology and detailed design, P.L.; analysis tools and resources, G.L.; writing—original draft preparation, P.L.; project administration, P.L.

Funding: This research was funded by National Natural Science Foundation of China, grant number 51705041 and the Education Department of Sichuan Province, China, project number 18ZB0034.

Conflicts of Interest: The authors declare no conflict of interest.

References

1. Tang, D.M.; Deng, J.M. Research and design of performance testing device for hydraulic control valve. *Mach. Des. Manuf.* **2011**, *6*, 57–59. [CrossRef]
2. Xie, B.Q. Experimental Study on Dynamic Characteristics of Phase Regulation of Variable Valve Timing Mechanism. Master's Thesis, Chongqing University, Chongqing, China, 2012.
3. Yao, M. Hydraulic System Testing of OCV and Actuator. Master's Thesis, Wuhan University of Technology, Wuhan, China, 2012.
4. Li, Z. The Online Testing System for Oil Control Valve and Actuator of Variable Valve Timing Engine. Master's Thesis, Wuhan University of Technology, Wuhan, China, 2012.
5. Lanza, G.; Stoll, J.; Koelmel, A.; Peter, S. Flexible Production Lines for Series Production of Automotive Electric Drives. In Proceedings of the 2nd International Electric Drives Production Conference (EDPC), Nuremberg, Germany, 15–18 October 2012.
6. Ren, Y. The application of industrial robot in flexible production line. *Autom. Instrum.* **2015**, *10*, 86–88. [CrossRef]
7. Ren, Z.G. Use situation of robot and development tendency analysis. *Aeronaut. Manuf. Technol.* **2015**, *13*, 106–108. [CrossRef]
8. Du, F.R. Design of the control system of Cartesian coordinate palletizing robot. *Mach. Build. Autom.* **2018**, *3*, 181–183. [CrossRef]
9. Zhou, Z. Research on Intelligent Test Methods of OCV. Master's Thesis, Wuhan University of Technology, Wuhan, China, 2014.
10. Ganbold, T.; Badarch, B.; Bat, B. An Investigation for Improving Power Performance by VVT Effect of Spark Ignition Engine by Wave Simulation. In Proceedings of the International Forum on Strategic Technology 2010, Ulsan, Korea, 13–15 October 2010; pp. 408–410.
11. Qiu, Y.H.; Perreault, D.J.; Keim, T.A.; Kassakian, J.G. Nonlinear system modeling, optimal cam design, and advanced system control for an electromechanical engine valve drive. *IEEE-ASME Trans. Mechatron.* **2012**, *17*, 1098–1110. [CrossRef]
12. Liu, Y.L.; Zhou, Z. Pattern recognition method of OCV testing. *Microcomput. Appl.* **2016**, *35*, 46–50. [CrossRef]
13. Sun, Y.; Wang, H.F.; Sun, D.H. A New Intermediate Locking VVT. China Patent ZL 201721658350, 6 July 2018.
14. Zhang, Y.D.; Jiang, J.G.; Liang, T. Structural Design of A Cartesian Coordinate Tooth-Arrangement Robot. In Proceedings of the 2011 International Conference on Electronic & Mechanical Engineering and Information Technology, Harbin, China, 12–14 August 2011.
15. Li, Z. Design and Research of A New Industrial Robot. Master's Thesis, Tianjin University, Tianjin, China, 2014.
16. Jin, J.Q.; Xu, Z.W.; Liu, C. Research on control system of Cartesian coordinate conveying robot. *Heavy Mach.* **2016**, *6*, 37–39. [CrossRef]
17. Liang, L.Q. Design and research of manipulator for automatic production line. *Modul. Mach. Tool Autom. Manu. Tech.* **2018**, *4*, 162–164. [CrossRef]
18. Wang, R.; Qu, H.W.; Sui, Y.Y.; Lv, J.L. Calibration of Cartesian Robot Based on Machine Vision. In Proceedings of the 2017 IEEE 3rd Information Technology and Mechatronics Engineering Conference (ITOEC), Chongqing, China, 3–5 October 2017.

19. Zhu, F.K.; Liu, Y.; Dong, H.; Meng, F.J.; Cong, M. Design and analysis for Cartesian robot cantilever structure. *Modul. Mach. Tool Autom. Manuf. Tech.* **2017**, *7*, 60–63. [CrossRef]
20. Wu, S.L.; Li, H.M.; Liu, N.C. Design of industry robot hand. *J. Shihezi Univ. (Nat. Sci.)* **2007**, *6*, 778–781. [CrossRef]
21. Zhang, H.Q. Application of pneumatic manipulator in automatic stamping production line design. *Public Commun. Sci. Technol.* **2016**, *14*, 176–177. [CrossRef]
22. Wu, D.; Li, J.; Qiao, M. Structure design and simulation analysis of pneumatic manipulator based on AMESim. *J. Guangxi Univ. Sci. Technol.* **2016**, *27*, 62–68. [CrossRef]

Article

Analysis of Vibration Plate Cracking Based on Working Stress

Zeyu Kang *, Gangjun Li, Fujun Wang, Huan Zhang and Rui Su

College of Mechanical Engineering, School of Chengdu Technological University, Chengdu 610071, China; lgjun1@cec.edu.cn (G.L.); wfjun1@cec.edu.cn (F.W.); zhuan1@cec.edu.cn (H.Z.); surui1988@126.com (R.S.)
* Correspondence: kangzeyu90@126.com

Received: 7 September 2018; Accepted: 10 October 2018; Published: 26 October 2018

Abstract: At present, vibroseis has become the major technique to achieve environmental protection and high efficiency in fossil fuel exploration. During such exploration, a vibrator transmits seismic waves to the surface. The waves are excited by continuously changing the load stress from the burden of weight of the vehicle and the vibrator's variable frequency load. This paper will apply a numerical simulation method to develop research on the analysis of vibration plate cracking based on working stress. Based on the structure and mechanism of vibroseis vibrator plate, a vibrator simulation model is built under system dynamics to develop research on the vibroseis plate load stress feature and gain distribution, and change pattern of the plate load stress. The results show that stress response around the upright welding of is high, and there is evident distortion in plate area, which matches the actual fracture position on the plate, and can be confirmed as a key area of plate fatigue.

Keywords: vibrator plate; working stress; simulated analysis

---

1. Introduction

At present, China's shallow oil and gas resources are gradually being exhausted, so oil and gas exploration is gradually moving to areas with complex topographies and harsh environments, such as the Gobi desert, mountainous areas, and the ocean [1]. In these areas of petroleum exploration and development, oil and gas exploration technology has been unable to meet the current reliability, safety, environmental protection, and efficient exploration acquisition requirements. Now, more efficient, environmentally friendly, and safe exploration technology is needed; therefore, the seismic method is also proposed. At present, seismic exploration signal sources cause significant damage to the environment [2]; however, vibroseis serves as a non-explosive source, which is not destructive to the environment or living creatures, as are the output of seismic signals generated by explosion sources. More importantly, though larger than the source of power in a short period of time to produce high-energy, vibroseis uses low power, and requires a long time for scanning excitation. During scanning, the continuous variation of the sinusoidal vibration waveform is known; vibroseis is therefore more accurate and controllable than explosion sources [3].

Vibroseis has the following advantages:

- The seismic signal excited by vibroseis is controllable. Vibroseis can be used to independently choose the appropriate excitation frequency width according to the working environment, thus improving the signal excitation quality. The direction of the output force is also controllable; vibroseis can reduce the energy loss in other directions and increase the signal-to-noise ratio (SNR) of the vibroseis excitation signal.
- Vibroseis uses continuous excitation signals. According to the requirements, vibroseis can be used to control the vibrator to continuously make contact with the earth within a few seconds,

and generate the required signal waveform. The repeated superposition of the signal can eliminate a large amount of random external interference, so as to obtain high-SNR data.

- Vibroseis is not destructive to the environment or organisms. Vibroseis is a low-power method which can efficiently complete the exploration of deep-stratum oil and gas reservoirs in cities, dykes, industrial areas, and other areas where it is inconvenient to use explosives.
- Vibroseis uses Combined Excitation Technology to effectively suppress the linear interference energy during operation.
- Vibroseis can be operated without drilling, which improves the mobility and reduces a lot of costs [4].

At present, vibroseis has become the first choice for oil and gas exploration in areas with harsh environments, complex topographies, and high environmental protection requirements. Wei established the finite element model of the vibrator plate and geo-coupling, and studied the influence of the plate and the geo-coupling on vibroseis performance [4,5]. Ding compared and analyzed the characteristics and performance of different vibrator plates, and put forward suggestions for their structural design [6]. Zhuang Juan derived a dynamical equation of high-frequency vibroseis, and analyzed the effect of plate area, plate mass, and soil medium density on the amplitude and frequency characteristics of the vibroseis output signal [7]. The literature above mainly studied the relationship between the macro characteristics of the plate and the output signal, but lacked microscopic and quantitative analysis.

At present, research on vibration plate performance mainly focuses on single frequencies. However, the excitation signal of vibroseis is a sinusoidal scanning signal, whose frequency increases linearly with time. In practice, the response performance of the plate at different scanning frequencies is different. The difference in plate response affects the stability of vibrator response performance at full frequency, and restricts the development of vibroseis in terms of broadband and precision. Therefore, it is necessary to carry out multi-frequency response analysis of the plate in order to master the response change rule at different scanning frequencies, so as to provide guidance for the structural design of the plate.

2. Structure and Working Principle of the KZ-28 Vibroseis Vehicle

Vibroseis mainly consists of a vibrator body and a vibrator. The vibrator is installed in the central position of the vibrator car, as shown in Figure 1.

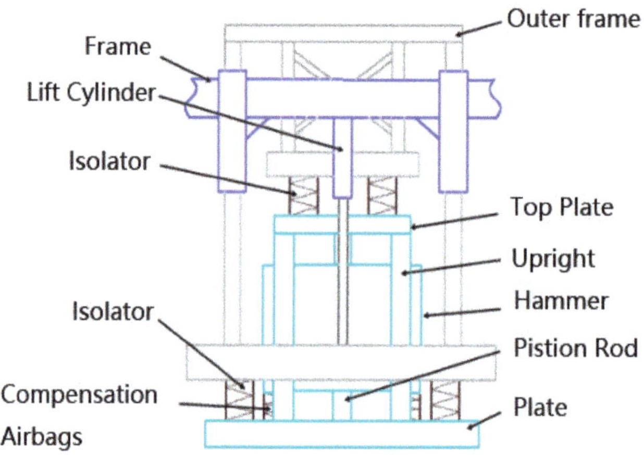

Figure 1. Schematic diagram of vibrator structure.

When a vibroseis vehicle moves, the lifting hydraulic cylinder lifts the vibrator and separates it from the ground [8]. As shown in Figure 2, at the beginning of the process, the lifting hydraulic cylinder transfers the vibrator and jacks up the car body, and the whole weight of the vibroseis vehicle acts upon the roof and plate of the vibrator through a vibration isolation air bag. When the vibroseis is working, the servo valve opens. The high-pressure hydraulic oil produced by the hydraulic system alternately enters the upper and lower cavity between the hammer and the piston rod, causing the hammer to move up and down. The reaction of hydraulic oil produced by the piston rod is passed to the tablet, and the signals produced by the vibrators are also passed to the earth, thereby exciting seismic waves [9,10].

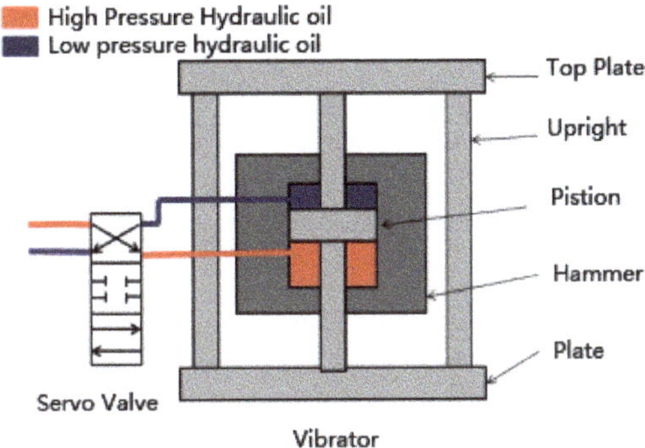

Figure 2. The working principle of a vibrator.

3. The Vibrator Cracking Problem

The vibrator is the key excitation device for the vibroseis seismic signal. The plate is an important component which connects the vibroseis and the ground; the column supports the car body and transfers the load. The column is fixed to the plate by welding.

In 2016, a group of vibroseis vehicles which had been sent to Saudi Arabia by Bureau of Geophysical Prospecting INC, experienced cracks in the plate weld during operation. As shown in Figure 3, the weld toe of the welding position between the vertical column and the plate was cracked (the position of the red line). At the time of the accident, the vehicle was exploring in the desert [10].

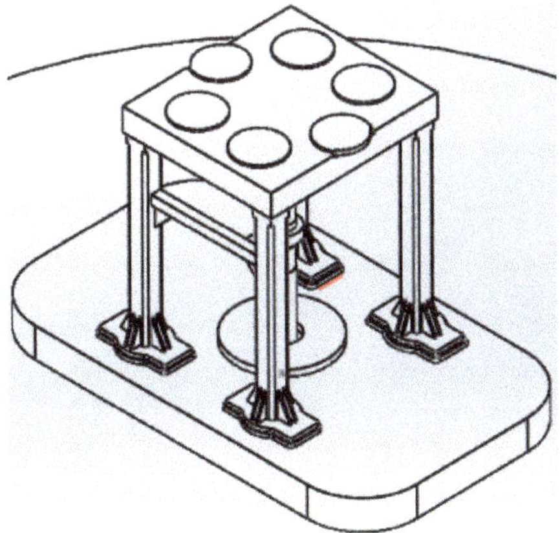

Figure 3. Schematic diagram of the weld cracking position.

It can be concluded by analyzing the causes of plate weld cracking that the weld structure is closed, and the deformation near the welding is not uniform, resulting in large tensile stress, which is one of the main reasons for the initiation of weld cracks.

When the vibrator is operating, under the effect of seismic wave excitation, the plate is affected by its own weight and by vibrating force at the same time; working stress is therefore generated. Additionally, the weld joint (especially in the weld toe) is prone to high tensile stress under the effect of vibrator working load, and fatigue cracks easily arise in this position [11,12].

4. Establishment of Finite Element Model for Working Stress Analysis of Vibrator Plate

By analyzing the structure and working characteristics of the KZ-28 vibroseis vibrator, we decided to use a simulation analysis to solve the problem of cracking. At present, there is a new method of analysis: Nguyen-Thanh et al. presented a new concurrent simulation approach to couple isogeometric analysis (IGA) with the mesh-free method for studying crack problems. The convergence rate of the present method is higher than that of the traditional method. However, the Gaussian integration of weak form is computationally expensive in the mesh-free sub-domain [13]. Therefore, we still choose the traditional method; the model has been established in ANSYS LS-DYNA, and the vibrator is simplified [14,15]:

- Simplify the hammer and its accessories without considering the weight of the hammer and the impact action when the hydraulic oil drives the hammer. Remove the hammer and its accessory parts in the analysis; the hammer force and dynamic hydraulic load are directly loaded into the vibrator model as the known load.
- Simplify the piston base of I-steel plate. According to the plate structure, the piston rod base is installed in the middle of the plate, which is used to connect the piston rod and I-steel plate. For the convenience of the analysis, this was simplified into solid cylindrical structure.
- The vibrator needs to be in contact with the ground during operation. Therefore, it is necessary to establish a ground model. However, the size cannot be unlimited; therefore, at the time of the earth model size selection, the size should be considered so as not to affect the deformation of the vibrator plate. Through repeated with different sizes of earth model, and considering the calculation time and accuracy, it was eventually determined that the optimal values were: diameter of the earth model, 5 m, and height, 2.5 m. The completed geometric model of the vibrator is shown in Figure 4.
- Apply a relatively fine hexahedral mesh to the grid setting, the junction area of flat slab and vertical column. The tetrahedral mesh is used in the grid as shown in Figure 5. For the sake of accuracy and efficiency, the grid far from the flat slab and vertical column is sparse [16].

In the analysis process, the size of the cell will cause grid-dependent numerical instability. In order to guarantee accuracy, the size of the flat cell was set to 6 mm, and the total number of grids was 119,988. The rest of the vibrator is not analyzed in detail, so the tetrahedral element with low calculation accuracy and fast calculation speed is selected for mesh division. The tetrahedral elements are 8–20 mm in size. The total number of grids after grid partitioning was 432,876. Since the vibrator plate is in contact with the ground, in order to improve the calculation accuracy, in the ground model a hexahedral mesh of equal size was adopted for the part that is in contact with the vibrator, while a larger tetrahedral mesh was adopted for the part that is not in contact with the vibrator. The total number of meshes after ground grid partitioning was 252,159. The material messages of the vibrator and earth are shown in Table 1.

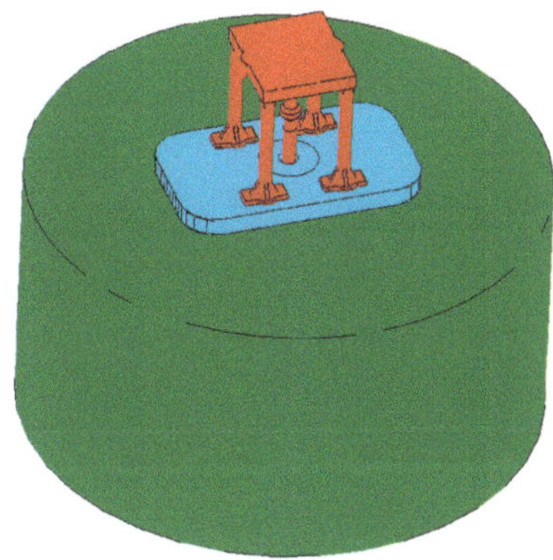

Figure 4. Vibrator geometric model.

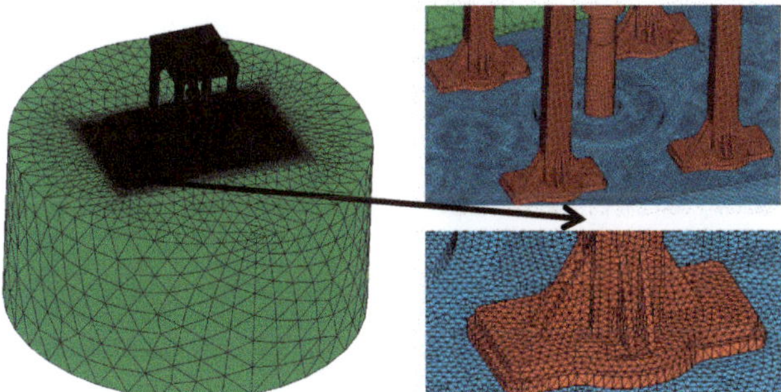

Figure 5. Vibrator geogrid.

Table 1. Material parameters of the vibrator model.

Parts	Materials	Density (kg/m^3)	Elastic Modulus (Pa)	Poisson's Ratio	Yield Strength (MPa)
Above-slab structure	45 steel	7890	2.09×10^{11}	0.269	355
Tablet	16 Mn	7850	2.12×10^{11}	0.310	345
Ground	rock	2600	5.5×10^{10}	0.270	–

Considering the actual situation, the simulated geodetic model is an elastic half-space, and the non-reflecting boundary conditions of the LS-DYNA software can simulate the effect of infinite earth. Therefore, the contact between the plate and the ground is defined as slide contact [17].

There are two kinds of loads on the vibrator, as described below:

(1) Static load (Figure 6 shows the loading diagram of static load).

The static load is mainly caused by the car body; its purpose is to ensure that the flat plate can stay close to the ground under the huge working load without deformation. The weight of the car body is 28 t, of which 90% uniformly acts on the shock absorber air pad of the roof, and the roof is transferred to the flat plate through the vertical column. The remaining 10% comes from the hammer [18,19].

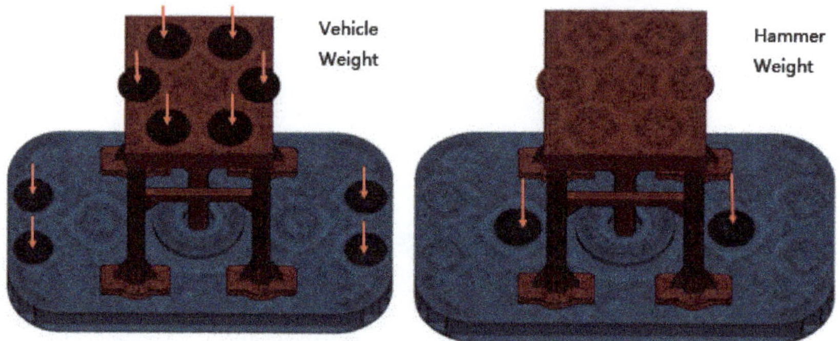

Figure 6. Vibrator loading diagram.

(2)   Dynamic load

The size and variation of dynamic loads are shown in Figure 7. The initial analyzed frequency in this study is 80 Hz, and the loading time is five cycles. The dynamic liquid pressure on the upper and lower end surfaces of the piston rod of the vibrator is consistent with the changing law of linear scanning signal. The rated peak pressure of hydraulic oil is 20 MPa. In the construction process, in order to ensure that the output signal does not generate large distortion, the hydraulic pressure is generally selected as 70–85% of the rated peak value. Therefore, 85% of the rated peak value was selected in this paper, that is, the peak value of liquid pressure was 17 MPa. Additionally, due to the instability of the system at the initial stage of loading, the data under the stable system (after three periods) were analyzed [20].

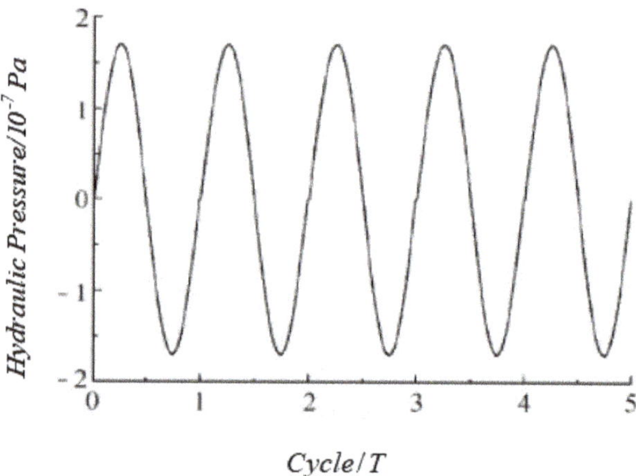

Figure 7. Dynamic load curve of vibrator.

5. Analysis of Working Stress of Vibrator Plate under Maximum Liquid Pressure

In Figure 8, in order to facilitate the study of the working stress of the vibrator plate, the plate area is divided into several parts for clarity of illustration.

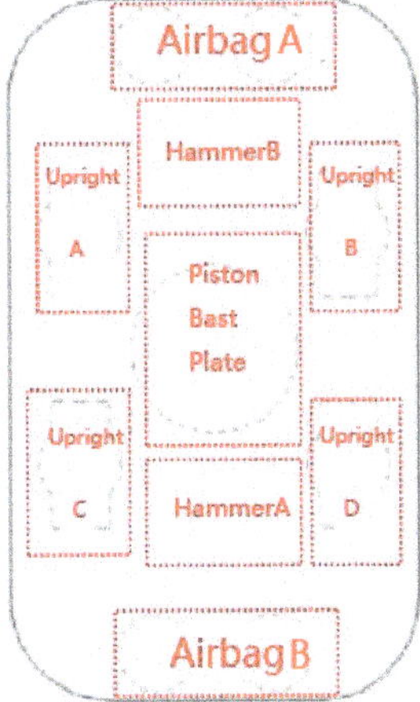

Figure 8. Regional location and labeling of the vibrator plate.

The stress distribution of the plate was extracted during a period when the liquid pressure reached its peak at the upper and lower ends of the piston rod. The results are shown in Figure 9.

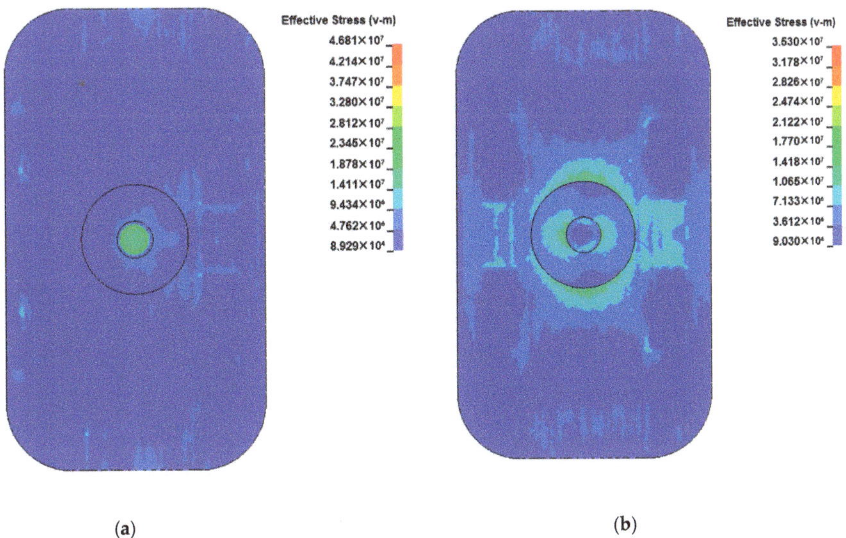

(a)   (b)

Figure 9. Stress distribution of plates. (**a**) plate stress distribution at the maximum of upper half cycle; (**b**) plate stress distribution at the maximum of lower half cycle.

- In the upper half of cycle when the liquid pressure reaches the maximum, the von Mises stress on the upper surface of the plate reaches the maximum value in the area of the piston rod; the maximum value is 25.3 MPa, the larger value around the long side of columns A and C is 16.4 and 12.2 MPa, respectively, and the larger value around the short edge of columns B and D is 10.1 and 18.6 MPa, respectively.
- In the second half cycle when the liquid pressure reaches the maximum, the von Mises stress on the upper surface of the plate reaches the maximum in the area of the piston rod; the maximum value is 28.2 MPa, which presents large values on the short edges of columns A, B, C, and D of 13.6, 9.5, 8.6, and 14.6 MPa, respectively.

- Based on the analysis of Figure 9, it can be found that the stress concentration and stress peak exist when the short edge weld of the column reaches its maximum value within one cycle of liquid pressure. The location of the peak area of stress concentration exists near the area of the piston rod and the pillar. When the liquid pressure of the vertical columns B and D reached the maximum in the upper and lower half periods, they all presented large values; the post short edge weld is the part of the welding crack in practical engineering.

For further analysis, the data for the joint stress in the direction of the welding seam along the short edge of the vertical column were extracted, and the stress–time curve was plotted, as shown in Figures 10 and 11.

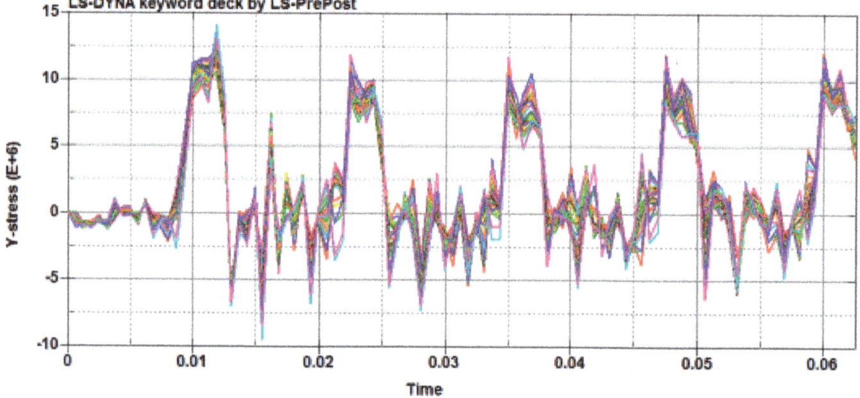

Figure 10. Longitudinal stress of weld joint.

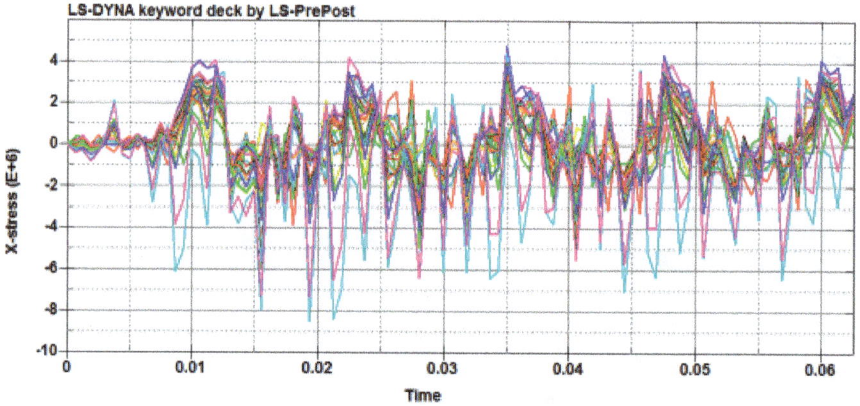

Figure 11. Transverse stress of weld joint.

The longitudinal and transverse stress of columns on the short edge weld in a cycle (T = 0.125 s) present the periodical change of tensile stress and compressive stress; the weld on each node of the change trend of stress is very close. As shown in Figures 10 and 11, the curve which is formed by the multiple nodes shows the same trend, so we can judge that the trends of these node stresses are almost the same. Therefore, the mid-point of the short edge of the welding seam was taken as being representative to analyze the variations of stress on the whole welding seam under one cycle, and the variation data for stress over time were extracted and are shown in Figure 12.

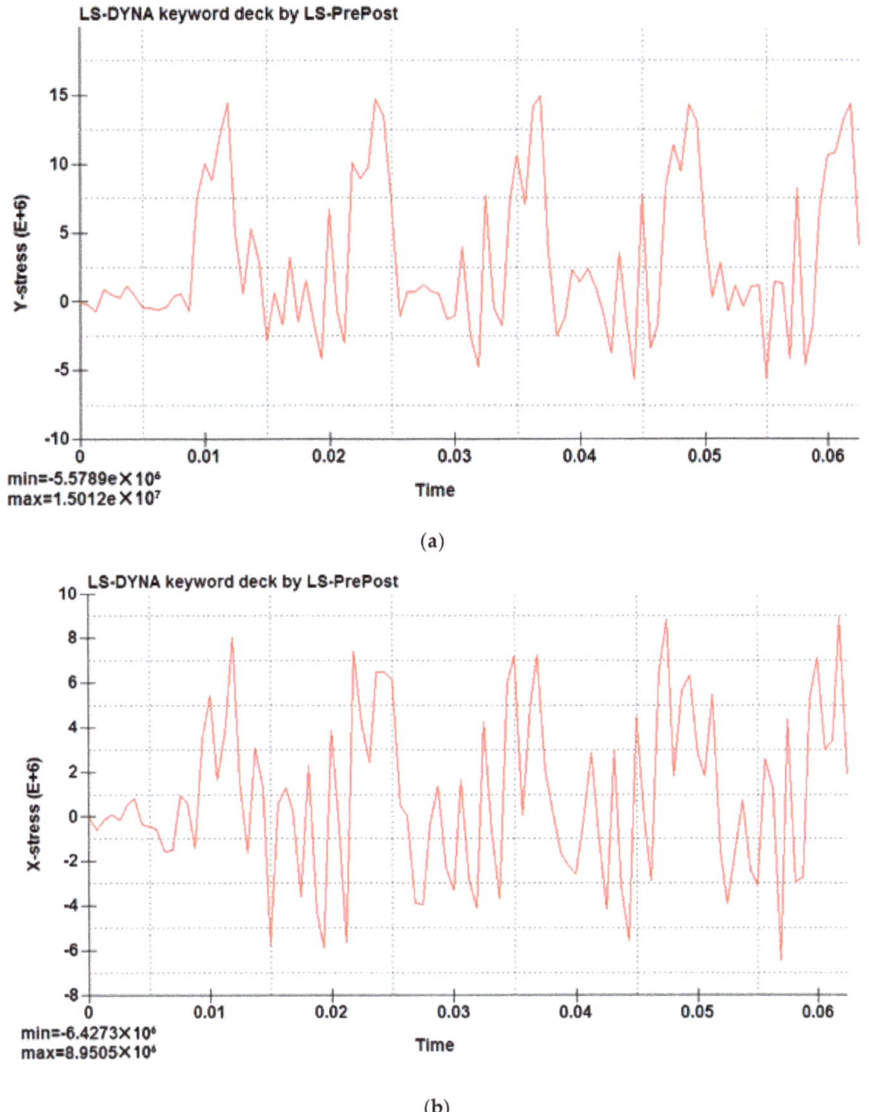

(a)

(b)

Figure 12. Schematic diagram of stress change at the mid-point of weld. (**a**) longitudinal stress. (**b**) transverse stress.

As shown in Figure 12, after the system is stable (the fifth cycle), the longitudinal stress on the weld mid-point in the first half cycle is converted from the tensile stress to compressive stress; the maximum value is 5.5 MPa. In the second half cycle, the stress is converted from compressive stress to tensile stress; the maximum value is 15.0 MPa. The transverse stress in the first half cycle is converted from tensile stress to compressive stress; the maximum value is 6.4 MPa. In the second half cycle, the stress is converted from compressive stress to tensile stress; the maximum value is 8.9 MPa. The overall trend shows that stress is present at the periodical change of tensile stress and compressive stress. Based on these observations, there is a peak stress in the weld.

6. Analysis of Plate Deformation under Maximum Liquid Pressure

The deformation distribution of the plate was extracted respectively when the liquid pressure reaches its peak at the upper and lower ends of the piston rod within one cycle. The results are shown in Figure 13.

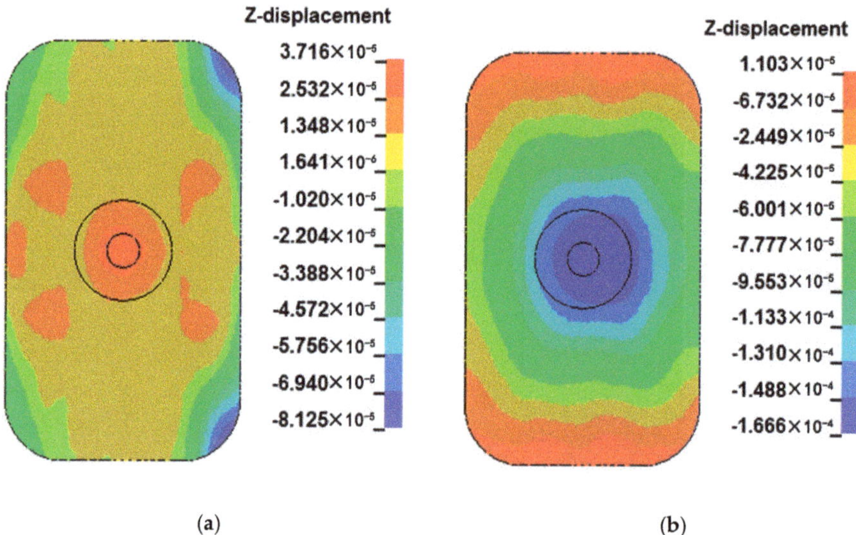

Figure 13. Schematic diagram of plate deformation. (**a**) plate deformation distribution at the upper half cycle maximum. (**b**) plate deformation distribution at the lower half cycle maximum.

- In the upper half of the cycle, when the liquid pressure reaches the maximum value, the whole plate deforms downward on the upper surface of the plate. The maximum value of deformation is 3.2×10^{-4} m near the piston rod; it deformed downward near the vertical column, showing a larger value. The edge of the plate has an upward deformation.
- In the second half of the cycle, when the liquid pressure reaches the maximum value, the whole plate deforms upward on the upper surface of the plate. The maximum value of deformation is 2.2×10^{-4} m near the piston rod, and the upward deformation near the vertical column presents a large value. The edge of the plate has downward deformation.

According to the deformation analysis, the whole of the plate is subjected to periodic deformation in the working process. The position of the short edge weld of the vertical column plate is shown in Figure 3. As a relatively fragile structure on the plate, the welding seam is prone to cracking under such conditions, which is consistent with the crack position observed in the welding seam.

For further analysis, the data for the mid-point of the short edge seam of the vertical column were extracted, and the displacement–time curve was drawn, as shown in Figure 14.

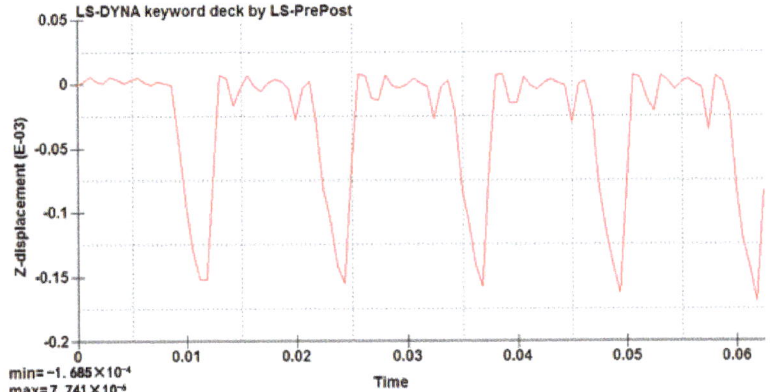

Figure 14. Displacement at the mid-point of weld.

After the system is stable (the fifth cycle), the displacement at the mid-point of the weld shows alternating changes with time in the positive and negative directions; the maximum value of the downward deformation is 7.74×10^{-6} m, and the that of the upward deformation in the lower half period is 1.69×10^{-4} m.

7. Analysis of Working Stress and Deformation of Plate at Different Frequencies under the Action of Liquid Pressure

The working load will lead to the deformation of the plate, and change its stress distribution, resulting in stress concentration, which is a hidden problem for the cracks in the plate's welding seam. Among them, the rated output and self-weight of vibroseis are restricted by the conditions. This does not usually change, and the vibrator as a source of multiple frequency output, frequency fluid pressure can be changed and controlled. With a working frequency for 5~125 Hz, we selected 50 Hz, 100 Hz, and 80 Hz, analyzed the stress and displacement of plate by using the method of simulation under three kinds of liquid pressure frequencies, and studied the influence of liquid pressure frequency on working stress and plate deformation.

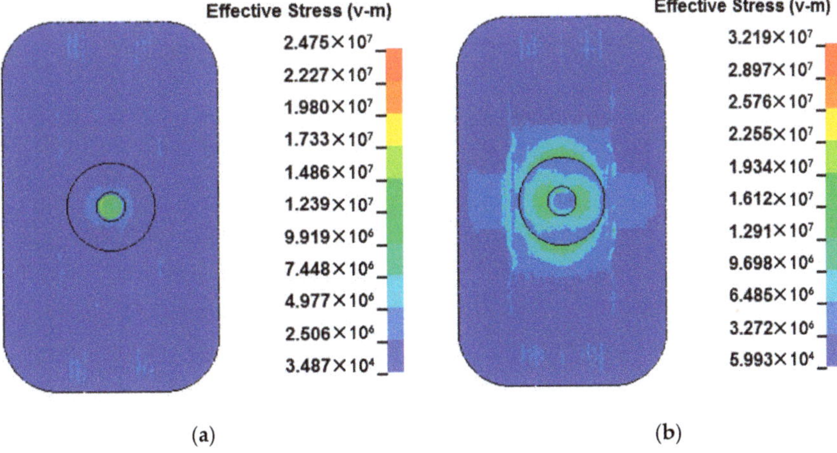

(a)   (b)

Figure 15. Distribution of plate stress under 50 Hz. (**a**) plate stress distribution at the maximum of upper half cycle. (**b**) plate stress distribution at the maximum of lower half cycle.

The stress distribution of the plate is obtained when the liquid pressure of 50 Hz reaches its peak at the upper and lower ends of the piston rod within a period. The results are shown in Figure 15.

- In the upper half of the cycle, when the liquid pressure reaches the maximum value, the von Mises stress on the surface of the plate reaches the maximum value in the piston rod area, 23.2 MPa. Larger values are presented near the long side of columns A, B, C, and D, respectively 12.9, 8.6, 7.5, and 7.8 MPa. Large values are presented at air springs A and B, respectively 4.6 and 3.8 MPa.
- In the second half of the cycle, when the liquid pressure reaches the maximum value, the von Mises stress on the upper surface of the plate reaches the maximum value in the piston rod area, 17.1 MPa, and larger values are displayed on the short edge of the vertical columns A, B, C, and D, respectively 9.4, 8.4, 10.1, and 11.9 MPa.
- Compare the stress distribution of the plate under 80 Hz liquid pressure to others. When the upper semi-periodic stress value under 50 Hz reaches the maximum, there is no stress concentration on the short edge of the column.

Similarly, the mid-point of the short weld edge was taken as being representative, and the data for the change of stress over time were extracted, as shown in Figure 16.

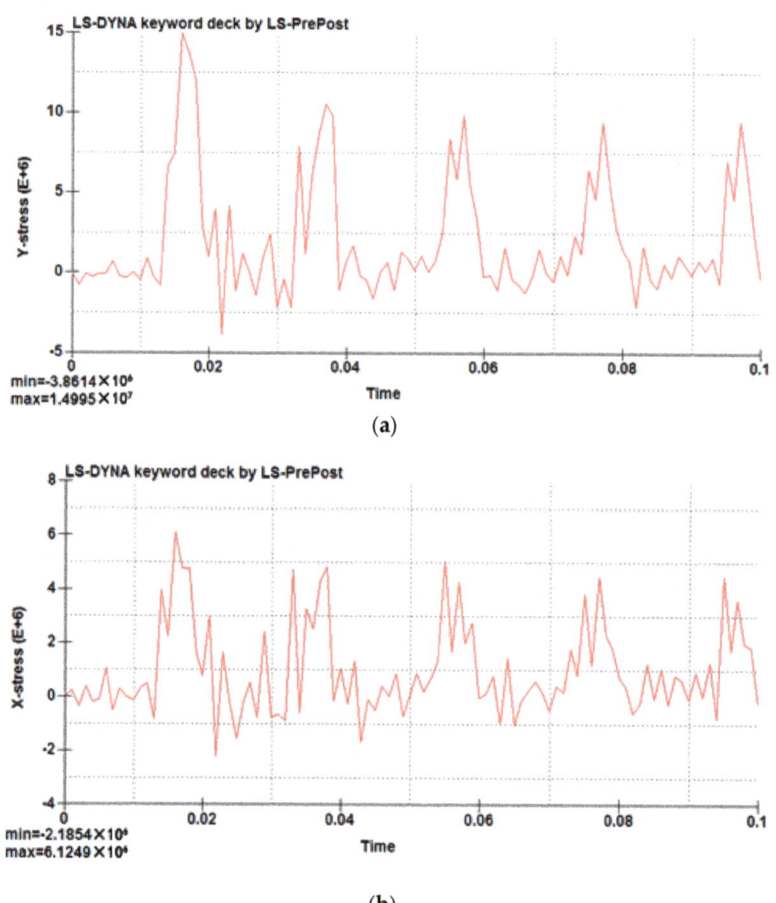

(a)

(b)

Figure 16. Variation of mid-point stress under 50 Hz. (**a**) longitudinal stress. (**b**) transverse stress.

As shown in Figure 16, after the system is stable (the fifth cycle), the longitudinal stress on the weld mid-point in the first half cycle is converted from tensile stress to compressive stress; the maximum value is 2.4 MPa. In the second half cycle, the stress is converted from compressive stress to tensile stress; the maximum value is 9.8 MPa. The transverse stress in the first half cycle is converted from tensile stress to compressive stress; the maximum value is 0.9 MPa. In the second half cycle, the stress is converted from compressive stress to tensile stress; the maximum value is 4.8 MPa. As an overall trend, stress is present during the periodical change of tensile stress and compressive stress. Based on this data, there is a peak stress in the weld.

The deformation distribution of 50 Hz liquid pressure on the plate in one cycle was extracted, as was the mid-point displacement–time curve of the welding seam. The results are shown in Figures 17 and 18.

- In the upper half of the cycle, when the liquid pressure reaches the maximum value, the whole plate deforms downward on the upper surface. The maximum value of the deformation is 2.8×10^{-5} m near the piston rod, while the downward deformation near the column presents a larger value, 8.8×10^{-6} m. The edge of the plate has upward deformation.
- In the second half of the cycle, when the liquid pressure reaches the maximum value, on the upper surface of the plate the deformation has a maximum value near the piston rod, 1.5×10^{-4} m, the upward deformation near the four vertical columns. The short edge of the plate deforms downward.
- By analyzing the deformation of the whole, under the effect of liquid pressure of 50 Hz, the deformation law of the whole plate in the working process is quite similar to that under 80 Hz, and the position of the short edge weld of the column plate shows up and down deformation alternately over the whole cycle.

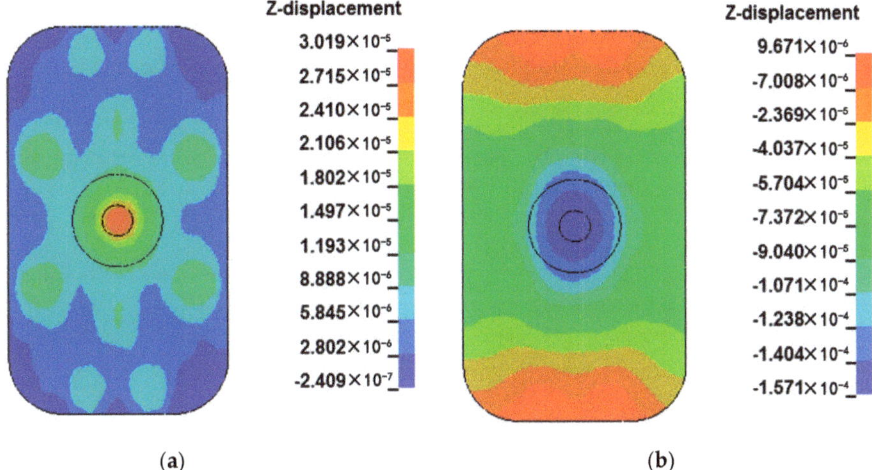

Figure 17. Deformation diagram under 50 Hz. (**a**) plate deformation distribution at the upper half cycle maximum. (**b**) plate deformation distribution at the lower half cycle maximum.

The data for the mid-point of the short edge seam of the vertical column were extracted, and the displacement–time curve was drawn, as shown in Figure 18.

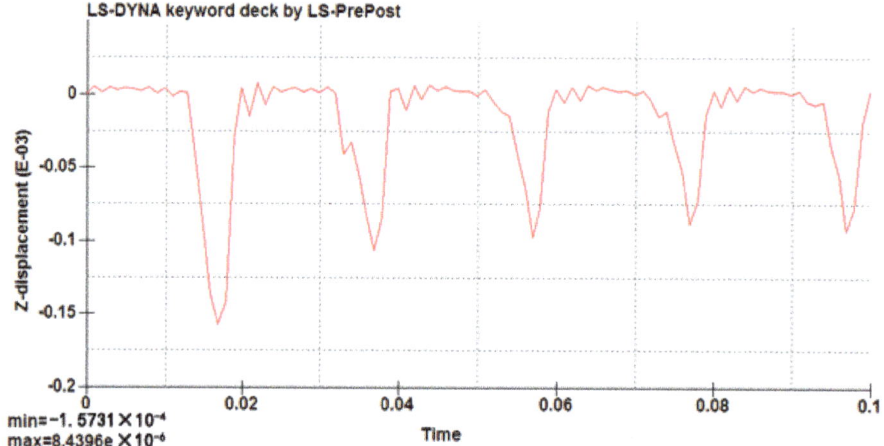

Figure 18. Displacement on mid-point of weld under 50 Hz.

After the system is stable (the fifth cycle), the displacement at the mid-point of the weld shows alternating changes with time in the positive and negative direction; the maximum value of downward deformation of the plate in the upper half of the cycle is 5.3×10^{-6} m. The maximum value of the upward deformation in the second half of the cycle is 9.1×10^{-5} m.

The stress distribution of the plate when the liquid pressure reaches its peak at the upper and lower ends of the piston rod within a period of 100 Hz was extracted, and the results are shown in Figure 19.

- In the upper half of the cycle when the liquid pressure reaches the maximum value, the von Mises stress on the upper surface of the plate reaches the maximum value in the piston rod area, 32.3 MPa, and the larger value is displayed near the vertical columns A, B, C, and D, respectively 10.1, 9.1, 13.4, and 9.6 MPa.
- In the second half of the cycle when the liquid pressure reaches the maximum value, the von Mises stress on the upper surface of the plate reaches the maximum value in the piston rod area, 33.6 MPa, and the larger value is shown at the long side of the vertical columns A and C, respectively 15.7 and 14.8 MPa, and the larger value is shown at the short edge of B and D, respectively 14.6 and 16.6 MPa.
- As for the stress distribution of the plate under 80 Hz liquid pressure, in the upper half of the cycle, the stress value under 100 Hz reaches the maximum. There is also a stress concentration near the column, while the stress distribution in the second half of the cycle is similar to that under 80 Hz.

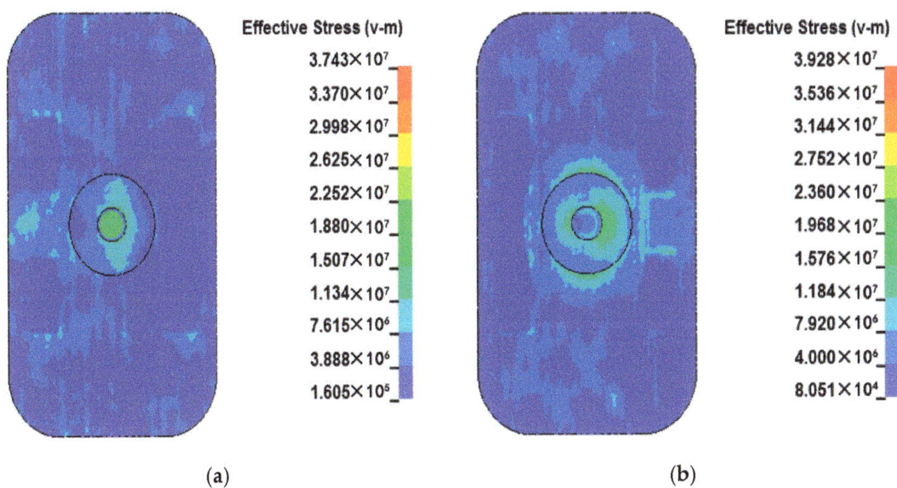

(a)   (b)

Figure 19. Distribution of plate stress under 100 Hz. (**a**) plate stress distribution at the maximum of upper half cycle. (**b**) plate stress distribution at the maximum of lower half cycle.

Similarly, the mid-point of the short weld edge was taken as being representative, and the data for the change of stress over time was extracted, as shown in Figure 20.

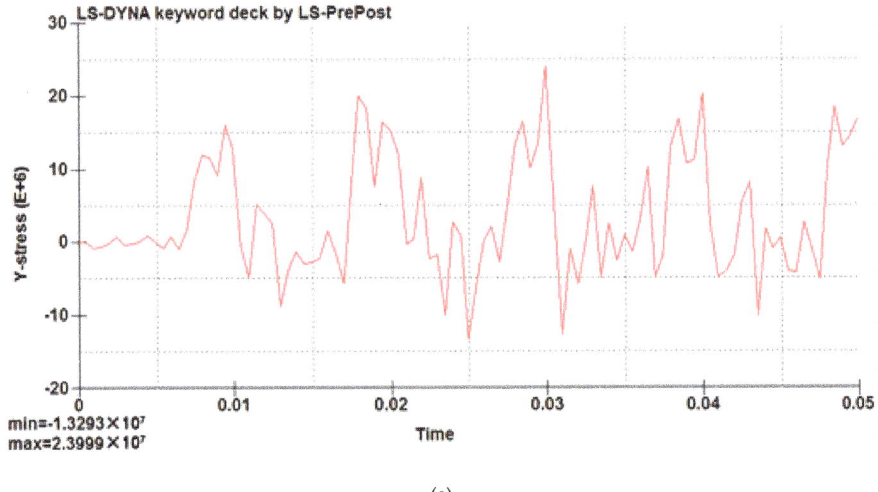

(a)

Figure 20. *Cont.*

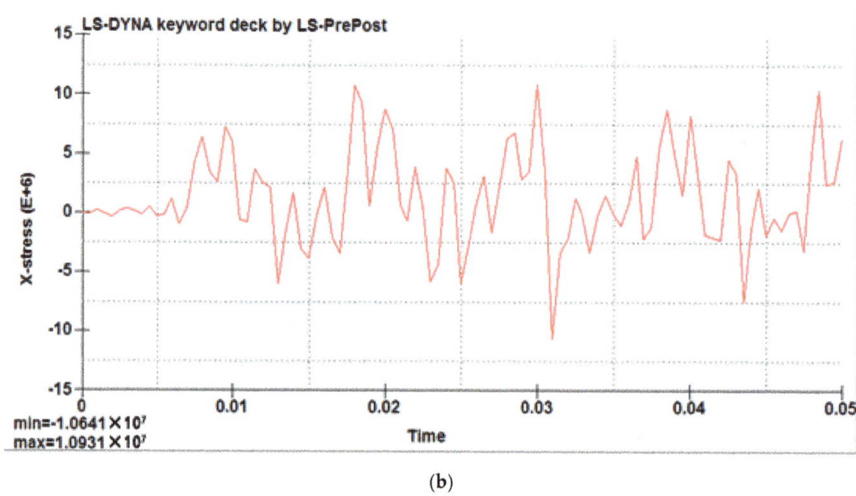

(**b**)

Figure 20. Variation of mid-point stress of welding seam under 100 Hz. (**a**) longitudinal stress. (**b**) transverse stress.

As shown in Figure 20, after the system is stable (the fifth cycle), the longitudinal stress on the weld mid-point in the first half cycle is converted from tensile to compressive stress; the maximum value is 10 MPa. In the second half cycle, the stress is converted from compressive stress to tensile stress; the maximum value is 18.6 MPa. The transverse stress in the first half cycle is converted from tensile stress to compressive stress; the maximum value is 7.43 MPa. In the second half cycle, the stress is converted from compressive stress to tensile stress; the maximum value is 10.04 MPa. In the overall trend, stress is present at the periodical change of tensile stress and compressive stress. Based on this data, there is a peak stress in the weld.

The deformation distribution of 100 Hz liquid pressure on the plate over one cycle was extracted, as was the time–displacement curve of the mid-point of the weld seam. The results are shown in Figures 21 and 22.

- In the upper half of the cycle, when the liquid pressure reaches the maximum value, the piston rod and the column are deformed downward and the rest of the parts are deformed upward on the upper surface of the plate. The maximum deformation is 3.4×10^{-5} m near the piston rod, and it is deformed downward near the vertical column, showing a larger value.
- In the second half of the cycle, when the liquid pressure reaches the maximum value, on the upper surface of the plate, the deformation takes on a maximum value near the piston rod, 1.4×10^{-4} m, showing a larger value near the four vertical columns. The short edge of the plate deforms downward.
- According to the global deformation analysis, under 100 Hz liquid pressure, the deformation law of the whole plate in the working process is similar to that under 80 Hz, and the position of short edge weld of the column plate shows up and down deformation alternately over the whole cycle.

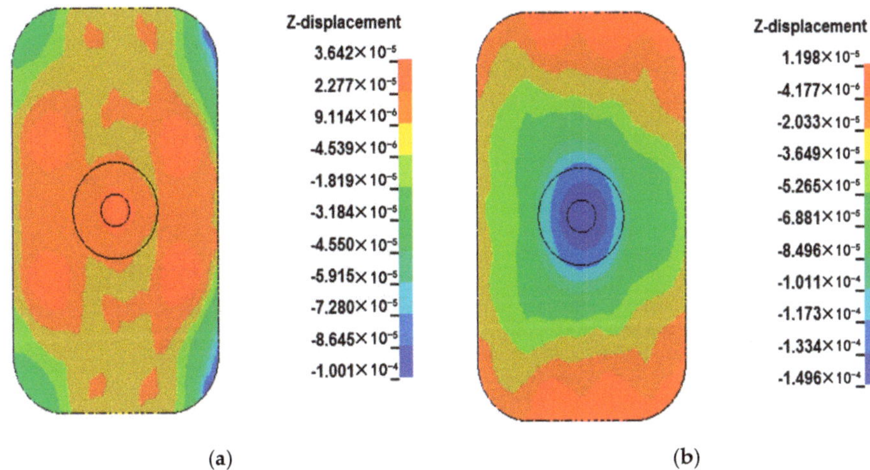

(a)   (b)

Figure 21. Deformation under 100 Hz. (**a**) plate deformation distribution at the upper half cycle maximum. (**b**) plate deformation distribution at the lower half cycle maximum.

The data for the mid-point of the short edge seam of the vertical column were extracted, and the displacement–time curve was drawn, as shown in Figure 22.

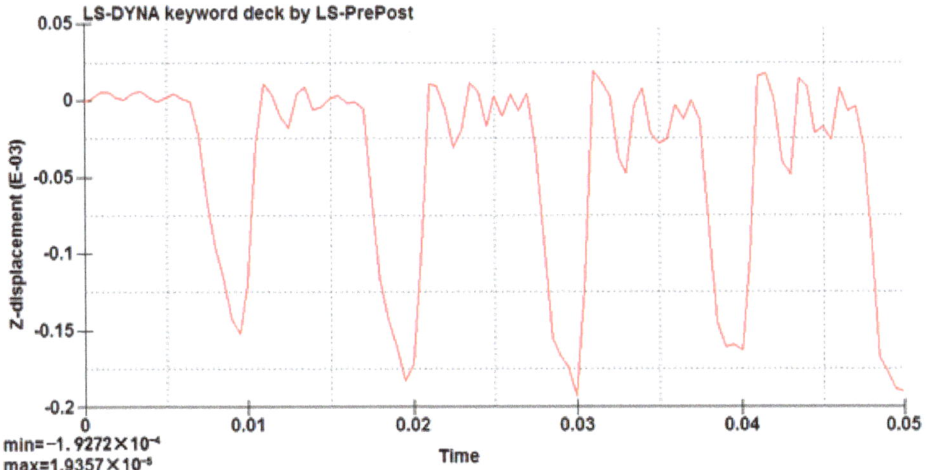

Figure 22. Displacement on mid-point of weld under 100 Hz.

After the system is stable (the fifth cycle), the displacement at the mid-point of the weld shows alternating changes with time in the positive and negative direction. The maximum value of downward deformation in the upper half cycle is 1.94×10^{-5} m, and that of upward deformation in the lower half period is 1.83×10^{-4} m.

The data were collated under the three frequencies, as shown in Table 2.

- The stress on the welding seam of the vertical column is mostly tensile stress, but is compressive stress in a small period.

- The maximum transverse stress and maximum longitudinal stress of the vertical welding seam of the vertical column increased with the rise in frequency, among which the tensile stress was clearly higher than the compressive stress.
- The peak value and variation amplitude of the deformation of the welding seam increased with the rise in liquid pressure frequency, and the deformation became more obvious with the rise in frequency.

Table 2. Stress and displacement of the mid-point of the weld at different frequencies.

	50 Hz	80 Hz	100 Hz
Maximum transverse tensile stress (MPa)	4.8	8.9	10
Maximum transverse compressive stress (MPa)	0.9	6.4	7.43
Maximum longitudinal tensile stress (MPa)	9.8	15.5	18.6
Maximum longitudinal compressive stress (MPa)	2.4	5.5	10
Maximum positive displacement ($\times 10^{-5}$ m)	9.1	16.9	18.3
Maximum negative displacement ($\times 10^{-5}$ m)	0.53	0.77	1.9

8. Summary

(1) The stress and displacement of the plate are greatly influenced by the liquid pressure frequency in the working frequency of the vibrator.

(2) Under the working frequency of the vibration device, the maximum stress in the weld area of the short edge of the vertical column increases with the increase in liquid pressure frequency.

(3) The welding zone of the short edge of the vertical column has a stress concentration under three working frequencies; there is a peak value, and most of the time, stress is tensile stress.

(4) The deformation of the welds on the short edge of the primary welding line alternates in positive and negative directions throughout the cycle, and the peak value and variation amplitude rise with the increase in liquid pressure frequency, and the deformation is more obvious.

(5) A stress peak exists in the short edge weld area, and the displacement deformation is produced continuously as the vibrator operates; this area can therefore be determined as being the most susceptible to plate cracking.

(6) Crack propagation is not discussed. In further research, we will use the coupling approach which integrates the mesh-free method and isogeometric analysis (IGA) for static and free-vibration analyses of cracks in thin-shell structures [21,22].

Author Contributions: Conceptualization, Z.K.; Methodology, Z.K.; Software, F.W.; Validation, G.L.; Formal Analysis, Z.K. and F.W.; Investigation, Z.K. and R.S.; Resources, H.Z. and G.L.; Data Curation, Z.K.; Writing-Original Draft Preparation, Z.K.; Writing-Review & Editing, R.S.; Visualization, H.Z.; Supervision, R.S.; Project Administration, G.L.; Funding Acquisition, G.L.

Funding: This research was funded by Basic research project of sichuan science and technology department grant number 18YYJC0217. The APC was funded by Gangjun Li.

Acknowledgments: I would like to thank my parents for their support and my mentor Zhiqiang Huang for their care and guidance.

Conflicts of Interest: The authors declare no conflict of interest.

References

1. Ma, Y.C. Research and application based on reflection wave seismic exploration. *Agric. Sci. Technol. Inf.* **2017**, *11*, 43–46.
2. Tong, X.Q.; Lin, J.; Jiang, T. Summary of development. *Prog. Geophys.* **2012**, *5*, 1912–1921.
3. Wei, Z.H.; Phillips, T.; Hall, F.; Michael, A. Fundamental discussions on seismic vibrators Basic overview on seismic vibrators. *Geophysics* **2010**, *75*, 13–25. [CrossRef]
4. Wei, Z.H. Modelling and modal analysis of seismic vibrator baseplate. *Geophys. Prospect.* **2010**, *58*, 19–32. [CrossRef]

5. Wei, Z.H. How good is the weighted-sum estimate of the vibrator ground force? *Lead. Edge* **2009**, *28*, 960–965. [CrossRef]

6. Ding, Z.G. Discussion on seismic vibration vibrator. *Equip. Geophys. Prospect.* **1998**, *4*, 36–38.

7. Zhuang, J.; Lin, J.; Wu, D.J. Analyzing coupling process of the earth-high frequency vibrator. *J. Changchun Univ. Sci. Technol.* **1999**, *29*, 184–187.

8. Tao, Z.F. Current situation and problems of vibroseis. *Equip. Geophys. Prospect.* **1995**, *5*, 11–26.

9. Zhuang, J.; Lin, J.; Wu, D.J.; Zhang, B.J.; Zhang, Z.S. Amplitude-frequency analysis of the coupling process of high frequency vibroseis and ground vibration. *J. Changchun Univ. Sci. Technol.* **1999**, *2*, 184–187.

10. Huang, Z.Q.; Ding, Y.P.; Tao, Z.F.; Hao, L.; Li, G. Research status and development direction of vibroseis vibrator baseplate at home and abroad. *Oil Field Equip.* **2015**, *6*, 1–5.

11. Tao, Z.F.; Su, Z.H.; Zhao, Y.L.; Ma, L. The latest development of low frequecy vibrator for seismic. *Equip. Geophys. Prospect.* **2010**, *20*, 1–5.

12. Claudio, B. Low-frequency vibroseis date with maximum displacement sweeps. *Lead. Edge* **2008**, *27*, 582–591.

13. Nguyen-Thanh, N.; Huang, J.; Zhou, K. An isogeometric-meshfree coupling approach for analysis of cracks. *Int. J. Numer. Methods Eng.* **2017**, *113*, 1630–1651. [CrossRef]

14. Mougenot, D. Land seismic: Needs and answers. *First Break* **2004**, *22*, 59–63.

15. Chapman, W.L.; Brown, G.L.; Fair, D.W. The Vibroseis System: A High Frequency Tool. *Geophysics* **1981**, *46*, 1657–1666. [CrossRef]

16. Walker, D. Harmonic resonance structure and chaotic dynamics in the earth-vibrator system. *Geophys. Prospect.* **1995**, *43*, 487–507. [CrossRef]

17. Gurbuz, B.M. Upsweep Signals with High Frequency Attenuation and Their Use in the Construction of VIBROSEIS Synthetic Seismograms. *Geophys. Prospect.* **1982**, *30*, 432–443. [CrossRef]

18. Sallas, J.J. Seismic vibrator control and down going P-wave. *Geophys. Prospect.* **1984**, *49*, 732–740.

19. Haines, S.S. Design and Application of an Electromagnetic Vibrator Seismic Source. *J. Environ. Eng.* **2006**, *11*, 9–15. [CrossRef]

20. Kalinski, M.E. Effect of Vibroseis Arrays on Ground Vibrations: A Numerical Study. *J. Environ. Eng.* **2007**, *12*, 281–287. [CrossRef]

21. Nguyen-Thanh, N.; Zhou, K. An adaptive extended isogeometric analysis based on PHT-splines for crack propagation. *Int. J. Numer. Methods Eng.* **2017**, *00*, 1–30.

22. Nguyen-Thanh, N.; Li, W.; Zhou, K. Static and free-vibration analyses of cracks in thin-shell structures based on an isogeometric-meshfree coupling approach. *Comput. Mech.* **2018**, *61*, 1–23. [CrossRef]

Review

A Systematic Review of Product Design for Space Instrument Innovation, Reliability, and Manufacturing

Kai-Leung Yung [1], Yuk-Ming Tang [1,2,*], Wai-Hung Ip [1,3] and Wei-Ting Kuo [1,2]

[1] Department of Industrial and Systems Engineering, The Hong Kong Polytechnic University, Hung Hom, Hong Kong; kl.yung@polyu.edu.hk (K.-L.Y.); wh.ip@polyu.edu.hk (W.-H.I.); david.kuo@polyu.edu.hk (W.-T.K.)

[2] Laboratory for Artificial Intelligence in Design, Hong Kong Science Park, New Territories, Hong Kong

[3] College of Engineering, University of Saskatchewan, Saskatoon, SK S7N 5A2, Canada

* Correspondence: yukming.tang@polyu.edu.hk

Abstract: The design and development of space instruments are considered to be distinct from that of other products. It is because the key considerations are vastly different from those that govern the use of products on planet earth. The service life of a space instrument, its use in extreme space environments, size, weight, cost, and the complexity of maintenance must all be considered. As a result, more innovative ideas and resource support are required to assist mankind in space exploration. This article reviews the impact of product design and innovation on the development of space instruments. Using a systematic literature search review and classification, we have identified over 129 papers and finally selected 48 major articles dealing with space instrument product innovation design. According to the studies, it is revealed that product design and functional performance is the main research focuses on the studied articles. The studies also highlighted various factors that affect space instrument manufacturing or fabrication, and that innovativeness is also the key in the design of space instruments. Lastly, the product design is important to affect the reliability of the space instrument. This review study provides important information and key considerations for the development of smart manufacturing technologies for space instruments in the future.

Keywords: space environment; space instrument; product design; performance; innovation; manufacturing

Citation: Yung, K.-L.; Tang, Y.-M.; Ip, W.-H.; Kuo, W.-T. A Systematic Review of Product Design for Space Instrument Innovation, Reliability, and Manufacturing. *Machines* **2021**, *9*, 244. https://doi.org/10.3390/machines9100244

Academic Editors: Zhuming Bi, Li Da Xu, Puren Ouyang and Pingyu Jiang

Received: 7 September 2021
Accepted: 17 October 2021
Published: 19 October 2021

Publisher's Note: MDPI stays neutral with regard to jurisdictional claims in published maps and institutional affiliations.

1. Introduction

Gold et al. [1] stated that space instruments are essential components for most space missions. The instruments help in gathering intelligence information, observing other planets, and monitoring the environment on earth. Providing the data to analysts and scientists on the ground, instruments are important for spacecraft in conducting regular structural verification (Garcia, [2]). During the launch operation of space missions, a mechanical environment that combines high and low frequencies, shock loads and vibrations, and high static acceleration, is generated. Each type of mechanical load must be simulated by analysis and tested to qualify the mechanical design. Some examples of space instruments include the supra thermal electrons and protons (STEP) instrument, which is constituted with other instruments such as the supra thermal ion spectrograph (SIS) and the energy particle detector (EPD) for solar orbiter spacecraft. Innovation, reliability, and product design [1] are essential for scientists to miniaturize space instruments. The size of the launch vehicle can reduce the weight of the instruments, allow the transition to smaller launch vehicles and provide accurate measurements from the space. Scientists have also modified the product designs of space instruments by creating completely new space instruments that enable previously impossible measurements. For instance, hyperspectral observations of the settings below the horizon and stars by visible imagers, spectrographic imagers, and ultraviolet imagers on the Midcourse Space Experiment Spacecraft. Tam [3] states that the best way to improve space instruments is by improving the technology through innovation, which may be in the form of designing space instruments with fewer components. Chau et al. [4] investigated the critical success factors (CSFs) for improving the management in manufacturing. Instead of using the traditional manufacturing method of using nuts and bolts to join complex systems and subsystems, new technologies such as Industry 4.0, 3D printing, and additive manufacturing could be used to produce complex yet monolithic structures, which do not require nuts and bolts. The process of innovation will help in reducing the number of pieces that might break down in case of a collision in orbit. Recently, a new method to design using "replicative" structures in different sizes and achieve required mechanical properties to manufacture with the minimum weight is investigated in [5]. In the study, manufacturing process parameters and design performance are analyzed with various examples.

Nevertheless, most of the smart manufacturing processes are mainly applied to traditional product development. Not many works are focused on developing smart manufacturing for space instruments and devices. This is because the space instrument usually consists of numerous factors such as size, weight, cost, extreme space environmental conditions, as well as a large number of components, high reliability, and stability, etc. With the recent advancement of Industry 4.0 and smart manufacturing technologies such as artificial intelligence, big data, augmented reality [6], and blockchain, it is important to extensively explore the product innovation, design, and reliability issues and CSFs in order to develop an optimal solution of smart manufacturing for space instrument. The main aim of the study is to explore how innovation and product design is essential in the development of space instruments and manufacturing. The study addresses the following key research questions.

RQ1: The research focuses on the development of space instruments;

RQ2: The key consideration factors that may influence the design of the space instrument.

This article presents the key contributions in the field. First, there is not much research focus on the design and development of a complex space instrument. Second, this study explores the influences of product innovation and design on the development of space instruments that are important to formulate the key consideration factors in the design and manufacturing of a complex space instrument. The key factors are important in formulating the smart manufacturing protocol for space instruments in the future. This is important to enhance the design and manufacturing efficiency of space products in the field.

2. Literature Review

2.1. Space Instrument

To design a space instrument or spacecraft to work in the space environment, three issues are important, making the design process very difficult, challenging, and exciting. The first one is that the complicated instrument work in a tough environment. This required high precision on the material selection and the physical mechanism. Moreover, the design and manufacture must have high precision to achieve the requirements for the best quality. Second, as the instruments will work remotely from the Earth, signal communication between the earth center and the instrument is a big concern. On the other hand, the design for the processes command, self-calibrate and operation are the other remotely issue. The third is the sensor. As there are many unknown environments in space, the sensor is the only reliable and detectable component for us to understand the situation. However, regarding the unknown environment, investing in a sensor to complete the mission is a big challenge. These issues are important in many processes of a space mission instrument such as space component replenishment. Yung et al. [7] proposed the multi-attribute fuzzy ABC classification to support the space components inventory decisions based on the tough situation of space missions.

All early and most current space programs are carried out or strongly dominated by governmental programs and choices. The reason for those monopolies is related to the high technical skill and knowledge required for developing space instruments and it is not worth it for a business to step into the industry. However, along with the mature environment of technology and the large developing margin in the deep space environment, such as mining, space travel, etc., more and more businesses are interested to enter the space market. The most outstanding example is the SpaceX program, developed by Elon Musk, the owner of the Tesla Company.

2.2. Product Development Process

2.2.1. Product Innovation

The exploitation and exploration of space have led to the emergence of new technologies in science including areas such as telecommunications, navigations, and medicine (van der Veen et al. [8]). Product innovation is the major goal of the space fairing nations to increase the capabilities of space technology to increase the benefits of space utilization. Over the generations, the space sector has focused mainly on advancing the technology conservatively, as well as innovation increments that are of low risks instead of disruptive, radical, and breakthrough innovations.

According to Popa et al. [9], the concept of innovation presents the ability to continuously make ideas and knowledge into new systems, processes, and products. Innovation can be divided into three pairs, radical innovation-incremental innovation, process innovation-product innovation, and technical innovation-administrative innovation. Incremental innovation refers to improving the existing processes, services, and products, radical innovation refers to the re-conceptualization of the products and process, and process innovation refers to the introduction of new elements to various processes. Administrative innovations refer to innovations that are related to the basic activities of the administrative processes as well as the management of those processes, and technical innovations refer to the products and technology innovations in the production process such as using Blockchain technology to enhance the traceability and trackability of the aerospace and aviation industries (Ho et al., [10]).

Van der Veen et al. [8] described disruptive technology as the kind of technology that emerges out of the niche market and dominates the market to the extent of disrupting the status quo of the market. Innovation is described as disruptive when it starts to appeal to the majority of users of the technology in the market. The technological capabilities of the space sector are steadily increasing due to the development and research efforts and the resulting space innovations. Tkatchova [11], however, indicated that innovation in space is different from other technological innovations due to the harsh environment experienced in space. The space environment makes it hard for space instruments to operate. According to the authors, the operating environment in space is determined by factors such as the microgravity environment, the high g-forces during the launch of the instruments, the vacuum environment, the temperature variations, extreme temperatures, and high-energy radiation. It is argued that space technologies are highly subject to the performance of the customer, which is similar to non-space technologies. The disruptive space technology differs from the other types of technologies in various ways: long development time with a high response time for new disruptive technologies, flight heritage, and market characteristics.

2.2.2. Disruptive Technologies in the Space Industry

Disruptive space technology is therefore a technology that changes the status of the space sector radically by having an alternative perceived performance mix, which fulfills the technical requirements of the user better than the previous technology (Van der Veen et al. [6]). The key difference between disruptive space technology and other space technologies is the fact that disruptive space technologies gain their relevance by outperforming the alternative performance mix that is valued by the customers of the niche market. In the space sector, there are various kinds of innovations to achieve the outperforming value.

A space elevator is an example of a disruptive innovation that has taken place in the space sector in the past years. According to Courtland [12], space elevators were proposed as a cheap alternative to costly rockets. The air elevator was considered a cheap alternative to transport cargo and humans into space.

The space elevator was designed to be made of a cable that was to be anchored to the surface of the earth and balanced by a counterweight in the space. On earth, the cable would have lasers that would beam power the climbers. The climbers would then crawl up the cable with their cargo to space. The technology has however stuck on the ground for years without progress. One of the main reasons the disruptive technology has not taken place is because the current materials are not strong enough to support the strain on the cable. Through carbon nanotubes have been found, it would be great news for the space elevators. Even with adequate materials, the concept of space elevators is still not achievable as it will still be highly unstable. This is because of the gravitational force from both the sun and moon as well as the pressure resulting from the solar wind. The solar wind and gravitational force would shake the cable causing the elevator to crash with other satellites. The author however recommends that thrusters are needed to keep the cable in line. Some of the significant negative effects expected to be caused by space elevators include sending a spacecraft to the wrong orbit, resulting in a slow crawl as compared to rocket launchers.

The motion of the cargo in the elevator will cause the cable to shake, which will either reduce or boost the velocity of the spacecraft exiting the elevator. The wobble could then send the spacecraft to the wrong orbit as well as damaging the elevator. The climbers in the space elevator also have to climb low to avoid creating large effects on the cable. Though slowing down the climbers can help minimize the effect, it will also slow down the trips to space. Ander Jorgensen of the New Mexico Institute of Mining and Technology indicated that building space elevators seem to be more complicated than originally expected.

2.2.3. Product Design

Product design is a situation or activity where people take industrial products as the main object for development and survival (Ren, [13]). The key to successful product design is an understanding of the end-user customer, the person for whom the product is being created. Khadke [14] stated that it is essential to consider the importance of technology innovation in product designs to avoid the destruction of key components as well as frequent redesign costs. Product designers attempt to solve real problems for real people by using both empathy and knowledge of their prospective customers' habits, behaviours, frustrations, needs, and wants.

Other than product design in normal practice, the product design process is much more complicated for space instruments. One of the reasons is the tough environments that the instrument needs to face. Another reason is the high accuracy of the product. There are a lot of trial-and-error processes during the product design stage. Moreover, there are many concerns not considered in earth products that are required to be included in the space instrument design. According to Meller [15], the product design for space instruments must have a low mass as well as high strength because of the hostility they face in the space environment. The product design used for creating the instruments should be able to avoid metal-to-metal contacts, must use liquid lubricants that are vacuum compatible as well as giving hardware error correction. In manufacturing space instruments, the manufacturers also have to incorporate latch-up protection circuits into the product design as well as radiative heat transfer mechanisms.

2.2.4. Reliability

The reliability of a space instrument is its ability to provide consistency in space and time or from different observers (Souza et al. [16]). According to the authors, reliability is one of the main quality criteria of an instrument regarding its ability to present aspects on homogeneity, equivalence, stability, and coherence. It refers to the equivalence, internal consistency, and stability of the space instrument. In space instruments, have the responsibility of ensuring the instruments are reliable for use in space. They should ensure that the onboard computers for the satellites are reliable as well as the infrastructure required for operating the instruments from the ground. According to the European Space Agency, there are no second chances in space missions hence reliability is a crucial aspect of space instruments. The current trend of increased autonomy of space systems and the unpredictable and rapid rate of technology change also poses new challenges to the reliability of the instruments. Reliability is therefore one of the main quality criteria of an instrument regarding its ability to present homogeneity, equivalence, stability, and coherence.

3. Methodology

3.1. Research Design

The study explores the relationships of innovation, product design, and reliability of space instruments. To do this, we conducted a review of previously published studies regarding space instruments and then analyzed the articles to investigate their findings. We systematically evaluate previous studies performed by different people to derive a conclusion about the research being carried out (Haidich [17]). The outcomes of analysis include a more precise estimate of the research body than any separate study, thus contributing to the collective analysis. The systematic review was carried out using PRISMA (Preferred Reporting Items for Systematic Reviews and Meta-analysis). According to Labaree [18], the PRISMA is designed to systematically summarize and evaluate the results from previous studies that meet the selection criteria of the research paper.

The selection of studies to be used is the first step in systematic analysis. According to Meline [19], the process involves the search of multiple databases to locate all studies that are potentially useful to determine the answers to the research questions. Secondary literature and data were used in the analysis. Secondary literature was composed of explanations and assessments from the primary result literature. The primary literature can be obtained, generalized, and summed up by researchers who can, later, generate new research. The studies used for the research originated from various databases that contained research papers related to the research topic. The research investigated papers that were published in English and incorporated search terms such as qualitative research and other terms related to the research topic. The Web of Science (WoS) database was used for investigation in this study. It is because the WoS is one of the widely used databases for research articles in academic disciplines. It also enables access to multiple databases that provide comprehensive citation data. We applied the following keywords in this study and select the articles until the end of September of 2021: "instrument" AND "space environment" AND "design".

This study mainly considers space instruments, the design of space instruments, and the space environment. However, the research does not indicate a specific timeline for the studies because we wish to acquire all the studies that had relevant information irrespective of the year it is published in order to explore the evolution of the design and manufacturing of space instruments.

3.2. Exclusion and Inclusion Criteria

The eligibility criteria specify the studies that will be included and those that will be excluded from the review (Meline [19]). The studies were selected and evaluated for eligibility based on their acceptability and relevance. The inclusion and exclusion criteria were guided by two questions: whether the study was acceptable for the analysis and whether the research was relevant to the purpose of the analysis. The full texts of the studies examined during the research were used to determine the study's trustworthiness and reliability. This study considered articles that were peer-reviewed and were written in the English language. After searching the terms based on the pre-defined keywords, the articles were later screened. The articles selected were those that met the inclusion criteria were retained within the study (Table 1).

Table 1. Inclusion and exclusion criteria.

Inclusion	Exclusion
Studies discussed space instruments and other topics relating to space exploration	Studies that did not discuss space instruments and other topics relating to space exploration
Studies that researched the space environment and the current trends in space exploration	Studies that failed to research the space environment and the current trends in space exploration
Journal articles that focus on the product designs or manufacturing of the space instruments	Journal articles that did not focus on the product designs or manufacturing of the space instruments
Journal articles published in the English language	Articles not published in the English language
Peer-reviewed articles	Articles that were not peer-reviewed

3.3. Sources of Information and Relevant Studies

This review followed the four-stage stream chart of PRISMA in looking for the investigations pertinent for the examination. PRISMA was utilized as it empowers to locate a wide scope of investigations of premium and suitable examinations for the exploration question (Moher et al. [20]). The four stages of PRISMA are recognizable proof, screening, qualification, and consideration of studies. The study used a single WoS database to search for relevant papers. The databases were utilized as they were considered to have increasingly centered data around the sort of studies and the researcher was searching for.

4. Results

To address the research questions of this study, the results are divided into several sections. The first section is to addresses the first research question on investigating the research focuses of the existing studies. Then, the research questions on the key consideration factors that may influence the design of the space instrument are discussed next. Further elaboration on the reviewed studies and the key consideration factors are elaborated in Sections 4.3 and 4.4, followed by the review on the product design and reliability.

4.1. Research Focuses

Figure 1 illustrates the overall systematic review process and the number of searched articles based on PRISMA. After searching the databases, 129 records were found. After removal of the similar and screening of the articles', and screened based on the inclusion and exclusion criteria, there were only 56 studies were left. Out of the 56 studies, 8 were excluded as the articles are not related to the instrument or product design nor manufacturing. After running through the inclusion and exclusion criteria and screening processes, only 48 remained for the final review. Any disagreement regarding the selection of the studies was resolved by keeping the research objectives as the focus. A systematic analysis was then conducted on the selected articles to extract information about the topic of the studies, the sample sizes, and the findings of the studies. Table 2 illustrates the summary of the research focuses of the articles on innovation, product design, instrument performance, and manufacturing. The articles were sort according to the last name of the first author. The instrument used in each study were also illustrated. It was found that most of the space instruments were applied in various outer space environments including orbit, spacecraft or space station, satellite or space telescope, lunar, mars, and mercury planetary missions, etc. Most of the research articles were focused on product design and instrument performance. It was also found that many of the product innovations were associated with the instrument product design, followed by the instrument fabrication and manufacturing of technologies. Innovation referred to the adoption of novel technologies and ideas, improvement of existing instruments using new techniques. In which, most of the instrument performance research focuses were related to the product design.

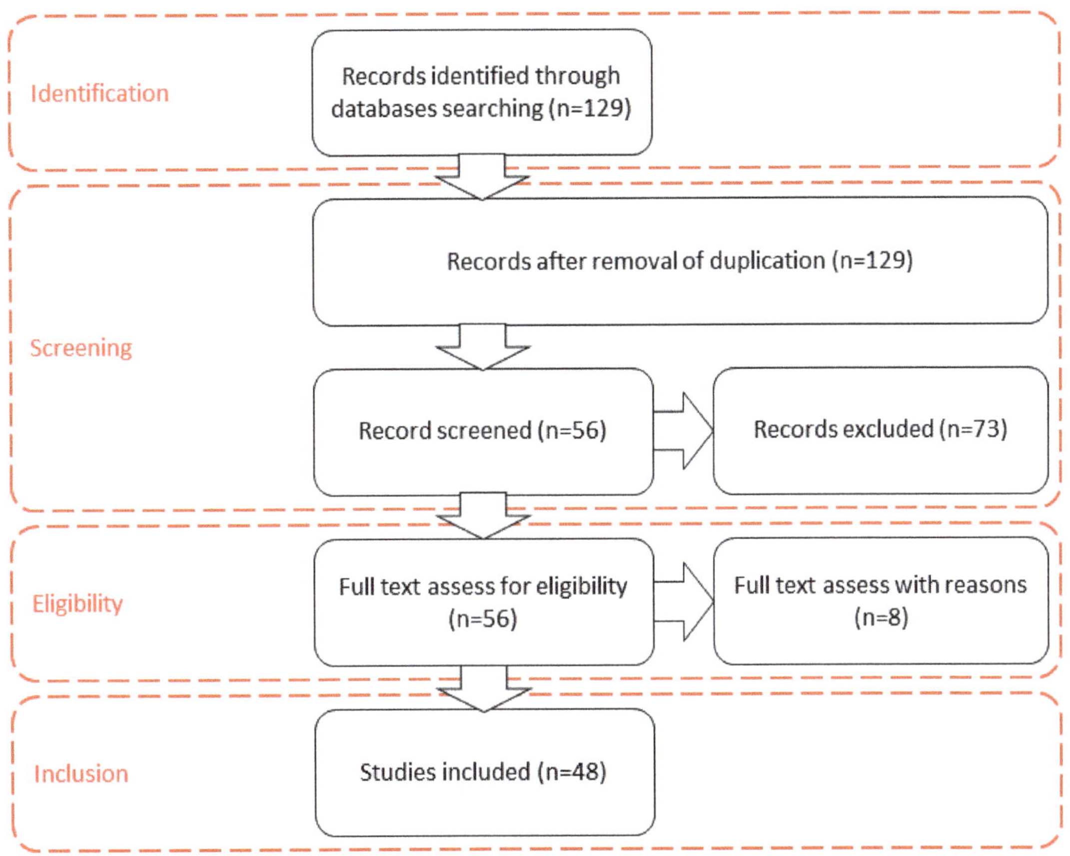

Figure 1. Systematic review on the instrument design of the space industry.

Table 2. Summary of the studies information and their research focus on innovation, product design, instrument performance, and manufacturing.

Authors	Year	Ref.	Instrument	Environment	Performance	Product Design	Innovation	Manufacturing
						Research Focus		
Barker	2018	[21]	Thermal Infrared	Lunar, Mercury		✓		✓
Biasotti	2020	[22]	Lunar Orbiter Laser Altimeter	Lunar	✓	✓		✓
Borgarelli	1998	[23]	Cassini Radar	Spacecraft		✓		
Bunce	2020	[24]	Imaging X-ray Spectrometer	Orbit	✓	✓		
Cavanaugh	2007	[25]	Mercury Laser Altimeter	Mercury	✓	✓		
Clark	2016	[26]	Energetic Charged Particle Detectors	Space		✓	✓	✓
Cress	2020	[27]	Falcon Solid-State Energetic Electron Detector	Orbit	✓	✓		
Delkowski	2021	[28]	Optical and Radar Instrument	Space			✓	✓

Table 2. Cont.

Authors	Year	Ref.	Instrument	Environment	Research Focus			
					Performance	Product Design	Innovation	Manufacturing
Dichter	1998	[29]	Compact Environmental Anomaly Sensor	Spacecraft		✓		
Dichter	2015	[30]	Gene Expression Measurement Module	Space	✓	✓	✓	
Dickie	2017	[31]	Micromachined Plasma Spectrometer	Satellites	✓	✓		✓
Dou	2017	[32]	Proton Microprobe	Space		✓		
Gilbert	2010	[33]	X-rays Space Telescopes	Space Telescope		✓		✓
Godet	2009	[34]	X- and Gamma-Ray Sensor	Space	✓			
Goldsten	2007	[35]	Gamma-Ray and Neutron Spectrometer	Spacecraft		✓		
Hall	2017	[36]	Charge-Coupled Device	Space	✓	✓		
Han	2016	[37]	Differential Electrostatic Space Accelerometer			✓	✓	✓
Hsiao	2010	[38]	Radiatively Cooled Instrument	Space	✓	✓		
Hu	2014	[39]	Scanning Fabry-Perot Interferometer	Space Station		✓		
Hudson	2007	[40]	Differential Electrostatic Accelerometer	Orbit	✓	✓		
Koehn	2002	[41]	Fast Imaging Plasma Spectrometer	Mercury		✓	✓	
Koga	3002	[42]	Neutron Monitor	Space Station		✓	✓	
Krebs	2005	[43]	Mercury Laser Altimeter	Mercury		✓		
Lepri	2017	[44]	Fast Imaging Plasma Spectrometer	Mercury		✓		
LIFE	2019	[45]	Charged Particle Instruments	Orbit	✓	✓	✓	
Lindstrom	2018	[46]	Environmental Anomaly Sensor	Space		✓		
Ling	2019	[47]	Space Welding Technology	Space			✓	✓
Liu	2021	[48]	Mass Spectrometers	Space			✓	
Lopes	2021	[49]	Radiometers	Space	✓	✓		
MacDonald	2006	[50]	Magnetospheric Plasma Analyzer	Satellites		✓		
Magnes	2020	[51]	Space Weather Magnetometer	Orbit		✓		
Mauk	2017	[52]	Energetic Particle Detector Instruments	Jupiter	✓	✓		
Moretti	2010	[53]	Magneto-Optical Filter-based system	Space	✓			
Ostgaard	2019	[54]	X- and Gamma-Ray Sensor	Space Station	✓	✓		

Table 2. *Cont.*

Authors	Year	Ref.	Instrument	Environment	Research Focus			
					Performance	Product Design	Innovation	Manufacturing
Rothkaehl	2011	[55]	Plasma-Wave Complex	Space Station	✓	✓		
Sadrozinski	2002	[56]	Gamma-ray Large Area Space Telescope	Space		✓		
Schlemm	2007	[57]	X-ray Spectrometer	Mercury		✓		
Soli	1995	[58]	Proton-spectrometer	Spacecraft, Satellite		✓		
Swinyard	2000	[59]	Moderate-Resolution Imaging Spectroradiometer (MODIS) Instrument	Orbit	✓	✓		
Thuillier	1992	[60]	Michelson Interferometer	Satellites	✓	✓		
Warren	2017	[61]	Differential Electrostatic Space Accelerometer	Space Station		✓		✓
Wei	2013	[62]	X-ray Detector and Energetic Particle Detectors	Space			✓	
Wesolek	2005	[63]	Microwave Sounder Instrument	Space		✓	✓	✓
Wise	1995	[64]	Materials in Devices as Superconductors	Spaceflight	✓	✓		✓
Wright	2013	[65]	Thermal Hyperspectral Imager	Space		✓		
Xiong	2019	[66]	Optical Thin Films	Space		✓		✓
Zanoni	2016	[67]	Doped Germanium Photoconducting Detectors	Space	✓			
Zurbuchen	2016	[68]	Plasma sensors	Space			✓	

4.2. Key Consideration Factors on Instrument Design

The results of the studies are summarized in Table 3. The summaries for the selected papers are given in terms of objectives and the research of the key consideration factors in the space environment. The summaries of the 48 studies are summarized in the table as per the guidelines provided by Arksey et al. [19]. The key consideration factors of the instrument in the space environment can be divided into categories including the design and performance considerations. Design consideration refers to the key factors and parameters considered in the space instrument design in order to suit the extreme space environment, such as materials, duration, size, power consumption, weight can perform its designed functions in long space travel. Performance consideration focuses on whether the designed and manufactured instrument can achieve and maintain certain design functions and accuracy under harsh space weather conditions and long-term operations. The performance considers the accuracy of measurement on the collected data and signals.

Table 3. Research objectives and the key consideration factors in the space environment.

Author	Ref.	Aims and Objectives	The Key Considerations in Space
Barker	[21]	Measured changes in the laser characteristics and obtain data to understand the laser behavior and refine the instrument pointing model	Long-term laser behavior
Biasotti	[22]	Describe the design, with the preliminary phonon dynamics simulation, the fabrication, of first demonstration model	Sensitivity
Borgarelli	[23]	Development of a passive mode, implemented to measure Titan's surface emissivity	Reduced mass, low available room, low power consumption, severe environmental conditions, specific thermal control and on-ground test accessibility
Bunce	[24]	The design, performance, scientific goals and operations plans of the mercury imaging X-ray spectrometer	Design, material, size
Cavanaugh	[25]	Describes the instrument design, prelaunch testing, calibration, and results of postlaunch testing.	Performance
Clark	[26]	Review the Puck Energetic Particle Detector (EPD) design, its heritage, unexpected results from these past missions and future advancements	Review paper
Cress	[27]	Describes the design, development, and calibration of the Falcon Solid-state Energetic Electron Detector (FalconSEED)	Geosynchronous environment
Delkowski	[28]	Develop manufacturing methods for next generation of advanced composites for space instrument	Materials (composites)
Dichter	[29]	Designed an instrument to measure the local space radiation environment.	Small, lightweight, and low power
Dichter	[30]	Describe the design and novel features of the instruments and discuss their calibration program	Accurate measurements
Dickie	[31]	Design, manufacture, and characterization of a new frequency selective surface (FSS) structure	Performance
Dou	[32]	A systematic investigation of the ion beam optics to optimize the design for the Harbin system	Design optimization
Gilbert	[33]	Demonstrate an optimized design of a linear-electric-field time-of-flight technology that can be used to obtain a high signal to noise	Signal to noise, size or complexity
Godet	[34]	Study the instrument background and sensitivity of the coded-mask camera	Optimise the performances
Goldsten	[35]	Overview the gamma-ray and neutron spectrometer and describes its science and measurement objectives, the design and operation of the instrument, the ground calibration effort, and early in-flight data.	Thermal behavior, performance
Hall	[36]	Optimise the device design to suffer minimum impact from radiation damage effects	Radiation
Han	[37]	Describe the design and capability of the differential accelerometer to test weak space acceleration	Electrostatic suspension, electrostatic motor
Hsiao	[38]	Design and fabrication of optical thin films for remote sensing instruments	Optical stability
Hu	[39]	Investigate the instrument design to measure the mesospheric and thermospheric wind velocities	Mesospheric and thermospheric wind velocities
Hudson	[40]	This paper presents the current design of the accelerometer, specifically the critical areas for the instrument design, integration, and final performance requirements.	Accurate measurements

Table 3. *Cont.*

Author	Ref.	Aims and Objectives	The Key Considerations in Space
Koehn	[41]	Discuss the design and prototype tests of the fast-imaging plasma spectrometer (FIPS) deflection system	Lightweight, fast, and have a very large field of view
Koga	[42]	Discuss the results of the engineering model (EM) and its properties	Particle and plasma
Krebs	[43]	Develop the mercury laser altimeter	Space-flight environmental tests
Lepri	[44]	Discuss an adaptation of the fast-imaging plasma spectrometer (FIPS) for the measurement of negatively charged particles.	Design modification
LIFE	[45]	Developed an automated, miniaturized, integrated fluidic system for in-situ measurements of gene expression in microbial samples	Biological validation
Lindstrom	[46]	Design a new sensor compact environmental anomaly sensor risk reduction (CEASE-RR) for anomaly attribution	Calibration and planned flight experiment, radiation environment
Ling	[47]	Carry out the environmental adaptability design and analysis	Mechanical property and the thermal environment
Liu	[48]	Research on the effects of the space environment on the welding technology	Microgravity, vacuum conditions, and temperature differences
Lopes	[49]	Understand how each component interferes with sensitivity and response time of the instrument depending on its design, material, volume, and thermal contact.	Thermal behavior, design, material, size, performance effect
MacDonald	[50]	Extrapolate the background response to the inner magnetosphere, a highly relevant instrument design parameter for future missions to this region.	Response to the inner magnetosphere
Magnes	[51]	Describes the magnetometer instrument design, discusses the ground calibration methods and results.	Avoiding strict magnetic cleanliness requirements, dynamic stray fields
Mauk	[52]	Describe the science objectives of the Jupiter Energetic Particle Detector Instruments (JEDI), the science and measurement requirements, the challenges that the JEDI team had in meeting these requirements, the design and operation of the JEDI instruments, their calibrated performances, the JEDI inflight and ground operations, and the initial measurements of the JEDI instruments in interplanetary space	Performances
Moretti	[53]	Present a low-cost, low-weight instrument, thus particularly fit to space applications, capable of providing stability and sensitivity of signals on long-term observations.	Stability and sensitivity of signals on long-term observations
Ostgaard	[54]	Describe the scientific objectives, design, performance, imaging capabilities and operational modes of the modular X- and gamma-ray sensor (MXGS) instrument.	Instrument performance, imaging capabilities
Rothkaehl	[55]	Design of the instrument for monitoring the electromagnetic ecosystem for space weather purpose	Ionospheric plasma property and artificial noises
Sadrozinski	[56]	The Gamma-ray Large Area Space Telescope (GLAST) instrument designed for high sensitivity, high precision gamma-ray detection in space.	High sensitivity, high precision gamma-ray detection
Schlemm	[57]	Summarizes XRS's science objectives, technical design, calibration, and mission observation strategy.	X-ray
Soli	[58]	Presents radiation dosimetry results from the radiation and reliability assurance experiments on the Clementine Spacecraft and Interstage Adapter Satellite.	Performance
Swinyard	[59]	Discuss the performance of the ten doped germanium photoconducting detectors on the infrared space observatory long wavelength spectrometer	Performance

Table 3. *Cont.*

Author	Ref.	Aims and Objectives	The Key Considerations in Space
Thuillier	[60]	Performances of the WINDII, a Michelson interferometer used to observe wind and temperature in the upper mesosphere and thermosphere are shown and analyzed.	Performance
Warren	[61]	Describes the design, build, calibration, and initial measurements from a new laboratory instrument	Performance
Wei	[62]	Presents the special technologies applied, for the solar X-ray spectrometer, and the first pre-flight calibration results	Solar X-ray and energetic charged particles
Wesolek	[63]	Design, fabrication, simulation, and testing of the instrument front end that consists of a collimator, parallel plate energy analyzer, and energy selector mask	Small-scale, energy analysis
Wise	[64]	Describes the design, fabrication, and testing of the primary subsystems of the instrument.	Critical superconductive properties
Wright	[65]	Describe the rationale for the project, the instrument design, and the quality of the data	Mass, volume, and power constraints
Xiong	[66]	Overview the calibration algorithms, operational activities, on-orbit performance, remaining challenges, and potential improvements.	Performance
Zanoni	[67]	Investigate the performance of a radiatively cooled instrument	Performance, thermal behavior
Zurbuchen	[68]	Review the innovation triggers in the context of the design literature and with the help of two case studies	Review

4.3. Product Innovation and Design

For the innovation concept of product design in a tough space environment and mission, the findings of the study indicated that there is a relationship between innovation, product design, and manufacturing of the space instruments. Figure 2 illustrates the number of articles showing the relationship between design, manufacturing, product innovation. Most of the reviewed articles demonstrated the relationship between instrument design and fabrication. Most of the instrument innovation and related to the instrument design.

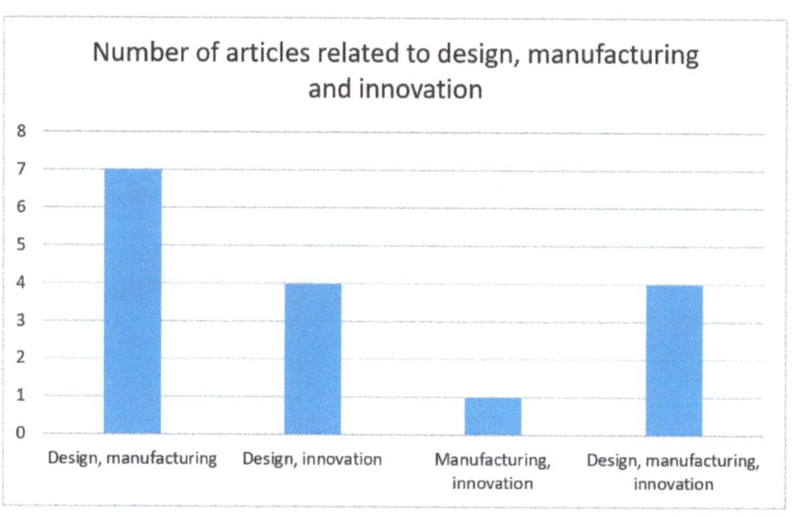

Figure 2. The number of articles demonstrates the relationship between design, manufacturing, product innovation.

Jiao et al. [69] indicated that the process of outgassing in space is a unique phenomenon in space instruments that can cause negative impacts on scientific exploration missions, high-voltage devices, and spacecraft optical systems. According to the authors, to mitigate the negative impact caused by outgassing, there is a need to develop a transient and long-term physical model of outgassing. This would be by developing new testing methods by combining the outgassing tests with the outgassing compound analysis, as well as on the improvement of the existing product design and manufacturing technology.

Dichter et al. [30] describe the next generation of GOES satellites will include a new suite of charged particle instruments. The design and novel features of the instruments and discuss their calibration program in terms of accuracy of on-orbit measurements. The innovation of the instrument development made significant improvements not only in the operational measurement of the space environment but also in the overall performance of the instrument covering a wider range of measuring abilities and lower power consumption compare with the previous version of the instrument.

Koehn et al. [41] discussed the design and prototype tests of the fast-imaging plasma spectrometer (FIPS) deflection system. The major piece of innovation is to improve the instrument to enable a larger instantaneous field of view. This novel design also enables a lightweight and fast product. Koga et al. [42] designed the engineering model (EM) and investigated its properties. A new neutron monitor instrument is designed to understand the particle acceleration mechanism at the solar surface. Life et al. [45] design a new biological system that can be deployed in near future for space missions. platforms other than the ISS to advance biological research in space. It can also prove useful for numerous terrestrial applications in the field. The novel instrument provided an automated, miniaturized, integrated fluidic system for biological validation.

4.4. Product Innovation and Manufacturing

As illustrated in Figure 2, the manufacturing of the instrument related to product innovation was usually associated with product design. Delkowski et al. [28] developed a new manufacturing method that was used to enhance polymer and composite structures in spacecraft. The novel approach of composite materials led to research and innovation over many decades. The new manufacturing of composite materials featuring 10–20 times greater resistance to cracking without affecting the stiffness of dimensionally stable structures.

Another research associated instrument innovation with the product design and the fabrication. In Clark et al. [26], new foil manufacturing processes were reviewed to discuss the association of high-voltage anomalies and the use of curved foils on recent Puck EPD designs. Han et al. [37] demonstrated the preliminary work on the development of the first instrument prototype. The space accelerometer is a newly designed instrument proposed to operate onboard China's space station. The new prototype was tested under a weak space acceleration. Modeling and simulation were performed to test the electrostatic suspension and electrostatic motor based on attainable space microgravity conditions. Noise evaluation was also performed to evaluate the performance of the instrument. This development confirmed several crucial fabrication processes and measurement techniques for the future design and development of space accelerometers.

Wesolek et al. [63] designed and fabricated a new version of a space environment that re-designed the one run in 2008. The redesigned system presented a lower cost, lower weight that fits space applications and long-term operations. The newly designed and fabricated instrument could provide stability and sensitivity of signals.

4.5. Product Design and Reliability

In the design of the space instrument, reliability is another key consideration concerned by the product designer. Indeed, the design and manufacturing of the instrument is usually relating the reliability of the product. For innovative product design improving the reliability of space instruments, García-Pérez et al. [2], found that the transient analysis performed on the STEP instrument provided accurate simulations of the shock environment.

The finite element method had higher confidence in the calculated results hence offering more information than the data obtained for the shock tests. Jiao et al. [69] also found that establishing a transient and long-term physical model of outgassing can help obtain the outgassing characteristics of different products. This shows that using innovation to develop new or improve the existing product designs helps increase the reliability of the space instruments.

According to Conscience et al. [70], improving the space instruments increases their reliability. The authors gave an example of the SOVAP instrument and how its improvement had increased its efficiency. According to the authors, the instrument has been improved by adding the bolometric oscillation sensor (BOS) in order to increase the time resolution. With the BOS, the SOVAP will be able to measure the albedo flux, the infrared flux of the Earth, and the solar irradiance with a smaller sampling period of ten seconds. Malandraki et al. [71] on the other hand conducted an experiment to compare the testing abilities of the space tool. The tool used microwave data that yielded no false alarms indicating that the product design of the instruments affected their reliability. Gold et al. [1] added that miniaturizing space instruments will help in improving the quality of the science from the instruments. The authors gave an example of the instrument of imageries which improved to include a version of the processing layer.

Jiggens et al. [72], found that the space radiation environment is an important factor for both astronauts and instruments. Other than the traditional shielding protection methods, the authors created a warning system for the solar particles event. The innovation in the study help to improve the reliability of the instrument by avoiding the large SPEs. Tam [3], raised the possibility of using new technology such as three-dimensional and additive manufacturing to replace the old manufacturing method which has a complex design and sub-system. The use of new manufacturing methods reduces the risk of pieces breaking off during collisions in the space environment.

Yung et al. [73] added that space instruments need to be designed in a such way that they perform reliably. The authors described an example of a new design of spacecraft made in 2011 that could provide both qualitative and quantitative measures of the composition of regolith. The SOPSYS however was designed in a such way that would enable it to the grid, sieve, transport, and measure samples of regolith in the absence of gravity. To increase its reliability, the instrument was developed with a reverse thread so that it would shroud any regolith that stuck in the mechanisms of the actuator. In this way, any stuck regolith would be pushed back to the grinding head. The author indicated that the new spacecraft provided an anti-jam solution that did not require additional mass hence increasing the reliability of space missions

5. Conclusions

In this study, we have explored the influences of innovation and reliability in the product design of space instruments. This was performed by conducting a review of previously published articles regarding space instruments and analyzing these articles to investigate their findings and review. PRISMA was used to search the articles systematically. The results in the study indicate that the product design of the space instrument was directly influenced by the innovation. This is because the space instrument is usually very complex and consists of many factors considering the complex situation of the deep space environment. On the other hand, the products are difficult to be found from the traditional design of a product. The study also found that the reliability of the instruments is directly influenced by the degree of innovation and product design of the space instruments. It was determined from the examples gathered in various literature sources that all innovation processes led to an improvement in the reliability of the instruments. This study is important to formulate the critical factors in the design and development of a space instrument that is important to develop the smart manufacturing protocol in the field in the future.

The current review focuses on articles about space instruments as well as product innovation, design. The criteria for inclusion are based on current trends in space exploration and product innovation. The keyword search focuses on the setting of the three themes mentioned above. However, product design and fabrication or manufacturing technologies are closely related but not included as one of the keywords in the search. Fabrication and manufacturing are not included because the keywords are too specific generating a small number of search results, particularly focusing on the space environment. As such, the screening processes have to be performed manually leading to less objective conclusions. In the future, more databases can be included in order to enhance the searching results and related articles. On the other hand, the product design is usually related to the reliability issue, particularly in space devices and instruments. Thus, a further review can be conducted to summarize whether product innovation, design, and reliability are correlated and affect the performances of space instruments. It is recommended that future research can also be made related to the performance and design of the space instrument. Lastly, instruments that are used in the space environment may include various interpretations such as near space, deep space, orbits, planetary missions, etc. These keywords may also be included in future review studies.

Funding: This research was funded by PolyU (project account code: ZG6W).

Informed Consent Statement: Not applicable.

Data Availability Statement: Not applicable.

Acknowledgments: We acknowledge the funding support from PolyU (project account code: ZG6W) for this research and support of the Laboratory for Artificial Intelligence in Design (Project Code: RP2-1), Hong Kong Special Administrative Region, in preparing this publication.

Conflicts of Interest: The authors declare no conflict of interest.

References

1. Gold, R.E.; Jenkins, R.E. Advanced space instruments. *Johns Hopkins APL Tech. Dig.* **1999**, *20*, 611–619.
2. García-Pérez, A.; Sorribes-Palmer, F.; Alonso, G.; Ravanbakhsh, A. Overview and application of FEM methods for shock analysis in space instruments. *Aerosp. Sci. Technol.* **2018**, *80*, 572–586. [CrossRef]
3. Tam, W. The Space Debris Environment and Satellite Manufacturing. Ph.D. Thesis, Walden University, Minnesota, MN, USA, 2015.
4. Chau, K.-Y.; Tang, Y.M.; Liu, X.; Ip, Y.-K.; Tao, Y. Investigation of critical success factors for improving supply chain quality management in manufacturing. *Enterp. Inf. Syst.* **2021**, 1–20. [CrossRef]
5. Calleja-Ochoa, A.; Gonzalez-Barrio, H.; de Lacalle, N.L.; Martínez, S.; Albizuri, J.; Lamikiz, A. A New Approach in the Design of Microstructured Ultralight Components to Achieve Maximum Functional Performance. *Materials* **2021**, *14*, 1588. [CrossRef]
6. Tang, Y.; Au, K.; Leung, Y. Comprehending products with mixed reality: Geometric relationships and creativity. *Int. J. Eng. Bus. Manag.* **2018**, *10*, 1847979018809599. [CrossRef]
7. Yung, K.L.; Ho, G.T.S.; Tang, Y.M.; Ip, W.H. Inventory classification system in space mission component replenishment using multi-attribute fuzzy ABC classification. *Ind. Manag. Data Syst.* **2021**, *121*, 637–656. [CrossRef]
8. Van Der Veen, E.J.; Giannoulas, D.A.; Guglielmi, M.; Uunk, T.; Schubert, D. Disruptive Space Technologies. *Int. J. Space Technol. Manag. Innov.* **2012**, *2*, 24–39. [CrossRef]
9. Popa, I.L.; Preda, G.; Boldea, M. A theoretical approach of the concept of innovation. Managerial Challenges of the Contemporary Society. *Proceedings* **2010**, *1*, 151–156.
10. Ho, G.; Tang, Y.M.; Tsang, K.Y.; Tang, V.; Chau, K.Y. A blockchain-based system to enhance aircraft parts traceability and trackability for inventory management. *Expert Syst. Appl.* **2021**, *179*, 115101. [CrossRef]
11. Tkatchova, S. *Space-Based Technologies and Commercialized Development*; IGI Global: Hershey, PA, USA, 2011.
12. Courtland, R. Space Elevator Trips Could be Agonizingly Slow. NewScientist. 2008. Available online: https://www.newscientist.com/article/dn16223-space-elevator-trips-could-be-agonisingly-slow/#:~{}:text=Space%20elevators%20have%20been%20proposed,a%20counter%2Dweight%20in%20space (accessed on 6 June 2020).
13. Ren, W. The Relations between Products Design and the Space Environment. *Asian Soc. Sci.* **2009**, *4*, 165. [CrossRef]
14. Khadke, K. Engineering Design Methodology for Planned Product Innovation. Ph.D. Thesis, Michigan Technological University, Michigan, MI, USA, 2007.
15. Meller, R. SpaceInstrument Instrument Development. 2010. Available online: https://www.mps.mpg.de/phd/space-instrument-development (accessed on 12 June 2020).
16. Souza, A.C.D.; Alexandre, N.M.C.; Guirardello, E.D.B. Psychometric properties in instruments evaluation of reliability and validity. *Epidemiol. Serviços Saúde* **2017**, *26*, 649–659. [CrossRef]
17. Haidich, A.B. Meta-analysis in medical research. *Hippokratia* **2010**, *14*, 29–37. [PubMed]

18. Labaree, R.V. Research Guides: Organizing Your Social Sciences Research Paper: Types of Research Designs. Retrieved 10 November 2020. Available online: https://libguides.usc.edu/writingguide/quantitative (accessed on 2 October 2021).

19. Meline, T. Selecting studies for systemic review: Inclusion and exclusion criteria. *Contemp. Issues Commun. Sci. Disord.* **2006**, *33*, 21–27. [CrossRef]

20. Moher, D.; Liberati, A.; Tetzlaff, J.; Altman, D.G. The PRISMA Group. Preferred Reporting Items for Systematic Reviews and Meta-Analyses: The PRISMA Statement. *PLoS Med.* **2009**, *6*, e1000097. [CrossRef] [PubMed]

21. Barker, M.K.; Sun, X.; Mao, D.; Mazarico, E.; Neumann, G.A.; Zuber, M.T.; Smith, D.E.; McGarry, J.F.; Hoffman, E.D. In-flight characterization of the lunar orbiter laser altimeter instrument pointing and far-field pattern. *Appl. Opt.* **2018**, *57*, 7702–7713. [CrossRef] [PubMed]

22. Biasotti, M.; Boragno, C.; Barusso, L.F.; Gatti, F.; Grosso, D.; Rigano, M.; Siri, B.; Macculi, C.; D'Andrea, M.; Piro, L. The Phonon-Mediated TES Cosmic Ray Detector for Focal Plane of ATHENA X-ray Telescope. *J. Low Temp. Phys.* **2020**, *199*, 225–230. [CrossRef]

23. Borgarelli, L.; Im, E.; Johnson, W.T.K.; Scialanga, L. The microwave sensing in the Cassini Mission: The radar. *Planet. Space Sci.* **1998**, *46*, 1245–1256. [CrossRef]

24. Bunce, E.J.; Martindale, A.; Lindsay, S.; Muinonen, K.; Rothery, D.A.; Pearson, J.; McDonnell, I.; Thomas, C.; Thornhill, J.; Tikkanen, T.; et al. The BepiColombo Mercury Imaging X-ray Spectrometer: Science Goals, Instrument Performance and Operations. *Space Sci. Rev.* **2020**, *216*, 1–38. [CrossRef]

25. Cavanaugh, J.F.; Smith, J.C.; Sun, X.; Bartels, A.E.; Ramos-Izquierdo, L.; Krebs, D.J.; McGarry, J.F.; Trunzo, R.; Novo-Gradac, A.M.; Britt, J.L.; et al. The Mercury Laser Altimeter Instrument for the MESSENGER Mission. *Space Sci. Rev.* **2007**, *131*, 451–479. [CrossRef]

26. Clark, G.; Cohen, I.; Westlake, J.; Andrews, G.B.; Brandt, P.; Gold, R.E.; Gkioulidou, M.; Hacala, R.; Haggerty, D.; Hill, M.E.; et al. The "Puck" energetic charged particle detector: Design, heritage, and advancements. *J. Geophys. Res. Space Phys.* **2016**, *121*, 7900–7913. [CrossRef]

27. Cress, R.; Maldonado, C.A.; Coulter, M.; Haak, K.; Balthazor, R.L.; McHarg, M.G.; Barton, D.; Greene, K.; Lindstrom, C.D. Calibration of the Falcon Solid-state Energetic Electron Detector (SEED). *Space Weather* **2020**, *18*, e2019SW002345. [CrossRef]

28. Delkowski, M.; Smith, C.T.; Anguita, J.V.; Silva, S.R.P. Increasing the robustness and crack resistivity of high-performance carbon fiber composites for space applications. *iScience* **2021**, *24*, 102692. [CrossRef]

29. Dichter, B.K.; McGarity, J.O.; Oberhardt, M.R.; Jordanov, V.T.; Sperry, D.J.; Huber, A.C.; Pantazis, J.A.; Mullen, E.G.; Ginet, G.; Gussenhoven, M.S. Compact Environmental Anomaly Sensor (CEASE): A novel spacecraft in-strument for in situ measurements of environmental conditions. *IEEE Trans. Nucl. Sci.* **1998**, *45*, 2758–2764. [CrossRef]

30. Dichter, B.K.; Galica, G.E.; McGarity, J.O.; Tsui, S.; Golightly, M.J.; Lopate, C.; Connell, J.J. Specification, Design, and Calibration of the Space Weather Suite of Instruments on the NOAA GOES-R Program Spacecraft. *IEEE Trans. Nucl. Sci.* **2015**, *62*, 2776–2783. [CrossRef]

31. Dickie, R.; Christie, S.; Cahill, R.; Baine, P.; Fusco, V.; Parow-Souchon, K.; Henry, M.; Huggard, P.G.; Donnan, R.S.; Sushko, O.; et al. Low-Pass FSS for 50–230 GHz Quasi-Optical Demultiplexing for the MetOp Second-Generation Microwave Sounder Instrument. *IEEE Trans. Antennas Propag.* **2017**, *65*, 5312–5321. [CrossRef]

32. Dou, Y.; Jamieson, D.N.; Liu, J.; Lv, K.; Li, L. The design of the 300 MeV proton microprobe system in Harbin. *Nucl. Instrum. Methods Phys. Res. Sect. B Beam Interact. Mater. Atoms* **2017**, *404*, 9–14. [CrossRef]

33. Gilbert, J.A.; Lundgren, R.A.; Panning, M.H.; Rogacki, S.; Zurbuchen, T.H. An optimized three-dimensional linear-electric-field time-of-flight analyzer. *Rev. Sci. Instrum.* **2010**, *81*, 53302. [CrossRef] [PubMed]

34. Godet, O.; Sizun, P.; Barret, D.; Mandrou, P.; Cordier, B.; Schanne, S.; Remoue, N. Monte-Carlo simulations of the background of the coded-mask camera for X- and Gamma-rays on-board the Chinese–French GRB mission SVOM. *Nucl. Instrum. Methods Phys. Res. Sect. A Accel. Spectrometers Detect. Assoc. Equip.* **2009**, *603*, 365–371. [CrossRef]

35. Goldsten, J.O.; Rhodes, E.A.; Boynton, W.V.; Feldman, W.C.; Lawrence, D.J.; Trombka, J.I.; Smith, D.M.; Evans, L.G.; White, J.; Madden, N.W.; et al. The MESSENGER Gamma-Ray and Neutron Spectrometer. *Space Sci. Rev.* **2007**, *131*, 339–391. [CrossRef]

36. Hall, D.J.; Skottfelt, J.; Soman, M.R.; Bush, N.; Holland, A. Improving radiation hardness in space-based Charge-Coupled Devices through the narrowing of the charge transfer channel. *J. Instrum.* **2017**, *12*, C12021. [CrossRef]

37. Han, F.; Liu, T.; Li, L.; Wu, Q. Design and Fabrication of a Differential Electrostatic Accelerometer for Space-Station Testing of the Equivalence Principle. *Sensors* **2016**, *16*, 1262. [CrossRef]

38. Hsiao, C.N.; Chen, H.P.; Chiu, P.K.; Cho, W.H.; Lin, Y.W.; Chen, F.Z.; Tsai, D.P. Design and fabrication of optical thin films for remote sensing instruments. *J. Vac. Sci. Technol. A Vac. Surf. Films* **2010**, *28*, 867–872. [CrossRef]

39. Hu, G.; Ai, Y.; Zhang, Y.; Zhang, H.; Liu, J. First scanning Fabry–Perot interferometer developed in China. *Chin. Sci. Bull.* **2014**, *59*, 563–570. [CrossRef]

40. Hudson, D.; Chhun, R.; Touboul, P. Development status of the differential accelerometer for the MICROSCOPE mission. *Adv. Space Res.* **2005**, *39*, 307–314. [CrossRef]

41. Koehn, P.L.; Zurbuchen, T.H.; Gloeckler, G.; Lundgren, R.A.; Fisk, L.A. Measuring the plasma environment at Mercury: The fast imaging plasma spectrometer. *Meteorit. Planet. Sci.* **2002**, *37*, 1173–1189. [CrossRef]

42. Koga, K.; Goka, T.; Matsumoto, H.; Muraki, Y.; Masuda, K.; Matsubara, Y. Development of the fiber neutron monitor for the energy range 15-100 MeV on the International Space Station (ISS). *Radiat. Meas.* **2001**, *33*, 287–291. [CrossRef]

43. Krebs, D.J.; Novo-Gradac, A.-M.; Li, S.X.; Lindauer, S.J.; Afzal, R.S.; Yu, A.W. Compact, passively Q-switched Nd:YAG laser for the MESSENGER mission to Mercury. *Appl. Opt.* **2005**, *44*, 1715–1718. [CrossRef]
44. Lepri, S.T.; Raines, J.M.; Gilbert, J.A.; Cutler, J.; Panning, M.; Zurbuchen, T.H. Detecting negative ions on board small satellites. *J. Geophys. Res. Space Phys.* **2017**, *122*, 3961–3971. [CrossRef]
45. Peyvan, K.; Karouia, F.; Cooper, J.J.; Chamberlain, J.; Suciu, D.; Slota, M.; Pohorille, A. Gene Expression Measurement Module (GEMM) for space application: Design and validation. *Life Sci. Space Res.* **2019**, *22*, 55–67. [CrossRef]
46. Lindstrom, C.D.; Aarestad, J.; Ballenthin, J.O.; Barton, D.A.; Coombs, J.M.; Ignazio, J.; Johnston, W.R.; Kratochvil, S.; Love, J.; McIntire, D.; et al. The Compact Environmental Anomaly Sensor Risk Reduction: A Pathfinder for Operational Energetic Charged Particle Sensors. *IEEE Trans. Nucl. Sci.* **2017**, *65*, 439–447. [CrossRef]
47. MingchunLing; MaoxinSong; JinHong; FeiTao; PengZou; ZhenSun Design and Validation of Space Adaptability for Particulate Observing Scanning Polarization. *Chin. J. Lasers* **2019**, *46*, 0704002. [CrossRef]
48. Liu, X.; Dong, Q.; Wang, P.; Chen, H. Review of Electron Beam Welding Technology in Space Environment. *Optik* **2021**, *225*, 165720. [CrossRef]
49. Lopes, A.G.; Irita, R.T.; Berni, L.A.; Vilela, W.A.; Savonov, G.D.S.; Carlesso, F.; Vieira, L.E.A.; de Miranda, E.L. Simplified Thermal Model for Absolute Radiometer Simulation. *J. Sol. Energy Eng.* **2021**, *143*, 1–22. [CrossRef]
50. Macdonald, E.; Thomsen, M.; Funsten, H. Background in channel electron multiplier detectors due to penetrating radiation in space. *IEEE Trans. Nucl. Sci.* **2006**, *53*, 1593–1598. [CrossRef]
51. Magnes, W.; Hillenmaier, O.; Auster, H.-U.; Brown, P.; Kraft, S.; Seon, J.; Delva, M.; Valavanoglou, A.; Leitner, S.; Fischer, D.; et al. Space Weather Magnetometer Aboard GEO-KOMPSAT-2A. *Space Sci. Rev.* **2020**, *216*. [CrossRef]
52. Mauk, B.H.; Haggerty, D.K.; Jaskulek, S.E.; Schlemm, C.E.; Brown, L.E.; Cooper, S.A.; Gurnee, R.S.; Hammock, C.M.; Hayes, J.R.; Ho, G.; et al. The Jupiter Energetic Particle Detector Instrument (JEDI) Investigation for the Juno Mission. *Space Sci. Rev.* **2013**, *213*, 289–346. [CrossRef]
53. Moretti, P.F.; Berrilli, F.; Bigazzi, A.; Jefferies, S.M.; Murphy, N.; Roselli, L.; Di Mauro, M.P.; Mauro, M.P. Future instrumentation for solar physics: A double channel MOF imager on board ASI Space Mission ADAHELI. *Astrophys. Space Sci.* **2010**, *328*, 313–318. [CrossRef]
54. Østgaard, N.; Balling, J.E.; Bjørnsen, T.; Brauer, P.; Budtz-Jørgensen, C.; Bujwan, W.; Carlson, B.; Christiansen, F.; Connell, P.; Eyles, C.; et al. The Modular X- and Gamma-Ray Sensor (MXGS) of the ASIM Payload on the International Space Station. *Space Sci. Rev.* **2019**, *215*, 23. [CrossRef]
55. Rothkaehl, H.; Morawski, M.; Puccio, W.; Bergman, J.; Klimov, S.I. Diagnostics of Space Plasma on Board International Space Station—ISS. *Contrib. Plasma Phys.* **2011**, *51*, 158–164. [CrossRef]
56. Sadrozinski, H.F.W. Radiation issues in the Gamma-ray Large Area Space Telescope GLAST. *Nucl. Instrum. Methods Phys. Res. Sect. A* **2002**, *476*, 722–728. [CrossRef]
57. Schlemm, C.E.; Starr, R.D.; Ho, G.; Bechtold, K.E.; Hamilton, S.A.; Boldt, J.D.; Boynton, W.V.; Bradley, W.; Fraeman, M.E.; Gold, R.E.; et al. The X-ray Spectrometer on the MESSENGER Spacecraft. *Space Sci. Rev.* **2007**, *131*, 393–415. [CrossRef]
58. Soli, G.; Blaes, B.; Buehler, M.; Jones, P.; Ratliff, J.M.; Garrett, H. Clementine dosimetry. *J. Spacecr. Rocket.* **1995**, *32*, 1065–1070. [CrossRef]
59. Swinyard, B.; Clegg, P.; Leeks, S.; Griffin, M.; Lim, T.; Burgdorf, M. Space Operation and Performance of Doped Germanium Photo-Conducting Detectors in the Far Infrared: Experience from the ISO LWS. *Exp. Astron.* **2000**, *10*, 157–176. [CrossRef]
60. Thuillier, G.; Alunni, J.-M.; Roland, J.-J.; Brun, J.-F. Frequency-stabilized He-Ne laser for WINDII interferometer calibration on board the UARS-NASA satellite. *Opt. Eng.* **1992**, *31*, 567–573. [CrossRef]
61. Warren, T.; Bowles, N.E.; Hanna, K.D.; Thomas, I.R. The Oxford space environment goniometer: A new experimental setup for making directional emissivity measurements under a simulated space environment. *Rev. Sci. Instrum.* **2017**, *88*, 124502. [CrossRef] [PubMed]
62. Wei, F.; Wang, S.-J.; Liang, J.-B.; Wang, Y.; Zhang, S.-Y.; Jing, T.; Zhang, H.-X.; Zhang, B.-Q.; Leng, S. Next generation Space Environment Monitor (SEM) for FY-2 satellite series. *Chin. J. Geophys. Ics Chin. Ed.* **2013**, *56*, 1–11. [CrossRef]
63. Wesolek, D.M.; Hererro, F.A.; Osiander, R.; Darrin, M.A.G. Design, fabrication, and performance of a mi-cromachined plasma spectrometer. *J. Micro* **2005**, *4*, 041403.
64. Wise, S.A.; Amundsen, R.M.; Hopson, P.; High, J.W.; Kruse NM, H.; Kist, E.H.; Hooker, M.W. Design and testing of the midas spaceflight instrument. *IEEE Trans. Appl. Supercond.* **1995**, *5*, 1545–1548. [CrossRef]
65. Wright, R.; Lucey, P.; Crites, S.; Horton, K.; Wood, M.; Garbeil, H. BBM/EM design of the thermal hyperspectral imager: An instrument for remote sensing of earth's surface, atmosphere and ocean, from a microsatellite platform. *Acta Astronaut.* **2013**, *87*, 182–192. [CrossRef]
66. Xiong, X.; Angal, A.; Twedt, K.; Chen, H.; Link, D.; Geng, X.; Aldoretta, E.; Mu, Q. MODIS Reflective Solar Bands On-Orbit Calibration and Performance. *IEEE Trans. Geosci. Remote. Sens.* **2019**, *57*, 6355–6371. [CrossRef]
67. Zanoni, A.P.; Burkhardt, J.; Johann, U.; Aspelmeyer, M.; Kaltenbaek, R.; Hechenblaikner, G. Thermal performance of a radiatively cooled system for quantum optomechanical experiments in space. *Appl. Therm. Eng.* **2016**, *107*, 689–699. [CrossRef]
68. Zurbuchen, T.H.; Gershman, D.J. Innovations in plasma sensors. *J. Geophys. Res. Space Phys.* **2016**, *121*, 2891–2901. [CrossRef]
69. Jiao, Z.; Jiang, L.; Sun, J.; Huang, J.; Zhu, Y. Outgassing Environment of Spacecraft: An Overview. *IOP Conf. Series: Mater. Sci. Eng.* **2019**, *611*, 012071. [CrossRef]

70. Conscience, C.; Meftah, M.; Chevalier, A.; Dewitte, S. The space instrument SOVAP of the PI-CARD mission. In Proceedings of the SPIE Optical Engineering + Applications, San Diego, CA, USA, 21–25 August 2011.
71. Malandraki, O.E.; Crosby, N.B. The HESPERIA HORIZON 2020 Project and Book on Solar Particle Radiation Storms Forecasting and Analysis. *Space Weather* **2018**, *16*, 591–592. [CrossRef]
72. Jiggens, P.; Chavy-Macdonald, M.A.; Santin, G.; Menicucci, A.; Evans, H.; Hilgers, A. The magnitude and effects of extreme solar particle events. *J. Space Weather Space Clim.* **2014**, *4*, A20. [CrossRef]
73. Yung, K.L.; Lam, C.W.; Ko, S.M.; Foster, J.A. The Phobos-Grunt microgravity soil preparation system. *Acta Astronaut.* **2017**, *141*, 22–29. [CrossRef]

MDPI
St. Alban-Anlage 66
4052 Basel
Switzerland
Tel. +41 61 683 77 34
Fax +41 61 302 89 18
www.mdpi.com

Machines Editorial Office
E-mail: machines@mdpi.com
www.mdpi.com/journal/machines